DIFFERENTIAL EQUATIONS, BIFURCATIONS, AND CHAOS IN ECONOMICS

Series on Advances in Mathematics for Applied Sciences

Published:*

Vol. 71 Dissipative Phase Transitions
eds. P. Colli, N. Kenmochi and J. Sprekels

Vol. 72 Advanced Mathematical and Computational Tools in Metrology VII
eds. P. Ciarlini et al.

Vol. 73 Introduction to Computational Neurobiology and Clustering
by B. Tirozzi, D. Bianchi and E. Ferraro

Vol. 74 Wavelet and Wave Analysis as Applied to Materials with Micro or Nanostructure
by C. Cattani and J. Rushchitsky

Vol. 75 Applied and Industrial Mathematics in Italy II
eds. V. Cutello et al.

Vol. 76 Geometric Control and Nonsmooth Analysis
eds. F. Ancona et al.

Vol. 77 Continuum Thermodynamics
by K. Wilmanski

Vol. 78 Advanced Mathematical and Computational Tools in Metrology and Testing
eds. F. Pavese et al.

Vol. 79 From Genetics to Mathematics
eds. M. Lachowicz and J. Miękisz

Vol. 80 Inelasticity of Materials: An Engineering Approach and a Practical Guide
by A. R. Srinivasa and S. M. Srinivasan

Vol. 81 Stability Criteria for Fluid Flows
by A. Georgescu and L. Palese

Vol. 82 Applied and Industrial Mathematics in Italy III
eds. E. De Bernardis, R. Spigler and V. Valente

Vol. 83 Linear Inverse Problems: The Maximum Entropy Connection
by H. Gzyl and Y. Velásquez

Vol. 84 Advanced Mathematical and Computational Tools in Metrology and Texting IX
eds. F. Pavese et al.

Vol. 85 Continuum Thermodynamics
Part II: Applications and Examples
by B. Albers and K. Wilmanski

*To view the complete list of the published volumes in the series, please visit:
http://www.worldscibooks.com/series/samas_series.shtml

Series on Advances in Mathematics for Applied Sciences – Vol. 68

DIFFERENTIAL EQUATIONS, BIFURCATIONS, AND CHAOS IN ECONOMICS

Wei-Bin Zhang
Ritsumeikan Asia Pacific University, Japan

World Scientific
NEW JERSEY • LONDON • SINGAPORE • BEIJING • SHANGHAI • HONG KONG • TAIPEI • CHENNAI

Published by

World Scientific Publishing Co. Pte. Ltd.
5 Toh Tuck Link, Singapore 596224
USA office: 27 Warren Street, Suite 401-402, Hackensack, NJ 07601
UK office: 57 Shelton Street, Covent Garden, London WC2H 9HE

British Library Cataloguing-in-Publication Data
A catalogue record for this book is available from the British Library.

DIFFERENTIAL EQUATIONS, BIFURCATIONS, AND CHAOS IN ECONOMICS
Series on Advances in Mathematics for Applied Sciences — Vol. 68

ISBN-13 978-981-256-333-0
ISBN-10 981-256-333-4

Printed and bound in Great Britain by
Marston Book Services Limited, Oxfordshire

Preface

Application of differential equations to economics is a vast and vibrant area. Concepts and theorems related to differential equations appear everywhere in academic journals and textbooks in economics. One can hardly approach, not to mention digest, the literature of economic analysis without "sufficient" knowledge of differential equations. Nevertheless, the subject of applications of differential equations to economics is not systematically studied. The subject is often treated as a subsidiary part of (textbooks of) mathematical economics. Due to the rapid development of differential equations and wide applications of the theory to economics, there is a need for a systematic treatment of the subject. This book provides a comprehensive study of applications of differential equations to dynamic economics. We not only study analytical methods, but also provide applications of these methods for solving economic problems.

This book is a unique blend of the theory of differential equations and its exciting applications to economics. It is mainly concerned with ordinary differential equations. The book provides not only a comprehensive introduction to applications of theory of linear (and linearized) differential equations to economic analysis, but also studies nonlinear dynamical systems which have been widely applied to economic analysis in recent years. It provides a comprehensive introduction to most important concepts and theorems in differential equations theory in a way that can be understood by anyone who has basic knowledge of calculus and linear algebra. In addition to traditional applications of the theory to economic dynamics, it also contains many recent developments in different fields of economics. It is mainly concerned with how differential equations can be applied to solve and provide insights into economic dynamics. We emphasize "skills" for application.

The book is divided, according to dimensions of dynamic systems, into three parts. The first part deals with scalar differential equations; the second part studies planar differential equations; and the third part introduces higher-dimensional differential equations. Each part consists of three chapters. The first chapter of each part mainly deals with key concepts and main mathematical results related to linear (linearized) differential equations and their applications to economics. The second chapter mainly studies key concepts and (some of) main mathematical results related to nonlinear differential equations and their applications to economics. For illustration, the first two chapters tend to use simple (simplified) economic systems. The third chapter of each part introduces "complicated" (in terms of the number of variables and relationships among variables) economic models, applying the concepts and theorems from the previous two chapters. Most of the chapters include problems that help the reader from routine exercises through extensions of the models. Except conducting mathematical analysis of the economic models like most standard textbooks on mathematical economics, we use computer simulation to demonstrate motion of economic systems. A large fraction of examples in this book are simulated with Mathematica. Today, more and more researchers and educators are using computer tools to solve – once seemingly impossible to calculate even three decades ago – complicated and tedious problems.

The lively pace of research on differential equations and theoretical and empirical applications of differential equations to economics means that this book cannot cover all the important applications of differential equations to economics, not to mention the current development of differential equations, irrespective of the endeavors to provide a comprehensive study of the subject.

I would like to thank Editor E H Chionh for effective co-operation. I completed this book at the Ritsumeikan Asia Pacific University, Japan. I am grateful to the university's pleasant and co-operative academic environment. I take great pleasure in expressing my gratitude to my wife,

Gao Xiao, who has been wonderfully supportive of my efforts in writing this book in Beppu City, Japan. She also helped me to draw some of the figures in the book.

W.B. Zhang

Contents

Chapter 1

Differential Equations in Economics

Applications of differential equations are now used in modeling motion and change in all areas of science. The theory of differential equations has become an essential tool of economic analysis particularly since computer has become commonly available. It would be difficult to comprehend the contemporary literature of economics if one does not understand basic concepts (such as bifurcations and chaos) and results of modern theory of differential equations.

A differential equation expresses the rate of change of the current state as a function of the current state. A simple illustration of this type of dependence is changes of the Gross Domestic Product (GDP) over time. Consider state x of the GDP of the economy. The rate of change of the GDP is proportional to the current GDP

$$\dot{x}(t) = gx(t),$$

where t stands for time and $\dot{x}(t)$ the derivative of the function x with respect to t. The growth rate of the GDP is $\dot{x}/x$. If the growth rate g is given at any time t, the GDP at t is given by solving the differential equation. The solution is

$$x(t) = x(0)e^{gt}.$$

The solution tells that the GDP decays (increases) exponentially in time when g is negative (positive).

We can explicitly solve the above differential function when g is a constant. It is reasonable to consider that the growth rate is affected by many factors, such as the current state of the economic system, accumulated knowledge of the economy, international environment, and many other conditions. This means that the growth rate may take on a complicated form $g(x,t)$. The economic growth is described by

$$\dot{x}(t) = g(x(t),t)x(t).$$

In general, it is not easy to explicitly solve the above function. There are various established methods of solving different types of differential equations. This book introduces concepts, theorems, and methods in differential equation theory which are widely used in contemporary economic analysis and provides many simple as well as comprehensive applications to different fields in economics.

This book is mainly concerned with *ordinary differential equations*. Ordinary differential equations are differential equations whose solutions are functions of one independent variable, which we usually denote by t. The variable t often stands for time, and solution we are looking for, $x(t)$, usually stands for some economic quantity that changes with time. Therefore we consider $x(t)$ as a dependent variable. For instance, $\dot{x}(t) = t^2 x(t)$ is an ordinary differential equation. Ordinary differential equations are classified as autonomous and nonautonomous. The equation

$$\dot{x}(t) = ax(t) + b,$$

with a and b as parameters is an *autonomous differential equation* because the time variable t does not explicitly appear. If the equation specially involves t, we call the equation *nonautonomous* or *time-dependent*. For instance,

$$\dot{x}(t) = x(t) + \sin t,$$

is a nonautonomous differential equation. In this book, we often omit "ordinary", "autonomous", or "nonautonomous" in expression. If an equation involves derivatives up to and includes the ith derivative, it is called an *ith order* differential equation. The equation $\dot{x}(t) = ax(t) + b$ with a and b as parameters is a first order autonomous differential equation. The equation

$$\ddot{x} = 3\dot{x} - 2x + 2,$$

is a second order equation, where the second derivative, $\ddot{x}(t)$, is the derivative of $\dot{x}(t)$.[1] As shown late, the solution is

$$x(t) = A_1 e^{2t} + A_2 e^{t} + 1,$$

where A_1 and A_2 are two constants of integration. The first derivative $\dot{x}$ is the only one that can appear in a first order differential equation, but it may enter in various powers: $\dot{x}$, $\dot{x}^2$, and so on. The highest power attained by the derivative in the equation is referred to as the *degree* of the differential equation. For instance,

$$3\dot{x}^2 - 2x + 2 = 0$$

is a second-degree first-order differential equation.

1.1 Differential Equations and Economic Analysis

This book is a unique blend of the theory of differential equations and their exciting applications to economics. First, it provides a comprehensive introduction to most important concepts and theorems in differential equations theory in a way that can be understood by anyone who has basic knowledge of calculus and linear algebra. In addition to traditional applications of the theory to economic dynamics, this book also contains many recent developments in different fields of economics. The book is mainly concerned with how differential equations can be applied to solve and provide insights into economic dynamics. We emphasize "skills" for application. When applying the theory to economics, we outline the economic problem to be solved and then derive differential equation(s) for this problem. These equations are then analyzed and/or simulated.

Different from most standard textbooks on mathematical economics, we use computer simulation to demonstrate motion of economic systems. A large fraction of examples in this book are simulated with Mathematica. Today, more and more researchers and educators are using computer tools such as Mathematica to solve – once seemingly

[1] The *n*th *derivative* of $x(t)$, denoted by $x^{(n)}(t)$, is the derivative of $x^{(n-1)}(t)$.

impossible to calculate even three decades ago – complicated and tedious problems.

This book provides not only a comprehensive introduction to applications of linear and linearized differential equation theory to economic analysis, but also studies nonlinear dynamical systems which have been widely applied to economic analysis only in recent years. Linearity means that the rule that determines what a piece of a system is going to do next is not influenced by what it is doing now. The mathematics of linear systems exhibits a simple geometry. The simplicity allows us to capture the essence of the problem. Nonlinear dynamics is concerned with the study of systems whose time evolution equations are nonlinear. If a parameter that describes a linear system, is changed, the qualitative nature of the behavior remains the same. But for nonlinear systems, a small change in a parameter can lead to sudden and dramatic changes in both the quantitative and qualitative behavior of the system.

Nonlinear dynamical theory reveals how such interactions can bring about qualitatively new structures and how the whole is related to and different from its individual components. The study of nonlinear dynamical theory has been enhanced with developments in computer technology. A modern computer can explore a far wider class of phenomena than it could have been imagined even a few decades ago. The essential ideas about complexity have found wide applications among a wide range of scientific disciplines, including physics, biology, ecology, psychology, cognitive science, economics and sociology. Many complex systems constructed in those scientific areas have been found to share many common properties. The great variety of applied fields manifests a possibly unifying methodological factor in the sciences. Nonlinear theory is bringing scientists closer as they explore common structures of different systems. It offers scientists a new tool for exploring and modeling the complexity of nature and society. The new techniques and concepts provide powerful methods for modeling and simulating trajectories of sudden and irreversible change in social and natural systems.

Modern nonlinear theory begins with Poincaré who revolutionized the study of nonlinear differential equations by introducing the qualitative techniques of geometry and topology rather than strict analytic methods to discuss the global properties of solutions of these systems. He considered it more important to have a global understanding of the gross behavior of all solutions of the system than the local behavior of particular, analytically precise solutions. The study of the dynamic systems was furthered in the Soviet Union, by mathematicians such as Liapunov, Pontryagin, Andronov, and others. Around 1960, the study by Smale in the United States, Peixoto in Brazil and Kolmogorov, Arnol'd and Sinai in the Soviet gave a significant influence on the development of nonlinear theory. Around 1975, many scientists around the world were suddenly aware that there is a new kind of motion – now called chaos – in dynamic systems. The new motion is erratic, but not simply "quasiperiodic" with a large number of periods.[2] What is surprising is that chaos can occur even in a very simple system. Scientists were interested in complicated motion of dynamic systems. But only with the advent of computers, with screens capable of displaying graphics, have scientists been able to see that many nonlinear dynamic systems have chaotic solutions.

As demonstrated in this book, nonlinear dynamical theory has found wide applications in different fields of economics.[3] The range of applications includes many topics, such as catastrophes, bifurcations, trade cycles, economic chaos, urban pattern formation, sexual division of labor and economic development, economic growth, values and family structure, the role of stochastic noise upon socio-economic structures, fast and slow socio-economic processes, and relationship between microscopic and macroscopic structures. All these topics cannot be effectively examined by traditional analytical methods which are concerned with linearity, stability and static equilibria. Nonlinear dynamical theory has changed economists' views about evolution. For instance, the traditional view of the relations between laws and consequences - between cause and effect - holds that simple rules imply simple behavior, therefore complicated behavior must arise from

---

[2] In the solar system, the motion traveled around the earth in month, the earth around the sun in about a year, and Jupiter around the sun in about 11.867 years. Such systems with multiple incommensurable periods are known as quasiperiodic.

[3] For recent applications of nonlinear theory to economics, see Rosser (1991), Zhang (1991, 2005), Lorenz (1993), Puu (2000), and Shone (2002).

complicated rules. This vision had been held by professional economists for a long time. But it has been recently challenged due to the development of nonlinear theory. Nonlinear theory shows how complicated behavior may arise from simple rules. To illustrate this idea, we consider the *Ueda attractor*

$$\ddot{x} + 2\gamma\dot{x} + x^3 = F\cos t .$$

This is a simple dynamical system. When $k = 0.025$ and $F = 7.5$, as illustrated in Fig. 1.1.1, its behavior are "chaotic". The model with the specified parameter values does not exhibit any regular or periodic behavioral pattern. Chaos persists for as long as time passes.

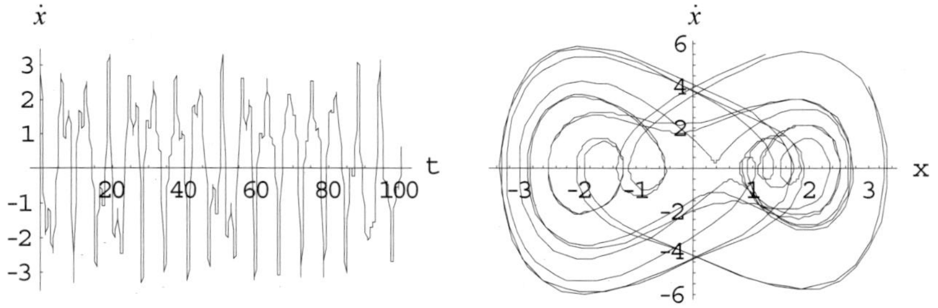

Fig. 1.1.1 Chaos of the Ueda attractor.

Another example is the Lorenz equations. The laws that govern the motion of air molecules and of other physical quantities are well known. The topic of differential equations is some 300 years old, but nobody would have thought it possible that differential equations could behave as chaotically as Edward N. Lorenz found in his experiments. Around 1960, Lorenz constructed models for numerical weather forecasting. He showed that deterministic natural laws do not exclude the possibility of chaos. In other words, determinism and predictability are not equivalent. In fact, recent chaos theory shows that deterministic chaos can be identified in much simpler systems than the Lorenz model.

The system of equations (with the parameter values specified) that Lorenz proposed in 1963 is

$$\dot{x} = 10(-x + y),$$

$$\dot{y} = 28x - y - xz,$$

$$\dot{z} = -\frac{8}{3}z + xy,$$

where x, y, and z are time-dependent variables.[4] We will come back to this system later. If we start with an initial state $(x_0, y_0, z_0) = (6, 6, 6)$, the motion of the system is chaotic, as depicted in Fig. 1.1.2. There are two sheets in which trajectories spiral outwards. When the distance from the center of such a spiral becomes larger than some particular threshold, the motion is ejected from the spiral and is attracted by the other spiral, where it again begins to spiral out, and the process is repeated. The motion is not regular. The number of turns that a trajectory spends in one spiral before it jumps to the other is not specified. It may wind around one spiral twice, and then three times around the other, then ten times around the first and so on.

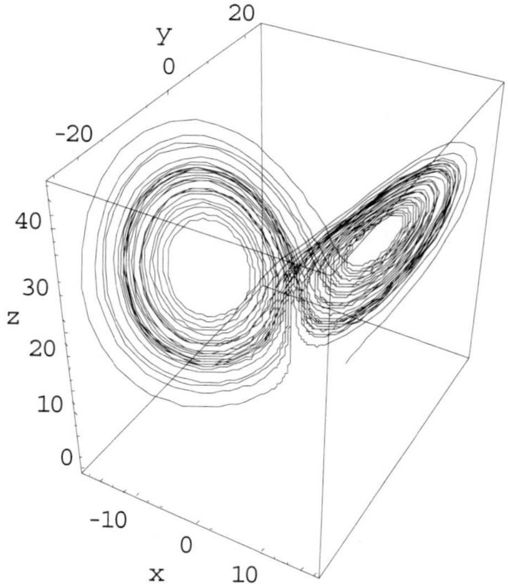

Fig. 1.1.2 The dynamics of the Lorenz equations.

Nonlinear dynamical systems are sufficient to determine the behavior in the sense that solutions of the equations do exist. But it is often

[4] A very thorough treatment of the Lorenz equations is given by Sparrow (1982).

impossible to explicitly write down solutions in algebraic expressions. Nonlinear economics based on nonlinear dynamical theory attempts to provide a new vision of economic dynamics: a vision toward the multiple, the temporal, the unpredictable, and the complex. There is a tendency to replace simplicity with complexity and specialism with generality in economic research. The concepts such as totality, nonlinearity, self-organization, structural changes, order and chaos have found broad and new meanings by the development of this new science. According to this new science, economic dynamics are considered to resemble a turbulent movement of liquid in which varied and relatively stable forms of current and whirlpools constantly change one another. These changes consist of dynamic processes of self-organization along with the spontaneous formation of increasingly subtle and complicated structures. The accidental nature and the presence of structural changes like catastrophes and bifurcations, which are characteristic of nonlinear systems and whose further trajectory is determined by chance, make dynamics irreversible.

Traditional economists were mainly concerned with regular motion of the economic systems. Even when they are concerned with economic dynamics, students are still mostly limited to their investigations of differential or difference equations to regular solutions (which include steady states and periodic solutions). In particular, economists were mainly interested in existence of a unique stable equilibrium. Students trained in traditional economics tend to imagine that the economic reality is uniquely determined and will remain invariant over time under "ideal conditions" of preferences, technology, and institutions. Nevertheless, common experiences reveal more complicated pictures of economic reality. Economic structures change even within a single generation. Economic systems collapse or suddenly grow without any seemingly apparent signs of structural changes.

1.2 Overview

This book presents the mathematical theory in linear and nonlinear differential equations and its applications to many fields of economics.

The book is for economists and scientists of other disciplines who are concerned with modeling and understanding the time evolution of nonlinear dynamic economic systems. It is of potential interest to professionals and graduate students in economics and applied mathematics, as well as researchers in social sciences with an interest in applications of differential equations to economic systems.

The book is basically divided into three parts - Part I concerns with one-dimensional differential equations; Part II concentrates on planar differential equations; Part III studies higher dimensional dynamical systems. Each part consists of three chapters – the first chapter is concentrated on linear systems, the second chapter studies nonlinear systems, and the third chapter applies concepts and techniques from the previous two chapters to economic dynamic systems of different schools.

Part I consists of three chapters. Chapter 2 deals with one-dimensional linear differential equations. Section 2.1 solves one-dimensional linear first-order differential equations. In Sec. 2.2, we examine a few special types of first-order equations, which may not be linear. Using the special structures of the equations, we can explicitly solve them. The types include separable differential equations, exact differential equations, and the Bernoulli equation. This section also examines the most well-known growth model, the Solow model and provides a few examples of applications. Section 2.3 is concerned with second-order differential equations. Section 2.4 gives general solutions to higher-order differential equations with continuous coefficients in time. Section 2.5 gives general solutions to higher-order differential equations with constant coefficients.

Chapter 3 is organized as follows. Section 3.1 introduces some fundamental concepts and theorems, such as equilibrium, trajectory, solution, periodic solution, existence theorems, and stability, about nonlinear differential equations. To avoid repetition in later chapters, the contents of this section are not limited to one dimension; they are valid for any finite dimensions. Section 3.2 states stability conditions of equilibria for scalar autonomous equations. We also apply the theory to two well-known economic models, the Cagan monetary model and the generalized Solow model with poverty traps. Section 3.3 introduces bifurcation theory and fundamental results for one-dimensional nonlinear systems. We examine saddle-node, transcritical, pitchfork, and cusp bifurcations. In Sec. 3.4, we demonstrate periodic solutions of one-dimensional second-order differential equations, using the Van der Pol equation and the Duffing equation as examples. Section 3.5 illustrates the energy balance method for examining periodic solutions. Section 3.6 introduces how to estimate amplitude and frequency of the periodic solutions examined in the previous section.

Chapter 4 applies concepts and theorems from the previous two chapters to analyze different models in economic model. Although the economic relations in these models tend to be complicated, we show that the dynamics of all these models are determined by motion of one-dimensional differential equations. Section 4.1 examines a one-sector growth model. As the economic mechanisms of this model will be applied in some other models in this book, we explain the economic structure in details. This section also applies the Liapunov theorem to guarantee global asymptotical stability of the equilibrium. Section 4.2 depicts the one-sector growth model proposed in Sec. 4.1 with simulation. Section 4.3 examines the one-sector-growth model for general utility functions. Section 4.4 examines a model of urban economic growth with housing production. In Sec. 4.5, we examine a dynamic model to how leisure time and work hours change over time in association with economic growth. Section 4.6 examines dynamics of sexual division of labor and consumption in association of modern economic growth. We illustrate increases of women labor participation as a "consequence" of economic growth as well as changes of labor market conditions. Section 4.7 introduces the Uzawa two-sector model. In Sec. 4.8, we re-examine the Uzawa model with endogenous consumer behavior. The models of this chapter show the essence of economic principles in many fields of economics, such as equilibrium economics (as a stationary state of a dynamic economics), growth theory, urban economics, and gender economics. The basic ideas and conclusions of this chapter require some books to explain, if that is possible. This also proves power of differential equations theory.

Part II consists of three chapters. Chapter 5 studies planar linear differential equations. Section 5.1 gives general solutions to planar linear first-order homogeneous differential equations. We also depict phase portraits of typical orbits of the planar systems. Section 5.2 introduces some concepts, such as positive orbit, negative orbit, orbit, limit set, and invariant set, for qualitative study. Section 5.3 shows how to calculate matrix exponentials and to reduce planar differential equations to the canonical forms. In Sec. 5.4, we introduce the concept of topological equivalence of planar linear systems and classify the planar linear homogeneous differential equations according to the concept. Section 5.5 studies planar linear first-order non-homogeneous differential equations. This section examines dynamic behavior of some typical economic models, such as the competitive equilibrium model, the Cournot duopoly model with constant marginal costs, the Cournot duopoly model with increasing marginal costs, the Cagan model with sluggish wages. Section 5.6 solves some types of constant-coefficient linear equations with time-dependent terms.

Chapter 6 deals with nonlinear planar differential equations. Section 6.1 carries out local analysis and provides conditions for validity of linearization. We also provide relations between linear systems and almost linear systems with regard to dynamic qualitative properties. This section examines dynamic properties of some frequently-applied economic models, such as the competitive equilibrium model, the Walrasian-Marshallian adjustment process, the Tobin-Blanchard model, and the Ramsey model. Section 6.2 introduces the Liapunov methods for stability analysis. In Sec. 6.3, we study some typical types of bifurcations of planar differential equations. Section 6.4 demonstrates motion of periodic solutions of some nonlinear planar systems. Section 6.5 introduces the Poincaré-Bendixon Theorem and applies the theorem to the Kaldor model to identify the existence of business cycles. Section 6.6 states Lienard's Theorem, which provides conditions for the existence and uniqueness of limit cycle in the Lienard system. Section 6.7 studies one of most frequently applied theorems in nonlinear economics, the Andronov-Hopf Bifurcation Theorem and its applications in the study of business cycles.

Chapter 7 applies the concepts and theorems related to two-dimensional differential equations to various economic issues. Section 7.1 introduces the IS-LM model, one of the basic models in contemporary macroeconomics and examines its dynamic properties.

Section 7.2 examines an optimal foreign debt model, maximizing the life-time utility with borrowing. In Sec. 7.3, we consider a dynamic economic system whose construction is influenced by Keynes' General Theory. Applying the Hopf bifurcation theorem, we demonstrate the existence of limit cycles in a simplified version of the Keynesian business model. Section 7.4 examines dynamics of unemployment within the framework of growth theory. In particular, we simulate the model to demonstrate how unemployment is affected by work amenity and unemployment policy. In Sec. 7.5, we establish a two-regional growth model with endogenous time distribution. We examine some dynamic properties of the dynamic systems. Section 7.6 models international trade with endogenous urban model formation. We show how spatial structures evolve in association of global growth and trade. In Sec. 7.7, we introduce a short-run dynamic macro model, which combines the conventional IS-LM model and Phillips curve. We also illustrate dynamics of the model under different financial policies. Section 7.8 introduces a growth model with public inputs. The public sector is treated as an endogenous part of the economic system. The system exhibits different dynamic properties examined in the previous two chapters.

Chapter 8 studies higher-dimensional differential equations. Section 8.1 provides general solutions to systems of linear differential equations. Section 8.2 examines homogeneous linear systems with constant coefficients. Section 8.3 solves higher-order homogenous linear differential equations. Section 8.4 introduces diagonalization and introduces concepts of stable and unstable subspaces of the linear systems. Section 8.5 studies the Fundamental Theorem for linear systems and provides a general procedure of solving linear equations.

Chapter 9 deals with higher dimensional nonlinear differential equations. Section 9.1 studies local stability and validity of linearization. Section 9.2 introduces the Liapunov methods and studies Hamiltonian systems. In Sec. 9.3, we examine differences between conservative and dissipative systems. We examine the Goodwin model in detail. Section 9.4 defines the Poincaré maps. In Sec. 9.5, we introduce center manifold theorems. Section 9.6 applies the center manifold theorem and Liapunov theorem to a simple planar system. In Sec. 9.7, we introduce the Hopf bifurcation theorem in higher dimensional cases and apply it to a predator-prey model. Section 9.8 simulates the Loren equations, demonstrating chaotic motion of deterministic dynamical systems.

Chapter 10 applies the mathematical concepts and theorems of higher differential equations introduced in the previous two chapters to differential economic models. Section 10.1 examines some tâtonnement price adjustment processes, mainly applying the Liapunov methods. Section 10.2 studies a three-country international trade model with endogenous global economic growth. Section 10.3 extends the trade model of the previous section by examining impacts of global economic group on different groups of people not only among countries but also within countries. We provide insights into complexity of international trade upon different people. Section 10.4 examines an two-region growth model with endogenous capital and knowledge. Different from the trade model where international migration is not allowed, people freely move among regions within the interregional modeling framework. Section 10.5 introduces money into the growth model. We demonstrate the existence of business cycles in the model, applying the Hopf bifurcation theorem. Section 10.6 guarantees the existence of limit cycles and aperiodic behavior in the traditional multi-sector optimal growth model, an extension of the Ramsey growth model. Section 10.7 proposes a dynamic model with interactions among economic growth, human capital accumulation, and opening policy to provide insights into the historical processes of Chinese modernization. Analysis of behavior of this model requires almost all techniques introduced in this book.

As concluding remarks to this book, we address two important issues which have been rarely studied in depth in economic dynamical analysis, changeable speeds and economic structures. The understanding of these two issues are essential for appreciating validity and limitations of different economic models in the literature, but should also play a guarding role in developing general economic theories. We also include an appendix. App. A.1 introduces matrix theory. App. A.2 shows how to solve linear equations, based on matrix theory. App. A.3 defines some basic concepts in study of functions and states the Implicit Function Theorem. App. A.4 gives a general expression of the Taylor Expansion.

App. A.5 briefly mentions a few concepts related to structural stability. App. A.6 introduces optimal control theory.

Part I

Dimension One

Chapter 2

Scalar Linear Differential Equations

Consider a consumer of a bank. Let her money in the bank be $x(t)$. If interest is continuously compounded at an annual rate a, then the differential equation

$$\dot{x}(t) = ax(t),$$

describes the amount of money in the bank account over time. Here, $\dot{x}(t)$ stands for the derivative of the function $x(t)$ with respect to the variable t. It should be noted that the derivative is also represented by $dx(t)/dt$ or $x'(t)$.

The solution of this equation is $x(t) = x(0)e^{at}$ - the bank account grows exponentially without bound if the size of the original deposit, $x(0)$, is positive. This same equation models the dynamics of the population with a constant percent rate of growth a. The assumption of a constant a is referred to as *Malthus's law*, and the corresponding equation $\dot{x}(t) = ax(t)$ as the *Malthus equation*. The solution says that if a society follows Malthus's law (with a positive a), then its population will grow exponentially without bound. The society may suffer from poverty due to over population if its economic growth fails to meet the basic need of the rapidly increasing population.

The equation, $\dot{x}(t) = ax(t)$, has infinitely many solutions, each of the form $x(t) = Ae^{at}$, for a constant real number A. At t, we determine $A = x(0)$. The number $x_0 = x(0)$ is called the *initial value* of the function x. An *initial value problem* often consists of a differential equation together with enough initial values to specify a single solution. Hence, we say that the solution of the initial problem

$$\dot{x}(t) = ax(t), \quad x_0 = x(0),$$

is $x(t) = x_0 e^{at}$. Figure 2.0.1 shows the family of solutions of differential equations for various initial values x_0 with $a < 0$ and $a > 0$. Each choice of initial value x_0 determines a curve. This picture is called *flow* of the differential equation. The flow, $\varphi(t, x_0)$, of an autonomous differential equation is the function of time t and initial value x_0, which represents the set of solutions. Thus $\varphi(t, x_0)$ is the value at time t of the solution with initial value x_0.

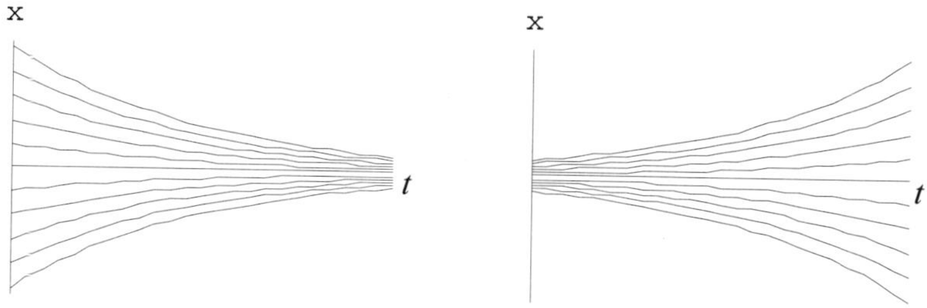

(a) $a < 0$; exponential decay (b) $a > 0$; exponential growth

Fig. 2.0.1 Solutions of $\dot{x} = ax$ with varied initial values.

The differential equation

$$\dot{x}(t) = ax(t),$$

contains a single dependent variable. An economic system often contains many dependent variables, such as outputs, capital stocks, money, and prices. The *dimension* of differential equations refers to the number of dependent variables in the system. In this section, there is one variable, which is a function of the independent variable t. If the derivative of the variable $x(t)$, denoted by $\dot{x}(t)$, $dx(t)/dt$, or $x'(t)$, is linear in x, we say that it is a *linear differential equation.*

This chapter is concerned with one-dimensional linear differential equations. Section 2.1 solves one-dimensional linear first-order differential equations. In Sec. 2.2, we examine a few special types of first-order equations, which may not be linear. Using the special structures of the equations, we can explicitly solve them. The types include separable differential equations, exact differential equations, and the Bernoulli equation. This section also examines the most well-known growth model, the Solow model and provides a few examples of applications. Section 2.3 is concerned with second-order differential equations. Section 2.4 gives general solutions to higher-order differential equations with continuous coefficients in time. Section 2.5 gives general solutions to higher-order differential equations with constant coefficients.

2.1 Scalar Linear First-Order Differential Equations

This section studies linear first-order differential equations, generally expressed as

$$\dot{x} + u(t)x = w(t), \tag{2.1.1}$$

where $u(t)$ and $w(t)$ are functions of t.

The homogeneous case with constant coefficient and constant term
First, we examine the homogeneous case of Eq. (2.1.1) when $u(t) = a$ and $w(t) = 0$. The solution of

$$\dot{x} + ax = 0, \tag{2.1.2}$$

is $x(t) = Ae^{-at}$, where A is an arbitrary constant. We have just examined this case in the previous section.

Example The Harrod-Domar model.[1]
The system is built on the hypothesis that any change in the rate of investment per year $I(t)$ will affect the aggregate demand and productivity of the economy. The demand effect of a change in $I(t)$ operates through the multiplier process. An increase in $I(t)$ will raise the rate of income flow per year $Y(t)$ by a multiple of the increment in $I(t)$. The agents regularly set aside some fairly predictable portion of its output for the purpose of capital accumulation. Since there is a single good, no question of changes in relative price can arise, nor can any questions of capital composition. Let us denote s a constant fraction of the total output flow that is saved and set aside to be added to the capital stock. For a predetermined s, the multiplier is $a = 1/s$.

[1] See Domar (1946) and Harrod (1948).

As $I(t)$ is the only expenditure flow that influences the rate of income flow, we have

$$\dot{Y}(t) = \dot{I}(t)/s\,.$$

The capacity effect of investment is reflected by the change in the rate of potential output the economy is capable of producing. The capacity-capital ratio is defined by $\rho \equiv \kappa(t)/K(t)$, where $\kappa(t)$ stands for capacity or potential output flow and ρ represents a (predetermined) constant capacity-capital ratio. The above equation implies that with a capital stock $K(t)$ the economy is potentially capable of producing an annual product κ. Taking derivatives of $\kappa(t) = \rho K(t)$ with respect to t yields

$$\dot{\kappa} = \rho\dot{K} = \rho I\,.$$

Here, equilibrium is defined as a situation, in which productive capacity is fully utilized, i.e., $Y(t) = \kappa(t)$. If we start initially from equilibrium, the requirement means the balancing of the respective changes in capacity and in aggregate demand; that is, $\dot{Y}(t) = \dot{\kappa}(t)$. The question is what kind of time path of investment $I(t)$ will keep the economy in equilibrium at all times. To answer this question, insert equations $\dot{Y}(t) = \dot{I}(t)/s$ and $\dot{\kappa} = I$ into $\dot{Y}(t) = \dot{\kappa}(t)$ to get

$$\dot{I} = s\rho I\,.$$

Therefore, the required path is given by the following solution of the above differential equation

$$I(t) = I(0)e^{\rho st}\,,$$

where $I(0)$ is the initial rate of investment. This implies that to maintain the balance between capacity and demand over time, the rate of investment flow must grow precisely at the exponential rate of ρs. Substituting $I(t) = I(0)e^{\rho st}$ into $\dot{K}(t) = I(t)$ yields $\dot{K} = I(0)e^{\rho st}$.

It is easy to check that the following function

$$K(t) = \frac{I(0)}{\rho s}e^{\rho st} + K(0) - I(0),$$

satisfies the above equation with initial capital stock $K(0)$. As $Y(t) = \rho K(t)$, we have

$$Y(t) = \frac{I(0)}{\hat{s}} e^{\rho st} + \rho(K(0) - I(0)).$$

We depict a solution to the system as in Fig. 2.1.1 for the following specified values of the parameters: $\rho = 0.5$, $s = 0.2$, with given initial values of K and I.

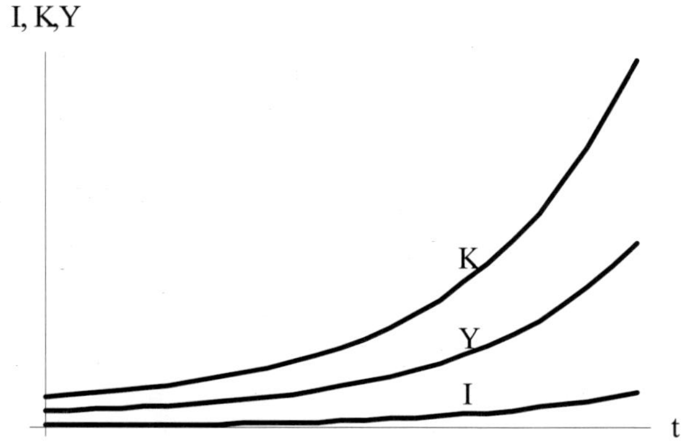

Fig. 2.1.1 The solution to the Harrod-Domar model.

The nonhomogeneous case with constant coefficient and constant term
A nonhomogeneous linear different equation with constant coefficient is generally given by

$$\dot{x} + ax = b, \quad b \neq 0. \tag{2.1.3}$$

In the case of $a = 0$, the solution is

$$x(t) = bt + A,$$

where A is an arbitrary constant. In the case of $a \neq 0$, the solution with known initial state $x(0)$ is

$$x(t) = \left(x(0) - \frac{b}{a} \right) e^{-at} + \frac{b}{a}.$$

Definition 2.1.1 A constant solution of the autonomous differential equation $\dot{x}(t) = f(x)$ is called an equilibrium of the equation.

It should be noted that this definition is also valid for nonlinear and higher dimensional problems. An equilibrium is a solution of $f(x)=0$. For $\dot{x}(t)=ax(t)$, the origin is an equilibrium solution. Evidently, $x=b/a$ is a solution to Eq. (2.1.3). It is an equilibrium point of the system.

Example We now consider dynamics of price of a single commodity. Suppose that the demand and supply functions for the commodity are

$$Q_d = a_1 - b_1 P, \quad Q_s = -a_2 + b_2 P, \quad a_j, b_j > 0, \qquad (2.1.4)$$

where Q_d and Q_s are respectively the demand and supply for price P and a_j and b_j are parameters. The market is in equilibrium when the demand equals supply, $Q_d = Q_s$. It is straightforward to show that if the price is

$$P^* = \frac{a_1 + a_2}{b_1 + b_2},$$

then the market is in equilibrium. Nevertheless, when the actual price deviates from P^*, then either the demand exceeds the supply or the supply exceeds the demand. We consider that in market price changes according to the relative strength of the demand and supply forces. For simplicity, assume that the rate of price changes with regard to time at t is proportional to the excess demand, $Q_d - Q_s$, that is

$$\dot{P}(t) = m(Q_d(t) - Q_s(t)), \quad m > 0.$$

Substituting Eqs. (2.1.4) into the above equation yields

$$\dot{P}(t) + m(b_1 + b_2)P = m(a_1 + a_2). \qquad (2.1.5)$$

This equation belongs to the type given by Eq. (2.1.3). Hence, its solution is

$$P(t) = (P(0) - P^*)e^{-m_0} + P^*, \quad m_0 \equiv m(a_2 + b_2) > 0. \qquad (2.1.6)$$

As m_0 is positive, we conclude that as $t \to +\infty$, $P(t) \to P^*$ (because $(P(0) - P^*)e^{-m_0} \to 0$). In the long term, the market mechanism will lead the market dynamics to its equilibrium position. Figure 2.1.2 depicts the price dynamics with different initial conditions. We see that if the initial price is above the equilibrium level, it decreases over time, and vice versa.

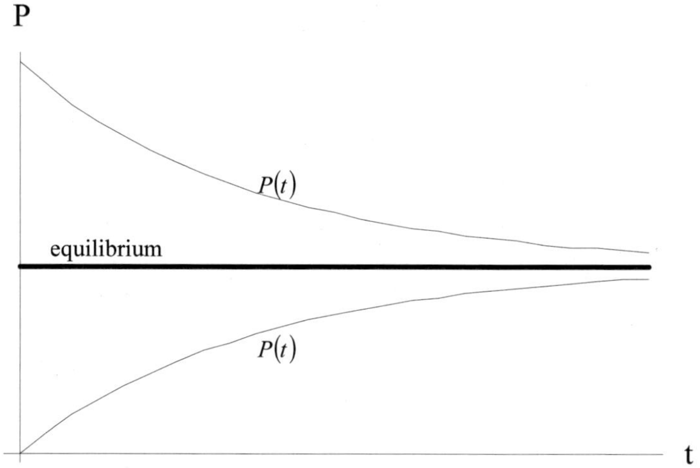

Fig. 2.1.2 Convergence towards the equilibrium price.

Example A simple expenditure model.
Consider a macroeconomic model for an open economy where prices are assumed constant. Expenditure, $E(t)$, is the sum of consumption expenditure, $C(t)$, investment expenditure, $I(t)$, government expenditure, G, and expenditure on net exports, $NX(t)$ (which is the difference between exports, X, and imports, $M(t)$, i.e., $NX = X - M$). We have

$$E = C + I + G + NX .$$

Assume the following relations among the economic variables

$$C = a + bY^d(t), \quad a > 0, \quad 0 < b < 1,$$

$$Y^d = Y - T,$$

$$T(t) = T_0 - \tau Y, \quad T_0 > 0, \quad 0 < \tau < 1,$$

$$I = I_0 + jY, \quad I_0, j > 0,$$

$$M = M_0 + mY, \quad 0 < m < 1, \tag{2.1.7}$$

where Y^d is the disposable income and T the net taxes. We assume that national income, $Y(t)$, adjusts continuously over time in response to the excess demand in the goods market, specified in the following way

$$\dot{Y} = \sigma(E - Y), \quad \sigma > 0 .$$

Substituting Eqs. (2.1.7) into the above equation yields

$$\dot{Y} = \sigma A - \sigma[1 - b(1 - \tau) - j + m]Y ,$$

where A is the sum of all autonomous expenditures

$$A \equiv a - bT_0 + I_0 + G + X - M_0 \ (> 0) .$$

The stability condition is straightforward.

Example The spread of disease.
The first application of differential equations to the study of epidemics and contagious disease was made by Daniel Bernoulli in 1760. His line of thought is illustrated by the following model.

The disease in question is smallpox. The disease is contagious, but confers completely immunity on anyone who has caught it and recovered. It is this last characteristic that makes vaccination so effective and finally made it possible to eradicate the disease. The Kangxi emperor (1654–1722), known as one of the greatest Chinese emperors in history, was chosen as the emperor in 1662 mainly because the boy had survived smallpox which was spreading in Peking at that time.

Bernoulli starts with a population of people at time $t = 0$. Suppose that at time t there are $x(t)$ people alive and $y(t)$ people who are alive and have not yet had smallpox. The model is

$$\dot{x} = -aby - d(t)x,$$

$$\dot{y} = -ay - d(t)y,$$

where a is the rate that the y-population are susceptible to the disease, b $(0 < b < 1)$ is the fraction of the y-population who get the disease and do not recover, and $d(t)$ is the death rate from all other diseases. Multiplying the first equation by $y(t)$ and the second by $x(t)$ and subtracting, we obtain

$$y\dot{x} - x\dot{y} = -aby^2 + axy .$$

This equation can be rewritten as

$$\frac{d(x/y)}{dt} = -ab + a\frac{x}{y}.$$

Thus the ratio $z = x/y$ satisfies the linear equation

$$\dot{z} = -ab + az.$$

The solution to the problem with an initial value $z(0) = 1$ is

$$z = b + (1-b)e^{at}.$$

Bernoulli estimated $a = b = 1/8$. After studying mortality tables, Bernoulli recommended vaccination.

The general case

Consider the general case

$$\dot{x} + u(t)x = w(t),$$

where $u(t)$ and $w(t)$ are functions of t. The solution of this equation is

$$x(t) = e^{-\int u\,dt}\left(A + \int we^{\int u\,dt}\,dt\right),$$

where A is an arbitrary constant.

Example For $\dot{x} + 2tx = 4t$, we solve

$$x(t) = e^{-\int 4t\,dt}\left(A + \int 4te^{\int 4t\,dt}\,dt\right) = e^{-2t^2}\left(A + \int 4te^{2t^2}\,dt\right)$$

$$= Ae^{-2t^2} + 1.$$

In particular, in the homogeneous case, i.e., $w = 0$, Eq. (2.1.1) becomes

$$\dot{x} + u(t)x = 0.$$

The solution of the equation is

$$x(t) = Ae^{-\int u\,dt}.$$

By this formula, for example, we solve $\dot{x} + t^2x = 0$ as

$$x(t) = Ae^{-\int t^2\,dt} = Ae^{-t^3}.$$

Exercise 2.1

1 According to Domar,[2] assume that income $Y(t)$ grows at a constant rate r. To maintain full employment, the budget deficit, $D(t)$, changes in proportion a to $Y(t)$, i.e., $\dot{D}(t) = aY(t)$. Show that

$$\frac{D(t)}{Y(t)} = \left(\frac{D_0}{Y_0} - \frac{a}{r}\right) + \frac{a}{r}.$$

2 Find solutions of the differential equations

(a) $\dot{x}(t) + x(t) = 4, \ \ x(0) = 0$;

(b) $\dot{x}(t) + 3x(t) = 2, \ \ x(0) = 4$.

3 Suppose that the demand and supply functions for the commodity are

$$Q_d = a_1 - b_1 P + \theta \dot{P},$$
$$Q_s = -a_2 + b_2 P, \ \ a_j, b_j > 0,$$

where Q_d and Q_s are respectively the demand and supply for price P and a_j, b_j, and θ are parameters. Suppose

$$\dot{P}(t) = m(Q_d(t) - Q_s(t)), \ \ m > 0.$$

(i) Find the time path $P(t)$ and the equilibrium price P^*. (ii) What restriction on the parameter θ would ensure that as $t \to +\infty$, $P(t) \to P^*$.

4 Solve

(i) $\dot{x} + 2tx = 0; \ x(0) = 2$;

(ii) $\dot{x} + 6tx = 5t^2; \ x(0) = 6$.

5 Solve the following Keynesian Cross Model and discuss the value of the income $Y(t)$ as $t \to +\infty$:

$$\dot{Y}(t) = \gamma(D(t) - Y(t)), \ \ \gamma > 0,$$

in which γ is a parameter and the aggregate demand $D(t)$ is given by

---

[2] Domar (1944).

$$D = C + I + G.$$

Consumption $C(t)$, investment $I(t)$, and the level of government spending $G(t)$ are respectively given by: (1) the consumption function,

$$C(t) = C_0 + cY(t), 1 > c > 0;$$

(2) the exogenous investment, $I(t) = \bar{I}$; and (3) the exogenous government expending, $G(t) = \overline{G}$, where C_0, c, $\vec{I}$, and $\overline{G}$ are constant.

2.2 A Few Special Types

This section solves a few special types of differential equations.

Separable differential equations
An ordinary differential equation that can be written in the form

$$\dot{x}(t) = M(t)N(x),$$

is called a *separable* differential equation. This equation can be solved by integrating

$$\int \frac{1}{N(x)} dx = \int M(t) dt.$$

Example Solve $\dot{x}(t) = tx - 4t$. Multiplying both sides by dt and $1/(x - 4)$ and then integrating, we get

$$\ln|x - 4| = \frac{t^2}{2} + C,$$

where C is an arbitrary constant.

Example Solve the equation $2t\,dx + x\,dt = 0$. Dividing the equation by tx yields

$$-\frac{2}{x} dx = \frac{1}{t} dt.$$

Integrating both sides, we have $\ln(yt^{1/2}) = c$. Hence, the general solution is $x = Ct^{-1/2}$.

Example Consider the initial value problem

$$\dot{x} = x^2, \quad x(0) = x_0 .$$

The problem has a solution

$$x(t) = \frac{x_0}{1 - x_0 t} .$$

The function, x^2, is "well-behaved", but the solution is defined on the interval $(-\infty, 1/x_0)$ for $x_0 > 0$, on $(-\infty, +\infty)$ for $x_0 = 0$, and $(1/x_0, +\infty)$ for $x_0 < 0$. The importance of this example is that the solution is not always defined on all of R and the interval of definition of the solution varies with the initial condition. Furthermore, the solution becomes unbounded as t approaches $1/x_0$.

Example Consider

$$\dot{x} = x^2, \quad x(0) = x_0, \quad \text{with } x \ge 0 .$$

A solution is $x(t) = \left(t + 2\sqrt{x_0}\right)/4$. If $x_0 = 0$, then there is also the solution which is identically zero for all t. Therefore, this initial value problem does not have a unique solution through x_0 at zero.

The above two examples demonstrate some of the difficulties in guaranteeing the existence and uniqueness of differential equations.

We now consider a type of equations which can be transformed to the separable form. We have

$$\frac{dx}{dt} = \frac{M(t,x)}{N(t,x)} .$$

Suppose that M and N are homogeneous of the same degree.[3] This means that we can rewrite the above function as

$$\frac{dx}{dt} = \frac{M(1,x/t)}{N(1,x/t)} .$$

Introducing $y(t) = x(t)/t$ transforms the equation to

---

[3] If a function f possesses the property $f(tx, ty) = t^{\alpha} f(x, y)$ for some real number α, then f is said to be a homogeneous function of degree α.

$$t\frac{dy}{dt} + y = \frac{M(1,y)}{N(1,y)}.$$

This is a separable equation in y and t.

Example We are now concerned with

$$\dot{x} = \frac{tx}{x^2 - t^2}.$$

Introducing $y = x/t$ transform this equation to

$$y + t\frac{dy}{dt} = \frac{y}{y^2 - 1}.$$

The general solution of this separable equation is

$$y^2 = \ln t^{-2} + C.$$

As $y(t) = x(t)/t$, we have

$$x^2 = \left(\ln t^{-2} + C\right)t^2.$$

Example The square law of military strategy.
Two opposing forces x_0 and y_0 soldiers, respectively, have a military encounter in an open field without cover or concealment. Each soldier can see and fire upon the opposing soldiers without hindrance, like in old-fashioned naval battles. The soldiers of a given force do not impede one another, so that if the size of the force is doubled, its effectiveness is also doubled. The effectiveness of the x force is measured by the rate of decrease $\dot{y}$ force and vice versa. If the soldiers of both forces have equal fighting skill, it is reasonable to establish

$$\dot{x} = -ky, \quad \dot{y} = -kx.$$

The general solution to the problem is

$$x^2 - y^2 = C,$$

where C is an arbitrary constant. The solution to the initial value problem is

$$x^2 - y^2 = x_0^2 - y_0^2.$$

A question is that if a smaller force, say y, is annihilated, what is the size of the larger force, x_1, by the end of the battle? The answer is given by $x_1 = \sqrt{x_0^2 - y_0^2}$ with $y = 0$. The number $\sqrt{x_0^2 - y_0^2}$ measures the numerical advantage of the superior force. This is one form of the square law, proposed by Lanchester in 1916. Figure 2.2.1 shows the dynamics when $k = 0.01$, $x_0 = 135$, and $y_0 = 120$.

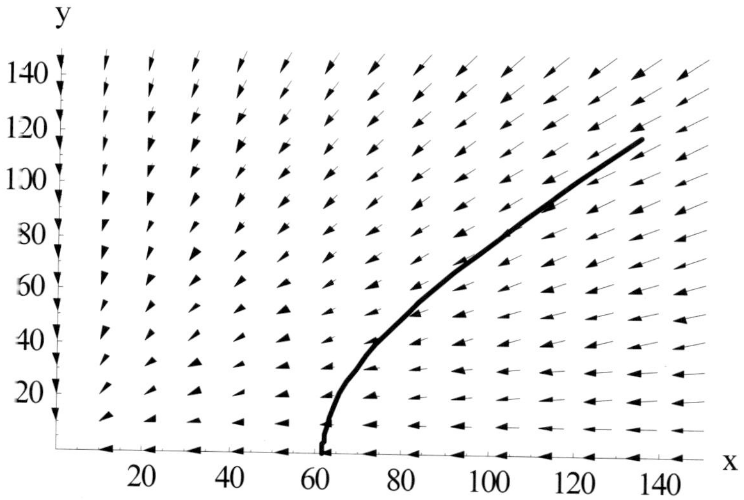

Fig. 2.2.1 A process of annihilating the enemy.

Example Find the demand function $Q = f(P)$ if point elasticity is -1 for all positive P. According to the definition of elasticity, we have

$$\frac{dQ}{dP}\frac{P}{Q} = -1.$$

Separating the variables and integrating, we have $Q = c / P$.

Exact differential equations

If $z = F(x, y)$ is a function of two variables with continuous first partial derivatives in a region R of the xy-plane, then its (total) differential is

$$dz = F_x dx + F_y dy.$$

Now if $F(x, y) = C$ where C is a constant, we have

$$F_x\,dx + F_y\,dy = 0\,.$$

In other words, given a one-parameter family of curves $F(x,y) = C$, we can generate a first-order differential equation by computing the total differential. Our problem is to turn the question around. For instance, given an equation

$$y^2dx + 2xydy = 0\,,$$

can we recognize that it is equivalent to the differential $d\left(xy^2\right) = 0$?

Definition 2.2.1 A differential expression

$$M(x,y)dx + N(x,y)dy = 0$$

is an *exact differential* in a region R of the xy-plane if it corresponds to the differential of some function $F(x,\ y)$. A first-order differential equation of the form

$$M(x,y)dx + N(x,y)dy = 0\,,$$

is said to be an *exact equation* if the expression on the left-hand side is an exact differential.

For instance,

$$x^2y^3dx + x^3y^2dy = 0\,,$$

is an exact equation because the left-hand side of the equation is an exact differential,

$$d\left(\frac{x^3y^3}{3}\right) = x^2y^3dx + x^3y^2dy\,.$$

The following theorem tells when a differential equation, $M(x,y)dx + N(x,y)dy = 0$ is exact.

Theorem 2.2.1 Suppose the functions M, N, M_y, and N_x are continuous on a connected region Ω. Then

$$M(x,y)dx + N(x,y)dy = 0$$

is an exact differential equation on Ω if and only if

$$M_y(x,y) = N_x(x,y)\,,$$

on Ω.

Proof: First, prove the necessity. If the expression

$$M(x,y)dx + N(x,y)dy$$

is exact, there exists some function $F(x, y)$ such that for all x and y in Ω

$$M(x,y)dx + N(x,y)dy = F_x dx + F_y dy.$$

Therefore

$$M(x,y) = F_x, \quad N(x,y) = F_y.$$

We thus have

$$\frac{\partial M}{\partial y} = \frac{\partial}{\partial y}\left(\frac{\partial F}{\partial x}\right) = \frac{\partial}{\partial x}\left(\frac{\partial F}{\partial y}\right) = \frac{\partial N}{\partial x}.$$

The equality of the mixed partials is a consequence of the continuity of the first partial derivatives of M and N.

We now show the sufficiency part of the theorem. We show that if $M_y(x,y) = N_x(x,y)$, we can construct a function $F(x,y)$ such that

$$F_x = M, \quad F_y = N.$$

To construct $F(x,y)$, we first integrate $F_x(x,y) = M(x,y)$ with respect to x and set

$$F(x,y) = \int M(x,y)dx + h(y). \tag{2.2.1}$$

The function $h(y)$ is the constant of integration. We thus have $F_x = M$. We now determine $h(y)$ so that $F_y = N$. Taking the partial derivatives on both sides of Eq. (2.2.1) with respect to y yields

$$F_y(x,y) = \int M_y(x,y)dx + h'(y).$$

From this equation and $F_y = N$ we solve

$$h'(y) = \int M_y(x,y)dx - N(x,y). \tag{2.2.2}$$

If

$$\int M_y(x,y)dx - N(x,y)$$

is independent of x, then we can integrate the right-hand side of Eq. (2.2.2) to solve $h(y)$. Since

$$\frac{\partial}{\partial x}\left\{\int M_y(x,y)dx - N(x,y)\right\} = M_y(x,y) - N_x(x,y) = 0,$$

we see

$$\int M_y(x,y)dx - N(x,y)$$

is independent of x.

Example Consider

$$\frac{dy}{dx} = \frac{e^x \cos y - 2xy}{e^x \sin y + x^2}.$$

It is straightforward to test that this equation is exact if we let

$$M = e^x \cos y - 2xy, \quad N = -e^x \sin y - x^2.$$

We calculate

$$F(x,y) = \int M(x,y)dx = \int \left(e^x \cos y - 2xy\right)dx$$

$$= e^x \cos y - x^2 y + h(y).$$

According to Eq. (2.2.2), we have

$$h'(y) = \int \left(-e^x \sin y - 2x\right)dx + \left(e^x \sin y + x^2\right) = 0.$$

This means that we can use any constant for $h(y)$. Hence,

$$F(x,y) = e^x \cos y - x^2 y + C_1,$$

where C_1 is an arbitrary constant. Therefore, the solution is $e^x \cos y - x^2 y = C$, where C is an arbitrary constant.

For $r(x,y)dx + s(x,y)dy = 0$, when $r_y(x,y) = s_x(x,y)$ is not held, sometimes we may transform the original differential equation to the form $M(x,y)dx + N(x,y)dy = 0$ which is exact. One method is to multiply the equation by $I(x,y)$ to create a new equation

$$Ir\,dx + Is\,dy = 0.$$

If we can find such a function I that $(Ir)_y = (Is)_x$, then the above procedure can be applied to solve

$$Ir\,dx + Is\,dy = 0.$$

Example Consider

$$\left(x^2 y + y^2\right)dx + \left(x^3 + 2xy\right)dy = 0.$$

This equation is not exact. To transform it into an exact form, we multiply the equation by $I(x,y) = x^m y^n$ where m and n are to be determined.[4] The condition of $(Ir)_y = (Is)_x$ is given by

$$(1+n)x^{2+m}y^n + (2+n)x^m y^{1+n}$$

$$= (3+m)x^{2+m}y^n + 2(1+m)x^m y^{1+n},$$

or

$$(-2+n-m)x^{2+m}y^n + (n-2m)x^m y^{1+n} = 0.$$

If

$$-2+n-m=0,\quad n-2m=0,$$

that is, $m=2$ and $n=4$, the new equation

$$\left(x^4y^5 + x^2y^6\right)dx + \left(x^5y^4 + 2x^3y^5\right)dy = 0,$$

is exact. Check that

$$\frac{x^5y^5}{5} + \frac{x^3y^6}{3} = C,$$

is the general solution.

Example Consider

$$\left(x^2 + y^2 + 1\right)dx + (xy + y)dy = 0.$$

This equation is not exact. The method with $I(x,y) = x^m y^n$ does not work for it. We might try to multiply the equation by a function $f(x)$ and try to find an f such that $(fr)_y = (fs)_x$. The condition $(fr)_y = (fs)_x$ is

---

[4] This is only one possible form for a try. The concrete form $x^m y^n$ often does not work.

$$f = (x+1)\frac{df}{dx}.$$

We have

$$f(x) = x + 1.$$

The new equation

$$(x+1)(x^2 + y^2 + 1)dx + (x+1)(xy + y)dy = 0$$

is exact. Check that the general solution is

$$\frac{x^4}{4} + \frac{x^2 y^2}{2} + \frac{x^2}{2} + \frac{x^3}{3} + xy^2 + x + \frac{y^2}{2} = C.$$

The Bernoulli equation

The following type of nonlinear equations is reducible to the linear form

$$\dot{x}(t) + u(t)x(t) = w(t)x^m(t), \tag{2.2.3}$$

where $m \neq 0, 1$. Introduce $z \equiv x^{1-m}$. Then Eq. (2.1.3) is written as

$$\dot{z}(t) + (1-m)u(t)z(t) = (1-m)w(t).$$

This belongs to the general type. The solution of this equation is

$$z(t) = e^{-(1-m)) \int u dt}\left(A + (1-m) \int we^{(1-m) \int u dt} dt \right).$$

Example The Solow model.[5]

We now examine the Solow model. It should be mentioned that mainly due to this model, Solow obtained the Nobel Prize in economics. We will explain the economic mechanism of this model in detail in Chap. 3.

When the production function takes the Cobb-Douglas form, the dynamics of per-capita capital $k(t)$ is governed by

$$\dot{k} = sk^{\alpha} - \delta k, \quad 0 < s, \delta, \alpha < 1,$$

where s, δ, and α are parameters. Introducing $z = x^{1-\alpha}$ transforms the Solow model into

$$\dot{z} + (1-\alpha)\delta z = (1-\alpha)s.$$

---

[5] See Solow (1956) and Swan (1956).

Its solution is

$$z(t) = \left(z(0) - \frac{s}{\delta} \right) e^{-(1-\alpha)\delta t} + \frac{s}{\delta}.$$

Substituting $z = x^{1-\alpha}$ into the above solution yields the final solution

$$k(t) = \left[\left(k(0)^{1-\alpha} - \frac{s}{\delta} \right) e^{-(1-\alpha)\delta t} + \frac{s}{\delta} \right]^{1/(1-\alpha)}.$$

We see that as

$$t \to +\infty, \; k(t) \to (s/\delta)^{1/(1-\delta)}.$$

We depict the Solow model in Fig. 2.2.2. Figure 2.2.2i shows the motion of $k(t)$ and $f(t)$. It can be seen that the growth rate gradually declines till the system approaches its long-term equilibrium. Figure 2.2.2ii depicts the mechanism of determination of the equilibrium. The per capita capital grows if it is on the left of the equilibrium value because $sf - \delta k$ is positive; $k(t)$ declines on the right of the equilibrium value.

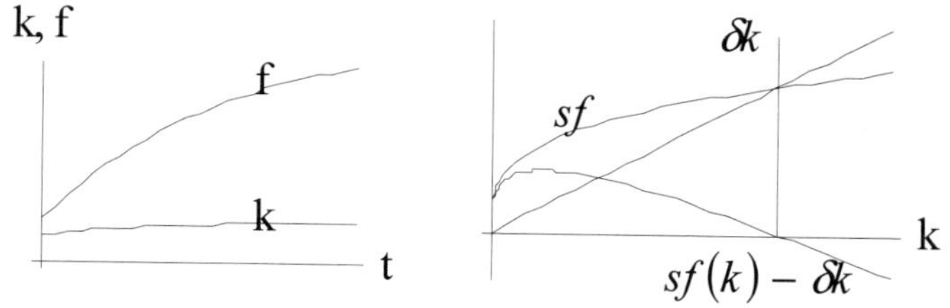

i) motion of k and f ii) equilibrium determination

Fig. 2.2.2 The Solow growth model.

Exercise 2.2

1 Find solution of $3tx^2\, dt - 6 \sin x\, dx = 0$.

2 Solve the *logistic differential equation* $\dot{N}(t) = aN(t)(1 - bN(t))$, where a and b are positive parameters and $N(t)$ is the population at time t.

3 Solve $\dot{x} = x \cos t - tx$ with $x(0) = 1$.

4 Solve the following equations

(i) $\left(x + te^{x/t}\right)dt - t\,dx = 0$;
(ii) $2xy\,dx + \left(x^2 + y^2\right)dy = 0$.

5 Solve the following equations
(i) $\left(\sin x + 3y^2e^x - 2xy\right)dx + \left(6ye^x - x^2\right)dy = 0, \; y(0) = 0$;
(ii) $2xy\,dx + \left(x^2 + y^2\right)dy = 0, \; y(1) = 0$.

6 Use $I = f(y)$ as an integrating factor to solve

$$\left(xy + x\right)dx + \left(x^2 + x^2 - 1\right)dy = 0.$$

7 Find the demand function $Q = f(P)$ if point elasticity is $-k$ for all positive P, where k is positive.

8 Given the following production function and parameters for the Solow growth model

$$f(k) = Ak^{\alpha}, \; A = 4, \; \alpha = 0.28, \; s = 0.15, \; \delta_k = 0.2, \; n = 0.03.$$

(i) Find the equilibrium of the dynamics

$$\dot{k} = sA\dot{k} - (n + \delta_k)k;$$

and (ii) determine whether the equilibrium is stable or unstable.

2.3 Second-Order Linear Differential Equations

We are now concerned with linear second-order differential equations with constant coefficients and constant term. Such an equation is generally given as

$$\ddot{x}(t) + a_1\dot{x}(t) + a_2x(t) = b, \tag{2.3.1}$$

where a_1, a_2, and b are constants. Corresponding to Eq. (2.3.1), we define the *characteristic equation* (or *auxiliary equation*)

$$\rho^2 + a_1\rho + a_2 = 0, \tag{2.3.2}$$

which has two *characteristic roots* given by

$$\rho_{1,2} = \frac{-a_1 \pm \sqrt{a_1^2 - 4a_2}}{2}.$$

We now give the solution according to the two roots.

Case 1 (distinct real roots) When $a_1^2 > 4a_2$, the two roots, ρ_1 and ρ_2 are real and distinct. The solution of Eq. (2.3.1) is $x(t) = A_1 e^{\rho_1 t} + A_2 e^{\rho_2 t} + x_0(t)$, where A_1 and A_2 are constants and

$$x_0(t) = \begin{cases} \frac{b}{a_2}, & a_2 \neq 0, \\ \frac{b}{a_1} t, & a_2 = 0, \ a_1 \neq 0, \\ \frac{b}{2} t^2, & a_1 = a_2 = 0. \end{cases} \tag{2.3.3}$$

Example It is straightforward to calculate the solution of

$$\ddot{x} + \dot{x} - 2x = -10$$

as

$$x(t) = A_1 e^t + A_2 e^{-2t} + 5.$$

If we have two initial conditions, for instance, $x(0) = 12$ and $\dot{x}(0) = -2$, then we can determine A_1 and A_2 with $A_1 = 4$ and $A_2 = 3$. The definite solution with the two initial conditions is

$$x(t) = 4e^t + 3e^{-2t} + 5.$$

Case 2 (two equal roots) When $a_1^2 = 4a_2$, the two roots, ρ_1 and ρ_2 are equal (to $-a_1/2$). The solution of Eq. (2.3.1) is

$$x(t) = (A_1 + A_2 t)e^{\rho t} + x_0(t),$$

where A_1 and A_2 are constants and $x_0(t)$ is defined in Eq. (2.3.3).

Example It is straightforward to calculate the definite solution to

$$\ddot{x} + 6\dot{x} + 9x = 27, \quad x(0) = 5, \ \dot{x}(0) = -5,$$

as $x(t) = (2 + t)e^{-3t} + 3$.

Case 3 (complex roots) When $a_1^2 < 4a_2$, the two roots, ρ_1 and ρ_2 are complex. Let

$$\rho_{1,2} = \alpha \pm \beta i, \ \alpha \equiv -\frac{a_1}{2}, \ \beta \equiv \frac{\sqrt{4a_2 - a_1^2}}{2}.$$

The solution of Eq. (2.3.1) is

$$x(t) = (A_1 \cos \beta t + A_2 \sin \beta t)e^{\alpha t} + x_0(t),$$

where A_1 and A_2 are constants and $x_0(t)$ is defined in Eq. (2.3.3).

Example We directly solve

$$\ddot{x} + 2\dot{x} + 17x = 34,$$

as

$$x(t) = (A_1 \cos 4t + A_2 \sin 4t)e^{-t} + 2.$$

Substituting the two initial conditions, $x(0) = 3$ and $\dot{x}(0) = 11$ into the above solution yields

$$A_1 + 2 = 3, \ -A_1 + 4A_2 = 11.$$

Hence, $A_1 = 1$ and $A_3 = 3$. The definite solution is

$$x(t) = (\cos 4t + 3\sin 4t)e^{-t} + 2.$$

Example We now consider an interaction of inflation and unemployment.[6] Denote the rate of inflation $p(t)$ (which is defined as $\dot{P}/P$, P being the price). The *expectations-augmented* version of the *Phillips relation* assumes the following relationship between the rate of inflation, the unemployment rate, $U(t)$, and the expected rate of inflation, $\pi(t)$

$$p = a - bU + h\pi, \ (0 < h \le 1), \tag{2.3.4}$$

where a, b, and h are parameters. The *adaptive expectations hypothesis* establishes a rule of the expected rate of inflation as follows

$$\dot{\pi} = j(p - \pi), \ 0 < j \le 1, \tag{2.3.5}$$

---

[6] The model and its solution are based on Chiang (1984: 535–538).

which states that if the actual rate of inflation exceeds the expected rate of inflation, then the expected rate of inflation tends to rise.

Denote the nominal money balance by M and its rate of growth by $m = \dot{M}/M$. The model contains a feedback from inflation to unemployment

$$\dot{U} = -k(m - p), \ \ k > 0, \tag{2.3.6}$$

where

$$m - p = (\dot{M}/M - \dot{P}/P),$$

is the rate of growth of real-money balance. The model consists of three equations, Eqs. (2.3.4), (2.3.5), and (2.3.6), with three variables, p, U and π. We now show that the dynamics can be described by a second-order linear differential equation.

First, substitute Eq. (2.3.4) into Eq. (2.3.5)

$$\dot{\pi} = j(a - bU) + j(h - 1)\pi \ .$$

Taking the above equation with respect to t, we have

$$\ddot{\pi} = - jb\dot{U} + j(h - 1)\dot{\pi} \ .$$

Substituting Eq. (2.3.4) into this then results in

$$\ddot{\pi} + (bk + j - jh)\dot{\pi} + jbk\pi = jbkm \ , \tag{2.3.7}$$

where we use

$$p = \frac{\dot{\pi}}{j} + \pi \ .$$

This equation belongs to the type of Eq. (2.3.1). Hence, we can solve Eq. (2.3.7). Once we solve $\pi(t)$, we determine

$$p = \frac{\dot{\pi}}{j} + \pi \, , \ \ U = \frac{a - p + h\pi}{b} \ .$$

We depict an interaction between inflation and unemployment with specified values of the parameters.[7]

[7] We specify: $a = 0.9$, $b = 0.1$, $h = 0.3$, $m = 2.9$, $j = 1$ and $k = 0.6$.

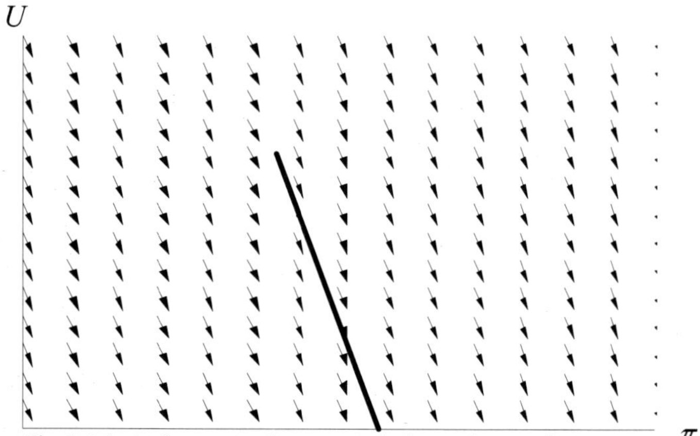

Fig. 2.3.1 An interaction between inflation and unemployment.

Example Denote $u(t)$ a utility function of wealth. At any wealth level, x, the *Arrow-Pratt measure of absolute risk aversion*, $\mu(t)$, equals $-u''(x)/u'(x)$. The function $\mu(t)$ is the percent rate of change of u' at x; it is a measure of concavity of the utility function u. We are interested in utility functions, which have constant risk aversion a. That is

$$-\frac{u''(x)}{u'(x)} = a \text{ , or } u''(x) + au'(x) = 0 .$$

This belongs to Eq. (2.3.2). The characteristic equation has two roots, 0 and $-a$. The solution of the differential equation is

$$u(x) = A_1 + A_2 e^{-at} ,$$

where A_1 and A_2 are constant.

Exercise 2.3

1 Solve

(a) $\ddot{x} + 4\dot{x} + 8x = 2, \ \ x(0) = 2, \ \ \dot{x}(0) = 2$;

(b) $\ddot{x} + 5x = 2, \ \ x(0) = 3, \ \ \dot{x}(0) = 5$.

2 Let the demand and supply functions for a single commodity be given by

$$Q_d(t) = 42 - 4P(t) - 4\dot{P}(t) + \ddot{P}(t),$$

$$Q_s(t) = -6 + 8P(t),$$

where P is the price. We assume that the demand is related to the price expectations, which are measured by $\dot{P}$ and $\ddot{P}$. Assuming market clearance at any point of time, that is, $Q_d(t) = Q_s(t)$, determine the paths of $P(t)$, $Q_d(t)$, and $Q_s(t)$ with $P(0) = 6$ and $\dot{P}(0) = 4$.

3 Let the three equations in the model of interactions between inflation and unemployment be given as follows

$$p = \frac{1}{6} - 3U + \pi,$$

$$\dot{\pi} = \frac{3}{4}(p - \pi),$$

$$\dot{U} = -\frac{1}{2}(m - p).$$

Determine the differential equation for π and find its general solution.

2.4 Higher-Order Linear Differential Equations

The previous sections solve first- and second-order linear differential equations. This section examines any nth order linear differential equations as

$$a_n(t)x^{(n)} + a_{n-1}(t)x^{(n-1)} + \ldots + a_0(t)x = h(t). \tag{2.4.1}$$

Throughout this section, it is assumed that $a_n, a_{n-1}, \cdots, a_0$ and h are continuous on an interval (a, b) and a_n is not zero for any t in (a, b). If $h(t) = 0$, the system is called *homogeneous*; otherwise it is called *nonhomogeneous*. If h is not zero, the equation

$$a_n(t)x^{(n)} + a_{n-1}(t)x^{(n-1)} + \ldots + a_0(t)x = 0, \tag{2.4.2}$$

is called the *corresponding homogeneous equation* of Eq. (2.4.1). If initial conditions on $x^{(n-1)}, x^{(n-2)}, \ldots, x'$ and x for Eq. (2.4.1) are known, then we call the system an *initial value problem*.

Theorem 2.4.1 Suppose $a_n, a_{n-1}, \dots, a_0$ and h are continuous on an interval (a, b) containing t_0 and a_n is not zero for any t in (a, b). Then the initial value problem has one and only one solution on the interval (a, b).

It can be shown that the solutions to an nth order homogeneous linear differential equation (2.4.2) on an interval where $a_n, a_{n-1}, \dots, a_0$ and h are continuous and a_n is not zero form a vector space of dimension n.[8] Hence, if we find a set of solutions to the homogeneous equation that forms a basis for the vector space, then a linear combination of these solutions is the general solution to Eq. (2.4.2). A set of n linearly independent solutions, $x_1, x_2, \dots, x_n$, each of which is a solution of an nth order homogeneous linear differential equation (2.4.2) is called a *fundamental set of solutions* for the homogeneous equation. If $x_1, x_2, \dots, x_n$ form a fundamental set of solutions (2.4.2), then the general solution to Eq. (2.4.2) is given by

$$x_H = \sum_{i=1}^{n} c_i x_i = x_c c, \qquad (2.4.3)$$

where c_i are constant, and

$$x_c \equiv [x_1, x_2, \dots x_n], \quad c \equiv [c_1, c_2, \dots c_n]^T.$$

We now show how to judge whether $x_1(t), x_2(t), \dots, x_n(t)$ is linearly dependent or independent. Suppose that we have n functions $x_1(t), x_2(t), \dots, x_n(t)$, each of which has $(n-1)$ derivatives on an interval (a, b). Consider a linear combination of these functions that is equal to the zero function

$$c_1 x_1(t) + x_2(t) + \cdots + c_n x_n(t) = 0.$$

Taking the derivatives of this equation $(n-1)$ times, we obtain a system of n equations

$$c_1 x_1^{(i)}(t) + x_2^{(i)}(t) + \cdots + c_n x_1^{(i)}(t) = 0, \quad i = 0, 1, \cdots, n-1,$$

or in the matrix form

[8] See Chap. 4 in Peterson and Sochacki (2002).

$$\begin{bmatrix} x_1(t) & x_2(t) & \cdots & x_n(t) \\ x_1'(t) & x_2'(t) & \cdots & x_n'(t) \\ \vdots & \vdots & \vdots & \vdots \\ x_1^{(n-1)}(t) & x_2^{(n-1)}(t) & \cdots & x_n^{(n-1)}(t) \end{bmatrix}_{nxn} \begin{bmatrix} c_1 \\ c_2 \\ \vdots \\ c_n \end{bmatrix} = 0 .$$

If there is some t in (a, b) for which this system has only the trivial solution, $c = 0$, then $x_1(t), x_2(t), ..., x_n(t)$ will be linearly independent. Having such a t is the same as having a t in (a, b) for which the $n \times n$ matrix in Eq. (2.4.4) is nonsingular. We call the determinant of this $n \times n$ matrix *the Wronskian* of the functions $x_1(t), x_2(t), ..., x_n(t)$, and denote by

$$W(x_1(t), x_2(t), ..., x_n(t)).$$

That is

$$W(x_1(t), x_2(t), \cdots, x_n(t)) = \begin{vmatrix} x_1(t) & x_2(t) & \cdots & x_n(t) \\ x_1'(t) & x_2'(t) & \cdots & x_n'(t) \\ \vdots & \vdots & \vdots & \vdots \\ x_1^{(n-1)}(t) & x_2^{(n-1)}(t) & \cdots & x_n^{(n-1)}(t) \end{vmatrix} .$$

Lemma 2.4.1 Suppose $x_1(t), x_2(t), ..., x_n(t)$ are functions each of which has $(n-1)$ derivatives on an interval (a, b). If the Wronskian of these functions is nonzero for some t in (a, b), then $x_1(t), x_2(t), ..., x_n(t)$ are linearly independent on (a, b).

Example Consider three functions, e^t, $\sin t$ and $\cos t$ on $(-\infty, +\infty)$. It is straightforward to calculate

$$W(e^t, \sin t, \cos t) = 2e^t .$$

Since at $t = 0$, $W \neq 0$, we conclude that e^t, $\sin t$ and $\cos t$ on $(-\infty, +\infty)$ are linearly independent.

It should be noted that when $W = 0$ at $t = -\infty$. Lemma 2.4.1 guarantees that if the Wronskian is nonzero for some t, then the functions are linearly independent. It does not mean that linearly independent functions have their Wronskian being nonzero for some t.[9]

Example We now show that for equation $x^{(3)} + x' = 0,\ 1, \cos 2t, \sin 2t$ is a fundamental set of solutions. Substituting respectively $1, \cos 2t, \sin 2t$ into the equation confirms that they are solutions to $x^{(3)} + x' = 0$. Since

$$W(1, \sin 2t, \cos 2t) = 8,$$

they indeed form a fundamental set of solutions. Hence the general solution to $x^{(3)} + x' = 0$ is

$$c_1 + c_2 \cos 2t + c_3 \sin 2t .$$

Lemma 2.4.2 Suppose $x_1(t), x_2(t), ..., x_n(t)$ are functions each of which has $(n-1)$ derivatives on an interval (a, b). If the Wronskian of these functions is zero for some t in (a, b), then $x_1(t), x_2(t), ..., x_n(t)$ are linearly dependent on (a, b).

Theorem 2.4.2 Suppose that $x_1(t), x_2(t), ..., x_n(t)$ form a fundamental set of solutions to Eq. (2.4.2) and that $x_p(t)$ is a solution to the nonhomogeneous system (2.4.1). Then every solution to Eq. (2.4.1) has the form

$$x = \sum_{i=1}^{n} c_i x_i + x_p = x_c c + x_p .$$

We call $x_p(t)$ a *particular solution* to the nonhomogeneous system.

Exercise 2.4

1 Find the Wronskian of $1, t, ..., t^{n-1}$ on $(-\infty, +\infty)$.

2 Show that $e^{rt}, xe^{rt}, \cdots, t^{n-1}e^{rt}$ are linearly independent on $(-\infty, +\infty)$.

---

[9] As shown in Chap. 2 in Peterson and Sochacki (2002), the Wronskian of the two functions, $f(x) = x^2$ and $g(x) = x|x|$ is zero for every t, and f and g are linearly independent on $(-\infty, +\infty)$.

2.5 Higher-Order Linear Differential Equations with Constant Coefficients

This section generalizes the analysis in Sec. 2 to any finite dimension. When $a_n, a_{n-1}, \dots, a_0$ in

$$a_n x^{(n)} + a_{n-1} x^{(n-1)} + \dots + a_0 x = h(t) \tag{2.5.1}$$

are constant real numbers, the equation is called a *constant coefficient* nth order linear differential equation. To solve this equation, we first find the general solution to the homogeneous equation

$$a_n x^{(n)} + a_{n-1} x^{(n-1)} + \dots + a_0 x = 0. \tag{2.5.2}$$

To solve this equation, we try solutions of the form of $x(t) = e^{\rho t}$. Substituting $x(t) = e^{\rho t}$ into Eq. (2.5.2) yields

$$p(\rho) e^{\rho t} = 0,$$

where

$$p(\rho) = a_n \rho^n + a_{n-1} \rho^{n-1} + \dots + a_0. \tag{2.5.3}$$

Since $e^{\rho t} \neq 0$, for $p(\rho) e^{\rho t} = 0$ we must have $p(\rho) = 0$. We call the polynomial of degree n and $p(\rho)$ the *characteristic polynomial* or *auxiliary polynomial* for Eq. (2.5.2) and call the equation

$$p(\rho) = 0,$$

the *characteristic equation* or *auxiliary equation*. The following theorem gives necessary and sufficient conditions that all the roots of $p(\rho) = 0$ have negative real parts.

Lemma 2.5.1 (The Routh-Hurwitz theorem) Necessary and sufficient conditions for all the roots of the polynomial equation with real coefficients

$$a_n \rho^n + a_{n-1} \rho^{n-1} + \dots + a_0 = 0, \quad a_n > 0,$$

to have negative real parts is given by the simultaneous verification of the following inequalities

$$\Delta_i > 0, \quad i = 1, 2, \cdots, n,$$

where

$$\Delta_1 = a_{n-1}, \quad \Delta_2 = \begin{vmatrix} a_{n-1} & a_{n-3} \\ a_n & a_{n-2} \end{vmatrix}, \quad \Delta_3 = \begin{vmatrix} a_{n-1} & a_{n-3} & a_{n-5} \\ a_n & a_{n-2} & a_{n-4} \\ 0 & a_{n-1} & a_{n-3} \end{vmatrix}, \cdots,$$

$$\Delta_n = \begin{vmatrix} a_{n-1} & a_{n-3} & a_{n-5} & a_{n-7} & \cdots & 0 \\ a_n & a_{n-2} & a_{n-4} & a_{n-6} & \cdots & 0 \\ 0 & a_{n-1} & a_{n-3} & a_{n-5} & \cdots & 0 \\ 0 & a_n & a_{n-2} & a_{n-4} & \cdots & 0 \\ 0 & 0 & a_{n-1} & a_{n-3} & \cdots & 0 \\ \cdots & \cdots & \cdots & \cdots & \cdots & \cdots \\ 0 & 0 & 0 & 0 & \cdots & a_0 \end{vmatrix}.$$

Example For

$$a_3\rho^3 + a_2\rho^2 + a_1\rho + a_0 = 0, \quad a_3 > 0,$$

the conditions are

$$\Delta_1 = a_2 > 0, \quad \Delta_2 = \begin{vmatrix} a_2 & a_0 \\ a_3 & a_1 \end{vmatrix} = a_1a_2 - a_0a_3 > 0,$$

$$\Delta_3 = \begin{vmatrix} a_2 & a_0 & 0 \\ a_3 & a_1 & 0 \\ 0 & a_2 & a_0 \end{vmatrix} = a_0(a_1a_2 - a_0a_3) > 0.$$

Example Consider

$$x''' - 4x' = 0.$$

The characteristic equation for this equation is

$$\rho^3 - 4\rho = 0.$$

It has three distinct real roots, 0, 2, and -2. The corresponding solutions $e^{\rho t}$ are e^{0t}, e^{2t}, and e^{-2t}. The Wronskian of these three functions is

$$W(1, e^{2t}, e^{-2t}) = -16.$$

Hence, the three solutions are linearly independent. The general solution to $x''' - 4x' = 0$ is

$$x = c_1 + c_2 e^{2t} + c_3 e^{-2t} .$$

The above example shows that the functions, $e^{\rho_i t}$ ($i = 1, 2, 3$), corresponding to the distinct real roots of the characteristic equation have nonzero Wronskian and hence are linearly independent. The following theorem shows that this is true for the general case.

Theorem 2.5.1 If $\rho_1, \rho_2, \cdots, \rho_m$ ($m \le n$) are distinct real roots for the characteristic equation, $p(\rho) = 0$, of Eq. (2.5.2), then $e^{\rho_1}, e^{\rho_2}, \cdots, e^{\rho_m}$ are linearly independent solutions of Eq. (2.5.2). If $m = n$, the general solution to Eq. (2.5.2) is

$$\sum_{i=1}^{n} c_i e^{\rho_i t} .$$

The following theorem shows the case of characteristic equations with complex roots.

Theorem 2.5.2 If $a \pm bi$ are roots of the characteristic equation, $p(\rho) = 0$, of Eq. (2.5.2), then $e^{at} \sin bt$ and $e^{at} \cos bt$ are two linearly independent solutions of the differential equation.

Example Consider

$$x''' + x'' - x' + 15x = 0 .$$

The characteristic equation for this equation is

$$\rho^3 + \rho^2 - \rho + 15 = 0 .$$

As shown below, the cubic equation has three roots, -3, $1 + 2i$, and $1 - 2i$. The general solution is therefore

$$x = c_1 e^{-3t} + c_2 e^t \sin 2t + c_3 e^t \cos 2t .$$

We now provide the formula for solving a cubic equation

$$\rho^3 + p\rho^2 + q\rho + r = 0 . \qquad (2.5.4)$$

Under the transformation

$$x \equiv \rho + \frac{p}{3}, \quad a \equiv \frac{3q - p^2}{3}, \quad b \equiv \frac{2p^3 - 9qp + 27r}{27},$$

Eq. (2.5.4) becomes

$$x^3 + ax + b = 0.$$

This equation has three roots

$$x_1 = A + B, \quad x_{2,3} = -\frac{A + B}{2} \pm \left(\frac{A - B}{2}\right)\sqrt{-3},$$

where

$$s \equiv \frac{b^2}{4} + \frac{a^3}{27}, \quad A \equiv \left(-\frac{b}{2} + \sqrt{s}\right)^{1/3}, \quad B \equiv \left(-\frac{b}{2} - \sqrt{s}\right)^{1/3}.$$

We can classify the three solutions of the equation,

$$\rho^3 + p\rho^2 + q\rho + r = 0,$$

according to the value of s. In the case of $s < 0$, the three solutions are real and distinct; in the case of $s = 0$, the three solutions are real and at least two are equal; and in the case of $s > 0$, one solution is real and the other two are complex conjugate values.

We now show how to solve the differential equation (2.5.2) when its characteristic equation has repeated real roots or repeated complex roots. If the characteristic equation has a repeated root ρ of multiplicity m, except $e^{\rho t}$, we get $m - 1$ additional solutions by multiplying $e^{\rho t}$ by $t, t^2, \cdots, t^{m-1}$. Similarly, if $a \pm bi$ are roots of the characteristic equation of multiplicity m, then except $e^{at} \sin bt$ and $e^{at} \cos bt$, we get $2m - 2$ additional solutions by multiplying respectively $e^{at} \sin bt$ and $e^{at} \cos bt$ by $t, t^2, \cdots, t^{m-1}$.

Example The characteristic polynomial of a constant coefficient homogeneous linear differential equation is

$$(\rho - 3)(\rho - 1)^2(\rho^2 + 4\rho + 13)^2.$$

The roots of the characteristic equation are 3, 1, and $-2 \pm 3i$. The root 1 has multiplicity 2 and the complex roots $-2 \pm 3i$ have multiplicity 2. This leads to the general solution

$$x = c_1 e^{3t} + c_2 e^{t} + c_3 te^{t} + c_4 e^{-2t} \sin 3t + c_4 e^{-2t} \cos 3t$$

$$+ c_5 te^{-2t} \sin 3t + c_6 te^{-2t} \cos 3t .$$

We now examine Eq. (2.5.1) when $h(t)$ is nonzero. We learned how to solve Eq. (2.5.2). From Theorem 2.3.2, we know that the general solution to Eq. (2.5.1) can be obtained by adding together the general solution x_H to Eq. (2.5.2) and a particular solution x_p of Eq. (2.5.1). Since we learned how to find x_H, we now try to find x_p. Here, we introduce the *method of undetermined coefficients* for finding the solution.[10] This method is based on the following theorem when $h(t)$ takes on some special forms.

Theorem 2.5.3 Suppose that ρ is a root of multiplicity m of the following characteristic equation

$$p(\rho) = 0,$$

of Eq. (2.5.2). Let k be a nonnegative integer and A and B be constant. (1) If ρ is real and $h(t)$ in Eq. (2.5.1) takes the form of $At^k e^{rt}$, then the differential equation (2.5.1) has a particular solution of the form

$$\left(A_k t^k + A_{k-1} t^{k-1} + \cdots + A_1 t + A_0\right) t^m e^{\rho t} .$$

(2) If $\rho = a + ib$ is imaginary and $h(t)$ in Eq. (2.5.1) takes the form of

$$(A \cos bt + B \sin bt) t^k e^{at} ,$$

then the differential equation (2.5.1) has a particular solution of the form

$$\left(A_k t^k + \cdots + A_1 t + A_0\right) t^m e^{at} \cos bt +$$

$$\left(B_k t^k + \cdots + A_1 t + A_0\right) t^m e^{at} \sin bt .$$

It should be noted that although this theorem is stated for a root ρ of multiplicity m, its conclusions are valid even if $m = 0$, that is, even if ρ is not a root of

$$p(\rho) = 0 .$$

[10] Refer to Chap. 4 in Peterson and Sochacki (2002) for other methods.

Example Consider

$$x'' - 2x' - 3x = 9t^2 .$$

The general solution of $x'' - 2x' - 3x = 0$ is

$$x_H = c_1 e^{3t} + c_2 e^{-3t} .$$

To find a special solution, try

$$x_p = At^2 + Bt + C .$$

Differentiating this candidate solution and plugging them into the original equation yields

$$2A - 4At - 2B - 3At^2 - 3Bt - 3C = 9t^2 .$$

Since the left- and right-hand sides of this equation are equal for all t, the coefficients of each power of t must be equal

$$-3A = 9,$$

$$-4A - 3B = 0,$$

$$2A - 2B - 3C = 0.$$

Hence,

$$A = -3, \quad B = 4, \quad C = -\frac{14}{3}.$$

The general solution is

$$x = c_1 e^{3t} + c_2 e^{-3t} - 3t^2 + 4t - \frac{14}{3}.$$

Example Solve the initial value problem

$$x''' + 4x' = 2t + 3\sin 2t - 3t^2 e^{2t},$$

$$x(0) = 1,\ x'(0) = 0,\ x''(0) = -1. \tag{2.5.5}$$

The roots of the characteristic equation are 0 and $\pm\, i$. Therefore, the general solution to

$$x''' + 4x' = 0 ,$$

is

$$x_H = c_1 + c_2 \cos 2t + c_3 \sin 2t .$$

We now try to find a special solution to Eq. (2.5.5). According to Theorem 2.5.3, we incorporate $(At + B)t$ for the $2t$ term into x_p, $Ct\cos 2t + Dt\sin 2t$ for $3\sin 2t$, and $(Et^2 + Ft + G)e^{2t}$ for $-3t^2e^{2t}$. Substituting

$$x_p = (At + B)t + Ct\cos 2t + Dt\sin 2t + (Et^2 + Ft + G)e^{2t}$$

into Eq. (2.5.5) yields

$$4(2At + B) + (-8C\cos 2t - 8D\sin 2t) + 16Et^2e^{2t} +$$

$$(16F + 32E)te^{2t} + (12E + 16F + 16G)e^{2t} = 2t + 3\sin 2t - 3t^2e^{2t}.$$

Equating the coefficients of the same functions on both sides of the above equation yields

$$A = \frac{1}{4},\ B = 0,\ C = 0,\ D = -\frac{3}{8},\ E = -\frac{3}{16},\ F = \frac{3}{8},\ G = -\frac{15}{64}.$$

We thus obtained the general solution $x = x_H + x_p$ to Eq. (2.5.5). Substituting the initial conditions into $x_H + x_p$ enables one to solve

$$c_1 = \frac{19}{16},\ c_2 = \frac{3}{64},\ c_3 = \frac{3}{64}.$$

Exercise 2.5

1 Determine the general solutions to the following differential equations

(i) $x''' - 5x'' + 6x' - 2x = 0$;

(ii) $x''' + 4x'' - x' - 4x = 0$;

(iii) $x^{(4)} + 2x''' + x'' = 0$;

(iv) $x''' + 4x'' - x' - 4x = 0$;

(v) $x''' - 2x' + 4x = 0$.

2 Solve the following initial value problems

(i) $x''' + 3x'' + 2x' = 0;\ x(0) = 1,\ x'(0) = 0,\ x''(0) = -1$;

(ii) $x''' - 5x'' + 3x' + 9x = 0;\ x(0) = 0,\ x'(0) = -1,\ x''(0) = 1$;

(iii) $x''' - x'' - 6x' = 0;\ x(0) = 1,\ x'(0) = 0,\ x''(0) = 2$.

3 Solve the following differential equations

(i) $x''' + 3x'' + 2x' = 3e^t$;

(ii) $x''' - 5x'' + 3x' + 9x = 5t^2 + \sin t$;

(iii) $x''' - x'' - 6x' = 4te^{2t}$.

Chapter 3

Scalar Nonlinear Differential Equations

We have already discussed some one-dimensional nonlinear differential equations in the previous chapter. We examined how to solve them, based on the special structures of these equations. We now use a simple model to illustrate the importance of nonlinear relations. Section 2.1 examined the Malthus equation, $\dot{N}(t) = aN(t)$, where $N(t)$ is the population at time t and a is a positive parameter. Such a population growth may be valid for a short time, but as shown in Fig. 2.1.1, it clearly cannot go on forever. There are limitations of natural resources that prevent population from limitlessly growing. The logistic growth model, which is defined by

$$\dot{N}(t) = aN(t)(1 - bN(t)),$$

takes account of checking effects of natural resources upon population growth. This differential equation is also known as the *logistic differential equation*. This equation is nonlinear because of the term $-aN^2$. It can be solved analytically by separating variables. This chapter is to introduce some other methods that are geometric in nature and that can quickly reveal some important qualitative properties of solutions.

In fact, the logistic differential equation is not proper for describing interactions between economic growth and population dynamics. From a long-term perspective, resources are changeable. To analyze how income affects population growth, Haavelmo suggested the following extension of Malthus' system[1]

$$\dot{N}(t) = aN(t)\left(1 - \frac{bN(t)}{Y(t)}\right), \quad a, b > 0,$$

[1] See Haavelmo (1954) or Zhang (1991).

$$Y = AN^{\beta}, \quad A > 0, \quad 0 < \beta < 1, \tag{3.0.1}$$

in which Y is real output. Substituting

$$Y(t) = AN(t)^{\beta}$$

into Eq. (3.0.1) yields

$$\dot{N}(t) = N\left(a - a_1 N^{1-\beta}\right), \tag{3.0.2}$$

where $a_1 \equiv ab / A$. The growth law is a generalization of the familiar logistic form widely used in biological population and economic analysis. To simulate the model, we specify the parameters as

$$a = 0.05, \quad b = 0.4, \quad \beta = 0.6, \quad A = 1.4. \tag{3.0.3}$$

From the initial condition $N(0) = 8$, we run the model for 100 years. Figure 3.0.1 depicts the motion of $N(t)$, $Y(t)$, and the income per capita, $y(t)$ $(\equiv Y(t)/N(t))$.

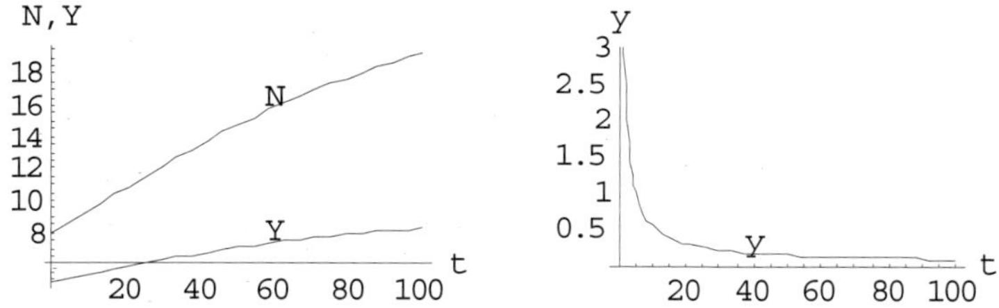

a) the population and the total product b) the income per capita

Fig. 3.0.1 The dynamics of a Malthusian economy.

As demonstrated in Fig. 3.0.1, the national product grows over time, but the income per capita (which is assumed to be the main factor for determining the level of consumption per capita) first decreases suddenly but afterwards steadily declines. This book introduces some economic mechanisms to avoid the decline of living standard – as observed in modern economic history, population growth may be associated with improvement in living standard. An obvious way to avoid the dismay implications of the Malthusian theory is to take account of technological change. At this initial stage, we specify the exogenous technological progress

$$A(t) = 1.4\exp(0.02t).$$

The productivity grows annually 2 percent. Under Eq. (3.0.3) except for the value of A, Fig. 3.0.1 shows the motion of the system. Comparing the two figures, we observe that with 2 percent annual technological change, the economy dramatically changes over 100 years. The economy is enlarged and the income per capita grows as the population grows.

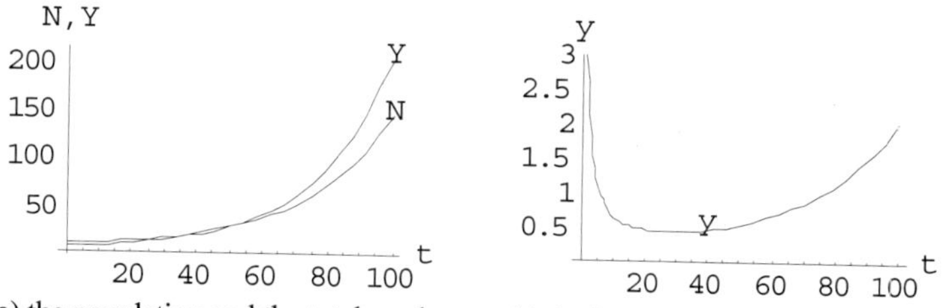

a) the population and the total product b) the income per capita

Fig. 3.0.2 The dynamics of the Malthusian economy with technological change.

3.1 Nonlinear Differential Equations

To avoid repeating concepts and theorems which are also valid for higher dimensions, this section examines systems of differential equations of any finite dimension.

Consider a dynamic system

$$\dot{x}(t) = f(x(t); \alpha), \tag{3.1.1}$$

where variables x, functions f, and parameters α

$$x = (x_1, x_2, \dots x_n)^T,$$

$$f = (f_1, f_2, \dots f_n)^T,$$

$$\alpha = (\alpha_1, \alpha_2, \dots \alpha_m)^T,$$

are vectors. We denote the solution of Eq. (3.1.1) by $\phi(t, x_0)$, where $x_0 \equiv x(0)$. Here, $\phi(t, x_0)$ is the m-parameter family of mappings:

$$\phi(t) := R^n \to R^n.$$

The two relationships, $\phi(0, x_0) = x_0$ and

$$\phi(t_1 + t_2, x_0) = \phi(t_2, \phi(t_1, x_0)),$$

hold. The set of points

$$\{\phi(t, x_0): -\infty < t < +\infty\} ,$$

is called the *trajectory through* x_0. The dynamical system (3.1.1) is (non-)linear if the vector field, $f(x;\alpha)$, is (non-)linear in x. The examination of the dynamical system usually includes determination of the equilibrium points, the periodic and quasi-periodic solutions, the chaotic behavior and the stability.

Definition 3.1.1 An equilibrium point x^* of an autonomous system is a constant solution of Eq. (3.1.1), that is, $x^* = \phi(t, x^*)$ for all t. In other words, x is an equilibrium point if $f(x) = 0$.

In general, the equilibrium point may not exist. As shown in Fig. 3.1.1, $\dot{x} = f_1(x)$ has no equilibrium point. If there exists a neighborhood about an equilibrium point x^*, in which there is no other equilibrium point, then x^* is called an *isolated equilibrium point*. In Fig. 3.1.1, $\dot{x} = f_3(x)$ has a (unique in this case) isolated equilibrium point. In Fig. 3.1.1, $\dot{x} = f_2(x)$ has nonisolated equilibrium points.

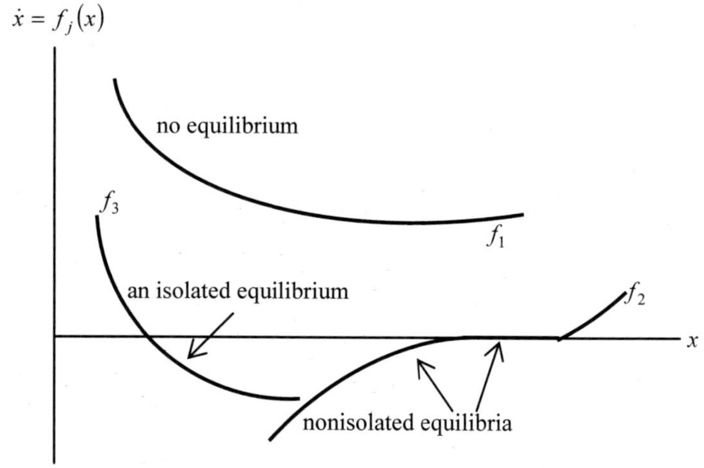

Fig. 3.1.1 Concepts of equilibrium points for $\dot{x} = f(x)$.

Definition 3.1.2 x^* is a periodic solution of the autonomous system, if for all t

$$\phi(t, x^*) = \phi(t + T, x^*),$$

for some minimum period $T > 0$.

Here, x^* is not unique since any point lying the periodic solution will satisfy the conditions. Changing x^* corresponds to changing the time origin. A periodic solution is isolated if it contains a neighborhood that possesses no other periodic solution. If the dynamical system is autonomous, an isolated periodic solution is referred as a *limit cycle.*

Definition 3.1.3 A quasi-periodic function is one that can be expressed as a countable sum of periodic functions

$$x(t) = \sum_i \varphi_i(t),$$

where $\varphi_i(t)$ has a minimum period T_i and frequency $\beta_i = 1/T_i$. Moreover, there exists a finite set of base frequencies $\{\beta_1^* \ \beta_2^*, \dots, \beta_p^*\}$ such that it is linearly independent and it forms a finite integral base for β_i.

There is no commonly accepted definition of chaos. From a practical point of view, chaos can be defined as bounded steady-state behavior that is not an equilibrium point, not periodic, and not quasi-periodic. In some sense, chaos may be defined as "stochastic behavior" in deterministic systems. The observable behavior of a dynamic system is called stochastic when the transition of the system from one state to another can only be given a probabilistic description as happens for truly random processes. Chaotic motion in continuous-time dynamical systems requires at least three-dimensional state space.

There are only a few types of differential equations which can be solved in "closed form", i.e., using elementary functions: the first order exact, linear, and homogeneous equations; the higher order linear equations with constant coefficients; the partial differential equations that are reducible to these by separation of variables. After this, one may find several methods: advanced theory, approximation of solutions on the computer by numerical analysis, and approximation of solutions by formulas.

Before stating an important theorem, we consider a function f on R^n. Let U be an open set in R^n. The function f on R^n is said to be *Lipschitz* on U if there exists a constant L such that

$$|f(x) - f(y)| \leq L|x - y|,$$

for all x and y in U. The constant L is called a *Lipschitz constant* for f. For instance, the function $f(x) = \sin x$ has Lipschitz constant $L = 1$. The primary fundamental theorem related to differential equations is known as the *Picard-Cauchy-Lipschitz theorem*, which is given as follows.

Theorem 3.1.1 Consider the system of equations

$$\dot{x}(t) = f(x(t), t).$$

Let the functions $f_j(x,t)$ satisfy the Lipschitz conditions in their arguments. Then there exists a unique solution $x = \phi(t, x_0)$ in the neighborhood of $t = t_0$ satisfying initial conditions $x_0 = x(0)$. Moreover, this solution is a continuous function of the initial conditions. If

$$\dot{x}(t) = f(x(t), t, \alpha),$$

where α is a parameter, and each $f_j(x,t,\alpha)$ also satisfies a Lipschitz condition uniformly in α in the neighborhood of α_0, and is continuous in α, then the same conclusions hold in the neighborhood of α_0. Moreover, the solution $\phi(t, x_0, \alpha)$ is a continuous function of α in this neighborhood.

The uniqueness implies that for a differential equation satisfying a Lipschitz condition, solutions to an initial value problem do not cross each other.

Example Consider the following equation[2]

$$\dot{x}(t) = x(t)^{2/3}, \quad x \in R, \quad t \in R.$$

---

[2] The example is from Arnold (1978: 14–15).

Let $t_0 = 0$ and $x_0 = 0$. Both $\phi(t) = 0$ and $\phi(t) = (t/3)^3$ are solutions to the initial value problem. The function $x(t)^{2/3}$ does not satisfy the Lipschitz condition near the origin.

Theorem 3.1.1 establishes the existence of a unique solution for a certain neighborhood of t_0. This means that the theorem only provides a local existence theorem.

Example Consider the following equation[3]

$$\dot{x}(t) = x(t)^2, \quad x \in R, \quad t \in R.$$

Let $t_0 = 1$ and $x_0 = 1$. Then $\phi(t) = -1/t$ is the solution to the initial value problem. On the other hand, the solution of this differential equation does not exist for $t_0 = 0$. This example illustrates that the global existence of a solution needs a careful study when one finds a local solution.

The concept of stability is defined as follows.

Definition 3.1.4 Consider the system

$$\dot{x}(t) = f(x(t), t).$$

The solution $\phi(t, x_0)$ defined in $[t_0, \infty]$ is *stable* if, for any given $\varepsilon > 0$, there exists $\delta > 0$ such that if x_0^* is any given vector satisfying $|x_0 - x_0^*| < \delta$, then the solution $\phi(t, x_0^*)$ with the initial conditions x_0^* exists in $[t_0, \infty]$ and satisfies

$$|\phi(t, x_0) - \phi(t, x_0^*)| < \varepsilon,$$

for all $t \geq t_0$.

The stability defined above is called *Lyapunov stability* or *solution stability*. It says that how small is the permitted deviation measured by ε, there exists a nonzero tolerance, measured by δ, in the initial conditions when the system is activated, allowing it to run satisfactorily.

---

[3] The example is adapted from Takayama (1994: 330–331).

Example The general solution of $\dot{x} = ax$ with $a < 0$ is $x(t) = x(0)e^{at}$. This solution is stable. As

$$\left|\phi(t, x_0) - \phi^*(t, x_0^*)\right| = \left|x_0 - x_0^*\right|e^{at}$$

we can choose $\delta = \varepsilon e^{-at_0}$ to guarantee the stability of the solution.

Another property of the dynamic system is that a particular solution may become less stable as time goes on. It is possible that a system's sensitivity to disturbance might increase indefinitely with time although it remains technically stable, the symptom being that $\delta(\varepsilon, t_0)$ decreases to zero as t_0 increases. If a solution is stable for $t \geq t_0$, and the δ of Definition 3.14 is independent of t_0, the solution is *uniformly stable* on $t \geq t_0$. Any stable solutions of an autonomous system are uniformly stable, since the system is invariant with respect to time transformation. Another property is asymptotic stability.

Definition 3.1.5 A solution $\phi(t, x_0)$ of

$$\dot{x}(t) = f(x(t), t)$$

is *asymptotically stable* if (a) it is stable and (b) there exists $\mu > 0$ such that if $\left|x_0 - x_0^*\right| < \mu$, then

$$\left|\phi(t, x_0) - \phi^*(t, x_0^*)\right| \to 0, \text{ as } t \to +\infty.$$

It is *globally asymptotically stable* if μ may be chosen arbitrarily large.

Obviously, the solution of $\dot{x} = ax$ with $a < 0$ is globally asymptotically stable.

In Figs. 3.1.2a and 3.1.2b (for a two-dimensional dynamics), the equilibrium is stable. This means that all solutions that start "sufficiently" close to the equilibrium stay close to it. It should be noted that the trajectory of the solution does not have to approach the equilibrium point as $t \to +\infty$, as illustrated in Fig. 3.1.2b. If an equilibrium point is asymptotically stable, trajectories that start "sufficiently close" to it must not only stay "close" but must eventually approach it as $t \to +\infty$. This is the case for the trajectory in Fig. 3.1.2a but not for the one in Fig. 3.1.2b. Obviously, asymptotic stability is a stronger property than stability.

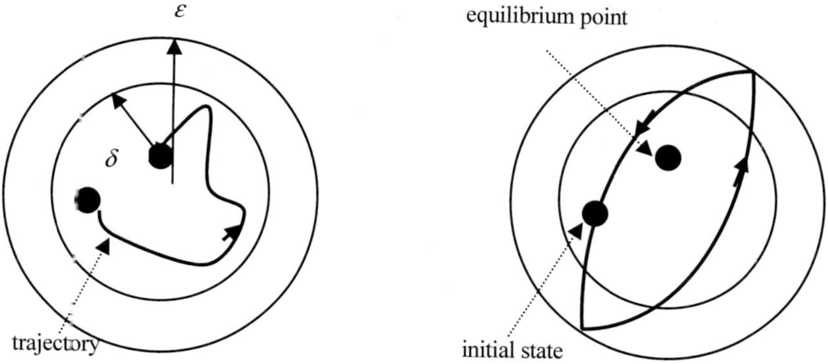

(a) asymptotic stability (b) stability

Fig. 3.1.2 Stability and asymptotic stability.

Definition 3.1.6 A solution $\phi(t,x_0)$ to $\dot{x}(t)=f(x(t),t)$ is *unstable* if, for any given $\varepsilon>0$ sufficiently small and any $\delta>0$, there is a solution $\phi^*(t,x_0^*)$ such that (a) $0<|x_0-x_0^*|<\delta$ and (b) $|\phi(t,x_0)-\phi^*(t,x_0^*)|>\varepsilon$ for some $t>t_0$.

The instability is also referred to as *Liapunov instability*.

Example The general solution of $\dot{x}=ax$ with $a>0$ is $x(t)=x(0)e^{at}$. As

$$|\phi(t,x_0)-\phi^*(t,x_0^*)|=|x_0-x_0^*|e^{at},$$

we see that if we can always choose t which satisfies $|x_0-x_0^*|e^{at}>\varepsilon$. Hence, the equation is unstable.

Definition 3.1.7 A solution $\phi_t(t,x_0)$ of a dynamical system, $\dot{x}=f(x,t)$ is *orbitally stable* if, for any given $\varepsilon>0$ there exists $\delta>0$ such that if $|x_0-x_0^*|<\delta$ then

$$\inf_{\theta\geq t_0}|\phi(t,x_0)-\phi^*(\theta,x_0^*)|<\varepsilon$$

for each $t > t_0$.

The concept of stability and orbital stability should not be confused. Figure 3.1.3 provides an example. Suppose that C and D are two orbits of different periods. Although the distance between them remains bounded for all times, the distance between two points 1 and 2 on the two orbits can increase in time owing to a phase shift induced by the difference between periods. Thus state 1 need not be stable, even though C is orbitally stable.

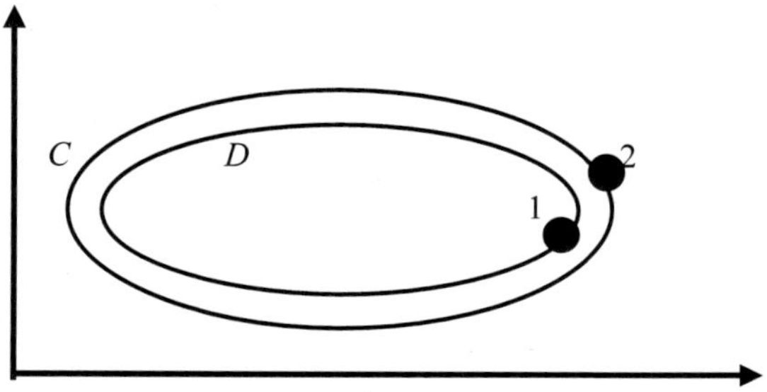

Fig. 3.1.3 Stability of equilibrium and orbital stability.

These definitions for $\dot{x} = f(x,t)$ are applicable to an autonomous system, $\dot{x} = f(x)$.

Definition 3.1.8 A periodic solution of a continuous-time dynamical system, in a neighborhood of which there are no other periodic solutions, is called a *limit cycle*.

Exercise 3.1

1 For the general two-dimensional equation

$$\begin{pmatrix} \dot{x}_1 \\ \dot{x}_2 \end{pmatrix} = \begin{pmatrix} a & b \\ c & d \end{pmatrix} \begin{pmatrix} x_1 \\ x_2 \end{pmatrix},$$

find a Lipschitz constant for the function Ax in terms of a, b, c, and d.

3.2 Stability of Equilibrium of Autonomous Equations

From the definitions in the previous section, we see that the stability of an equilibrium point x^* of the scalar differential equation

$$\dot{x} = f(x), \tag{3.2.1}$$

is a local property of the flow near the equilibrium. The following theorem tells that under certain conditions the stability properties of x^* can be determined from the linear approximation of the function f near x^*.

Theorem 3.2.1 Suppose that f is a C^1 function and x^* is an equilibrium point of $\dot{x} = f(x)$, that is, $f(x^*) = 0$. Suppose also that $f'(x^*) \neq 0$. Then the equilibrium point x^* is asymptotically stable if $f'(x^*) < 0$, and unstable if $f'(x^*) > 0$.

The linear differential equation $\dot{x} = f'(x^*)x$ is called the *linear variation equation* or the *linearization* of the vector field $\dot{x} = f(x)$ about its equilibrium point x^*. Since the equilibrium point $x = 0$ of $\dot{x} = f'(x^*)x$ is asymptotically stable if $f'(x^*) < 0$ and unstable if $f'(x^*) > 0$, we see that Theorem 3.2.1 asserts that if $f'(x^*) \neq 0$, the stability type of the equilibrium point x^* of $\dot{x} = f(x)$ is the same as the stability type of the equilibrium point at the origin of its linearized vector field. It should be noted that the condition $f'(x^*) \neq 0$ is important. An equilibrium point x^* of $\dot{x} = f(x)$ is called a *hyperbolic* equilibrium if $f'(x^*) \neq 0$. If $f'(x^*) = 0$, then x^* is called a *nonhyperbolic* or *degenerate* equilibrium point. When an equilibrium is degenerate, then its stability is determined by higher-terms in the Taylor expansion of the function $f(x)$ at x^*. For instance, the origin is an unstable equilibrium for $\dot{x} = x^2$, but it is an asymptotically stable equilibrium for $\dot{x} = -x^3$.

Example The Cagan monetary model.[4]
The Cagan model of hyperinflation is described by the pair of equations

$$m(t) - p(t) = -\alpha\pi(t), \quad \alpha > 0,$$

---

[4] Cagan (1956).

$$\dot{\pi} = \gamma(\dot{p} - \pi), \quad \gamma > 0,$$

where m is natural log of the nominal stock of money (M), p is natural log of price level (P), and π is the expected rate of inflation. In this model, output and real interest rate are fixed and wealth effects are ignored. The first equation describes continuous money market equilibrium, and the second equation expresses the usual form of the adaptive hypothesis about the expected rate of inflation. The coefficient

$$\alpha = \frac{1}{M}\frac{dM}{d\pi},$$

is interpreted as a semi-elasticity of demand. It measures the percentage change in the demand for money per percentage point change in the expected inflation rate. Supposing the nominal money stock remains constant, from the first equation we obtain, $\dot{p} = \alpha\dot{\pi}$. Substituting the first equation and $\dot{p} = \alpha\dot{\pi}$ into the second equation yields

$$\dot{p} = \frac{\gamma}{1 - \alpha\gamma}(m - p).$$

We conclude that the adjustment of prices is stable if and only if $\alpha\gamma < 1$. This relation emphasizes the trade-off between the semi-elasticity of the demand for money with respect to inflationary expectations and the rate of adaptation of inflationary expectations. A highly sensitive demand for money function is compatible with stability only if inflationary expectations adapt sufficiently slowly to past inflation rates.

Example Poverty traps generated in the Solow model.
King and Rebelo tried to explain poverty traps with the framework of neoclassical growth theory by using a utility function in which there is a subsistence level of per capita consumption and the elasticity of intertemporal substitution varies over time. [5] The model has two equilibria – a Solow-type and an unstable steady state at the level of the capital stock comparable with subsistence consumption.

---

[5] King and Rebelo (1993).

We now illustrate this approach based on Ros.[6] The model starts with a consumption function

$$c(t) = \left(y_0 - \delta_k k_0\right) + \bar{\xi}\left(y(t) - c_0\right), \tag{3.2.2}$$

where $c(t)$ and $y(t)$ are respectively per capita consumption and per capita income at time t, y_0 is subsistence income per capita, k_0 is the capital-labor ratio consistent with a subsistence level of income, δ_k the depreciation rate of capital, and $\bar{\xi}$ $(0 < \bar{\xi} < 1)$ is the propensity to consume out of nonsubsistence income. When income per worker is at the subsistence level of saving is equal to the depreciation of capital stock, i.e., $y_0 - c = \delta_k k_0$. The corresponding saving rate is

$$s(t) = \frac{y(t) - c(t)}{y(t)} = \hat{s}_0 - \frac{\delta^*}{y(t)}, \tag{3.2.3}$$

where $\hat{s}_0 \equiv 1 - \bar{\xi} > 0$, $\delta^* \equiv \hat{s}_0 y_0 - \delta_k k_0$. Here, $\hat{s}_0$ is the propensity to save out of nonsubsistence income. According to the definitions, the parameter δ^* may be either positive or negative. If $\hat{s}_0 y_0 > \delta_k k_0$, δ^* is positive, implying that as income rises, the saving rate rises. If $\hat{s}_0 y_0 < \delta_k k_0$, δ^* is negative, implying that as income rises, the saving rate falls. The saving rate is a nonlinear function of the level of income per capita. The saving rate rises with income per capita if the marginal propensity to consume out of nonsubsistence income is less than the average propensity to consume out of subsistence income. Otherwise, the saving rate tends to fall as income rises above the subsistence level. Substituting Eq. (3.2.3) into the fundamental equation

$$\dot{k} = s(t)y(t) - (n + \delta_k)k(t)$$

in k yields

$$\dot{k} = \hat{s}_0 f(k) - (n + \delta_k)k - \delta^*, \tag{3.2.4}$$

in which n is the fixed population growth rate. We call the above model the *generalized Solow model with poverty traps*. Check that this model can yield two equilibria as shown in Fig. 3.2.1 when $\delta^* > 0$.

[6] Ros (2000: 60–62).

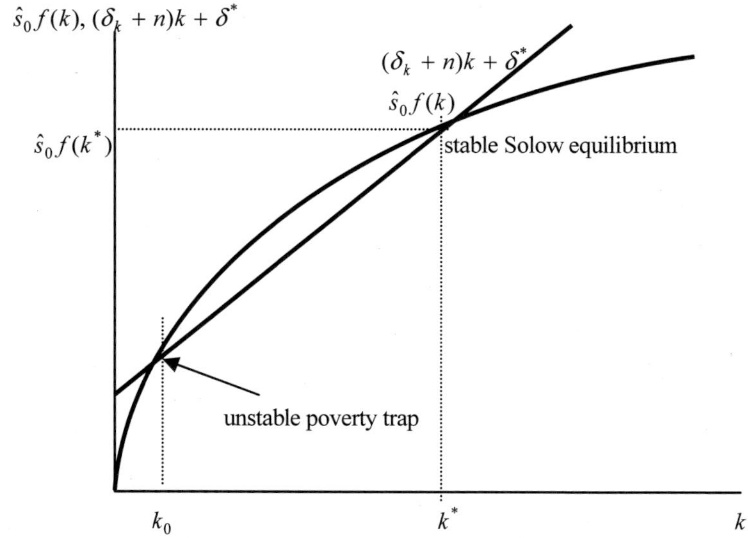

Fig. 3.2.1 Two equilibrium points with $\delta^* > 0$.

In the case of $\delta^* > 0$, the equilibrium at the high-k level is similar to the steady state in the Solow model. The other is at the subsistence level of income. This is often referred to as poverty trap, but it is a fortune one in the sense that it is unstable. The economic system will not stay there for long. Poverty may not be persistent under proper disturbances from, for instance, foreign aids and trade. This trap occurs because at low levels of capital-labor ratio income per capita is scarcely sufficient and savings fall below depreciation. If an economy is in this type of poverty, it is easy to start rapid development because once capital-labor ratio is higher than the level k_0, it will grow fast towards the stable equilibrium. Once it reaches the high level of living standard, it is trapped because this is a stable state.

The model has only one equilibrium – an unstable poverty when $\delta^* < 0$, i.e., $\hat{s}_0 y_0 < \delta_k k_0$. This situation occurs that the saving rate is too low or the subsistence income per capita is too low. If an economy is characterized by political and social instabilities, then it tends to have a low saving rate. Under such circumstances, the nation can hardly make any progress in economic development.

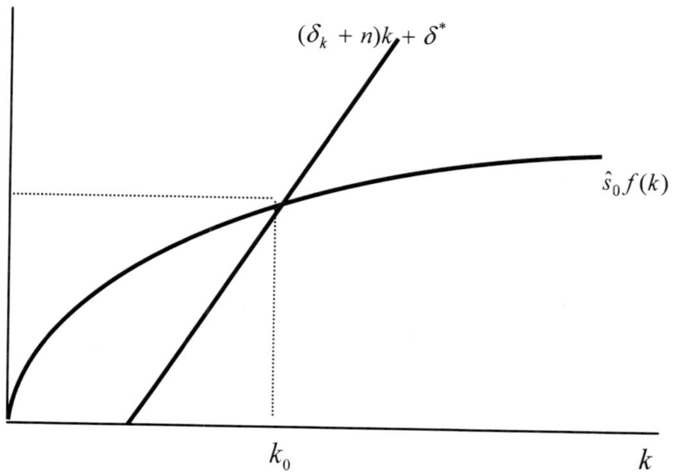

Fig. 3.2.2 A single poverty trap with $\delta^* < 0$.

Exercise 3.2

1 Determine the equilibrium points of the following scalar differential equations and determine their stability:

(a) $\dot{x} = 3x(1 - x)$;

(b) $\dot{x} = 2x^2 - x^3$;

(c) $\dot{x} = x - x^3 + 0.2$;

(d) $\dot{x} = 1 - \sin x$.

3.3 Bifurcations

Bifurcation theory studies possible changes in the structure of the orbits of a differential equation depending on parameters. There are two distinct aspects of bifurcation theory: static and dynamic. Static bifurcation theory is concerned with the changes that occur in the structure of the set of zeros of a function as parameters in the function are varied. In differential equations, the equilibrium solutions are the zeros of the vector field. Therefore, methods in static bifurcation theory are directly applicable.

Dynamic bifurcation theory is concerned with the changes that occur in the structure of solutions of differential equations as parameters in the vector field are varied. A change in the qualitative properties could mean a change in stability of the original system, and thus the system must assume a state different from the original design. In vague terms, the values of the parameters where this change takes place are called *bifurcation values*. Knowledge of the bifurcation values is necessary for a complete understanding of the system. Consider the following differential equation

$$\dot{x} = f(x, \lambda), \tag{3.3.1}$$

where x is defined in some space, λ represents a parameter vector, and f is a function vector which satisfies certain requirements. There may exist different types of solutions such as steady solutions, periodic solutions, sub-harmonic solutions, asymptotically quasi-periodic solutions, chaos, and so on.

Consider the equilibrium equations $f(x, \lambda) = 0$. Sometimes, derivatives of f are required, and it will always be assumed in this section that f has as many derivatives as necessary if this is not explicitly stated. We may consider equilibrium states as functions of the parameters. Usually multiple equilibria may exist for given values of the parameters. The basic question is to discuss how the equilibrium depends on the parameters. We now consider some specific examples to illustrate some of the key ideas from bifurcation theory.[7]

Example Hyperbolic equilibrium is insensitive.
Consider the linear differential equation

$$\dot{x} = \lambda - x = F(x, \lambda), \tag{3.3.2}$$

where λ is a real parameter. For $\lambda = 0$, we have

$$\dot{x} = -x.$$

We thus refer Eq. (3.3.2) as a *perturbation* of $\dot{x} = -x$. For all values of λ, there is a single hyperbolic equilibrium, which is asymptotically stable.

[7] The examples below in this section are from Sec. 2.1 in Hale and Koçak (1991).

Before generalizing this example, we need to state the implicit function theorem. The (static) bifurcation problem is equivalent to the study of the curves $f(x, \lambda) = 0$ and their singular points. The main tool for the proof of the existence is the implicit function theorem, which holds for vector-valued functions of multiple parameters.[8] For a one-dimensional problem, the theorem is stated as follows.

Lemma 3.3.1 (The implicit function theorem in R^1) Let $F(x^*, \lambda_0) = 0$ and f be C^1 in some open neighborhood of (x^*, λ_0). Then if $F_x \neq 0$ at (x_0, λ_0), there exists $\alpha, \beta > 0$ such that (i) the equation $F(x, \lambda) = 0$ has a unique solution $x = x(\lambda)$ with $x^* - \beta < x < x^* + \beta$ whenever $\|\lambda - \lambda_0\| < \delta$;[9] and (ii) $x_\lambda(\lambda)$ exists and

$$x_\lambda(\lambda) = -\frac{F_\lambda(x(\lambda), \lambda)}{F_x(x(\lambda), \lambda)}.$$

We knew that if f is a C^1 function with $f(0) = 0$ and $f'(0) \neq 0$, then the stability properties of the equilibrium point 0 of $\dot{x} = f(x)$ is determined by the linear approximation of the vector field near 0, that is, the higher order perturbations in the Taylor expansion of the vector field do not effect the qualitative structure of the flow near zero. But we did not discuss what will happen to the system if we make small perturbations.

To precisely answer this question, consider the perturbed differential equation

$$\dot{x} = F(x, \lambda),$$

where

$$F : R \times R^k \to R \,;\; (x, \lambda) \mapsto F(x, \lambda)$$

is a C^1 function satisfying

$$F(x, 0) = f(0), \quad F_x(0, 0) = f'(0) \neq 0.$$

As 0 is a hyperbolic equilibrium point of the differential equation $\dot{x} = F(x, \lambda)$ depending at $\lambda = 0$, then the conditions of the implicit

---

[8] The theorem is referred to App. A.3.

[9] Here, the "norm" is defined by $\|\lambda\| = \left(\sum_i \lambda_i^2\right)^{1/2}$.

function theorem are satisfied. This guarantees that the equation $F(x, \lambda) = 0$ may be solved locally as a function of the parameters

$$\lambda = (\lambda_1, \cdots, \lambda_k).$$

Furthermore, $F_x(x(\lambda), \lambda) \neq 0$ for $\|\lambda - \lambda_0\|$ sufficiently small. The stability type of the equilibrium $x(\lambda)$ of the perturbed equation is the same as the stability type of the equilibrium 0 of the unperturbed equation $\dot{x} = f(x)$. Thus, the qualitative structure of the flow does not change near the equilibrium. Consequently, there are no bifurcations in the neighborhood of the equilibrium. We thus conclude that the flow near a hyperbolic equilibrium point is insensitive to small perturbations of the vector field.

Example Saddle-node bifurcation.
The quadratic differential equation

$$\dot{x} = x^2 + \lambda = F(x, \lambda), \tag{3.3.3}$$

where λ is a real parameter and is a perturbation of $\dot{x} = x^2$. Note that the origin is a nonhyperbolic equilibrium point for $\lambda = 0$. We can easily determine the flow of Eq. (3.3.3) for all values of the parameter λ by leaving the original parabola $f(x, 0)$ fixed and vertically translating the x-axis by $-\lambda$. The resulting flows are depicted in Fig. 3.3.1. For all $\lambda < 0$ the system has two equilibrium points. For $\lambda = 0$, the system has a unique equilibrium point $x = 0$. For all $\lambda > 0$, there is no equilibrium point. If λ is varied, as long as it is negative, the number and the direction of the orbits remain the same; the only change is the shifting of the location of the equilibrium points. Similarly, for all positive λ there is only one orbit and its direction is from left to right. However, for $\lambda = 0$, regardless of how small an amount λ is varied, the number of orbits changes: there are two equilibrium points for any $\lambda < 0$, and none for any $\lambda > 0$.

For a scalar differential equation $\dot{x} = f(x)$, the equilibrium points and the sign of the function $f(x)$ between the equilibria determine the number of orbits and the direction of the flow on the orbits. We refer to the number of orbits and the direction of the flow on the orbits as the *orbit structure* of the differential equation or the *qualitative structure of the flow*.

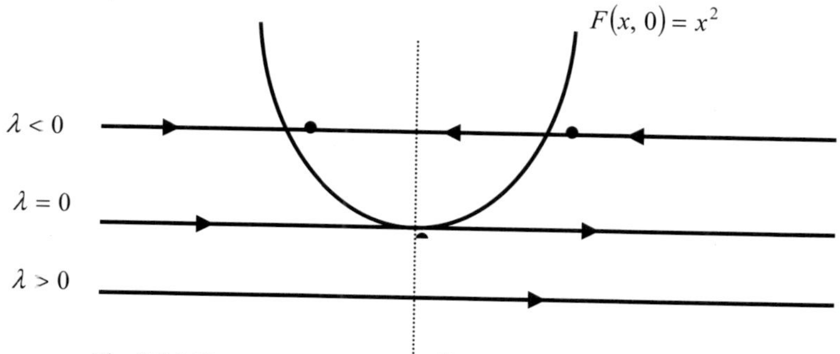

Fig. 3.3.1 Phase portraits of $\dot{x} = x^2 + \lambda$ for several values of λ.

Bifurcation theory is to study changes of the qualitative structure of the flow as parameters are varied. At a given parameter value, a differential equation is said to have *stable orbit structure* if the qualitative structure of the flow does not change for sufficiently small variations of the parameter. A parameter value for which the flow does not have stable orbit structure is called a *bifurcation value*, and the equation is said to be at a *bifurcation point*. In the above example, Eq. (3.3.3) has stable orbit structure for any $\lambda \neq 0$, but is at a bifurcation point for $\lambda = 0$. The bifurcation that Eq. (3.3.3) undergoes is called *saddle-node bifurcation*. This bifurcation has other names such as *limit point*, and *turning point*.

We may depict some of the important dynamic features in $\dot{x} = F(x, \lambda)$ depending on a parameter λ. This graphical method consists of drawing curves on (λ, x)-plane, where curves depict the equilibrium points for each of the parameters. A point lies on one of these curves if and only if $F(x, \lambda) = 0$. To represent the stability types of these equilibria, we label stable equilibria with solid curves and unstable equilibria with dotted curves. The resulting picture is called a *bifurcation diagram*. Figure 3.3.2 is the bifurcation diagram for

$$\dot{x} = x^2 + \lambda .$$

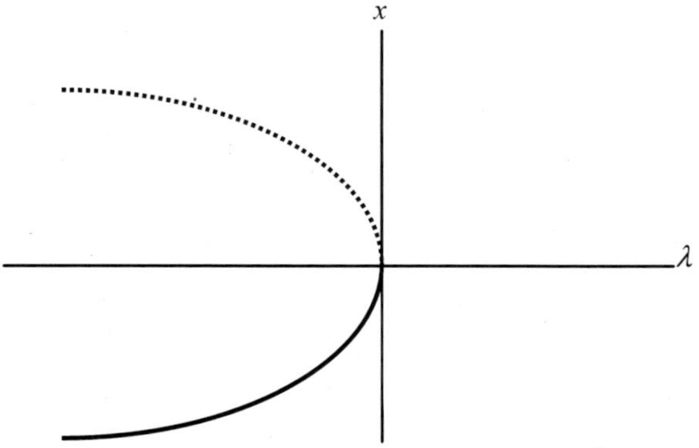

Fig. 3.3.2 Bifurcation diagram of $\dot{x} = x^2 + \lambda$.

The system $\dot{x} = \lambda - x^2$ can be considered in the same way. The analysis reveals two equilibria appearing for $\lambda > 0$.

Bifurcation diagrams are not entirely random. Different strata of bifurcation diagrams in generic systems exhibit similarities in different applications. To discuss this topic, we have to decide when two dynamical systems have qualitative similar or equivalent bifurcation diagrams. Before continuing, we introduce a concept, which defines "equivalence of two dynamical systems. Let us consider two dynamical systems

$$\dot{x} = f(x, \alpha), \quad x \in R^n, \quad \alpha \in R^m, \tag{A}$$

and

$$\dot{y} = f(y, \alpha), \quad y \in R^n, \quad \beta \in R^m, \tag{B}$$

with smooth right-hand sides and the same number of variables and parameters.

Definition 3.3.1 Dynamical system (A) is called *topologically equivalent* to a dynamical system (B) if (i) there exists a homeomorphism of the parameter space:

$$p : R^m \to R^m, \; \beta = p(\alpha);$$

and (ii) there is a parameter-dependent homeomorphism of the phase space

$$h_\alpha : R^n \to R^n, \; y = h_\alpha(x),$$

mapping orbits of the system (A) at parameter values α onto orbits of the system (B) at parameter values $\beta = p(\alpha)$, preserving the direction of time.

By definition, topologically equivalent parameter-dependent systems have equivalent bifurcation diagrams. A *homeomorphism* is an invertible map such that both the map and its inverse are continuous. In the above definition, we do not require the homeomorphism h_α to depend continuously on α. For this reason, the above definition is sometimes called *weak* or (*fiber*) topological equivalence.

In terms of the topological equivalences, the following theorems show that the system (3.3.3) is quite a "general" case, irrespective of its simplicity.[10]

Lemma 3.3.2 The system

$$\dot{x} = \lambda + x^2 + O(x^3)$$

is "locally" topologically equivalent near the origin to the system

$$\dot{x} = \lambda + x^2 + O(x^3).$$

Here, "locally" means that Definition 3.3.1 is modified for local behavior of the systems.

Theorem 3.3.1 (Topological normal form for the saddle-node bifurcation) Any generic scalar one-parameter system

$$\dot{x} = f(x, \lambda),$$

having the equilibrium $x = 0$ with $f_x(0, 0) = 0$ at $\lambda = 0$, is locally topologically equivalent near the origin to one of the following normal forms

$$\dot{y} = \beta \pm y^2 .$$

---

[10] See Sec. 3.2 in Kuznetsov (1998) for the following results related to saddle-node bifurcations.

Here, "genetic" means that the system satisfies a number of genericity conditions. These conditions have the form of inequalities:

$$N_i[f], \; i = 1, \cdots, s,$$

where each N_i is some algebraic function of certain partial derivatives of f with respect to x and α evaluated at $(x, \alpha) = (0, 0)$.[11]

Example Dynamics of labor market.[12]
Consider a simple partial-analytical model of the labor market. Let $l^s(w)$ and $l^d(w)$ be the supply and demand for labor, respectively, which both depend on the real wage, w. The change in the real wage rate is assumed to depend on the excess demand for labor in this market, i.e.

$$\dot{w} = \beta\left(l^d(w) - l^s(w)\right), \; \beta > 0.$$

Assume that the demand function is parametrized by μ and let

$$l^d(w, \mu) = \mu - bw, \quad \mu, b > 0.$$

Assume that the labor supply function reflects an inferiority such that it is bending backwards for high values of w, see Fig. 3.3.3.

Let $d^2 l^s(w)/dw^2 < 0$ for any $w > 0$ and $dl^s(w)/dw < 0$ for w greater than a value w_0. Introduce

$$f(w, \mu) = l^d(w, \mu) - l^s(w).$$

Let μ_0 be the value of μ such that $f(w_0, \mu_0) = 0$ and $\partial f / \partial \mu(w_0, \mu_0) = 0$. At μ_0, the demand and supply functions are tangent. Hence, the saddle-node bifurcation occurs in this labor market model.

Example Transcritical bifurcation.
Consider the differential equation

$$\dot{x} = \lambda x + x^2 = F(x, \lambda). \tag{3.3.4}$$

The origin is an equilibrium point for all values of λ. For $\lambda < 0$, the origin is asymptotically stable and there is another equilibrium point $x = -c$ which is unstable. The parameter value $\lambda = 0$ is a bifurcation

[11] See Sec. 3.2 in Kuznetsov (1998) for the definition.
[12] See Lorenz (1993: 88–89).

value at which the two equilibria coalesce at the origin, which is a nonhyperbolic unstable equilibrium point. For $\lambda > 0$, the origin is unstable and there is another equilibrium point $x = -c$ which is stable. The bifurcation that Eq. (3.3.4) undergoes is called *transcritical bifurcation*. We depict the bifurcation diagram in Fig. 3.3.4.

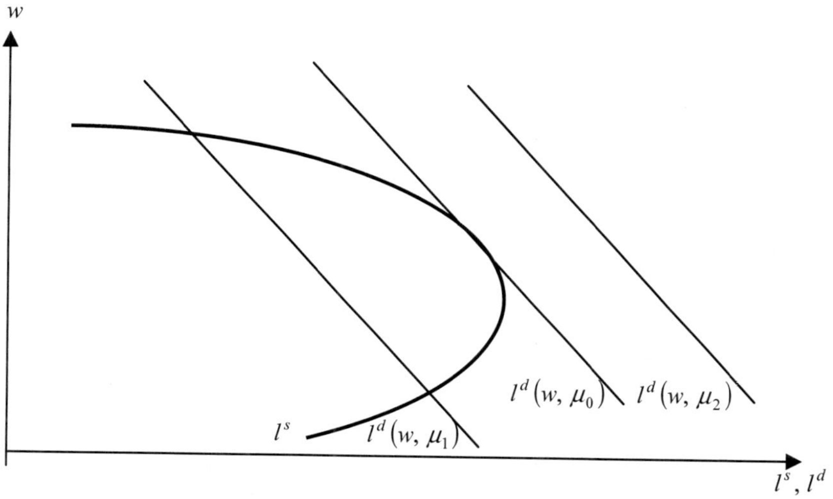

Fig. 3.3.3 Demand and supply of labor.

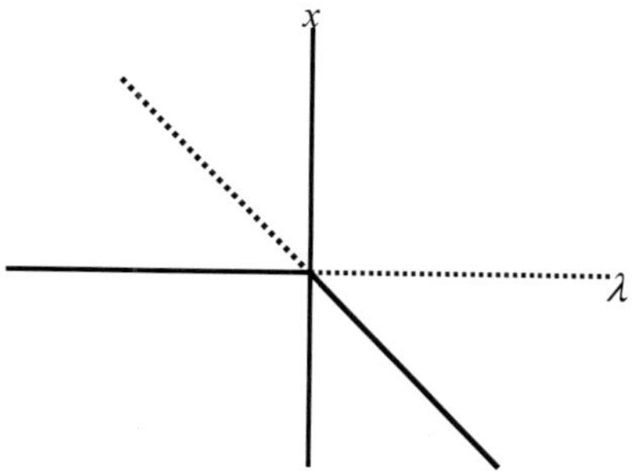

Fig. 3.3.4 Bifurcation diagram of $\dot{x} = \lambda x + x^2$.

Example Hysteresis.
The cubic differential equation used to illustrate hysteresis is

$$\dot{x} = -x^3 + x + \lambda = F(x, \lambda). \tag{3.3.5}$$

For $\lambda = 0$, the system is $\dot{x} = -x^3 + x$. It has three equilibrium points, $-1, 0, 1$ and it has stable orbit structure. The flow continues to have stable orbit structure for small values of the parameter, $-\lambda_0 < \lambda < \lambda_0$, where $\lambda_0 = 2/3\sqrt{3}$ is the local maximum value and $-\lambda_0$ is the local minimum value of $F(x, 0)$. For $\lambda = \lambda_0$ or $\lambda = -\lambda_0$, the equation is at a bifurcation point. The bifurcation diagram is shown in Fig. 3.3.5. The bifurcation that Eq. (3.3.5) undergoes is called *hysteresis*. It should be noted that the system experiences a jump at two different values of the parameter. The part in Fig. 3.3.5 that resembles a parallelogram is called the *hysteresis loop*.

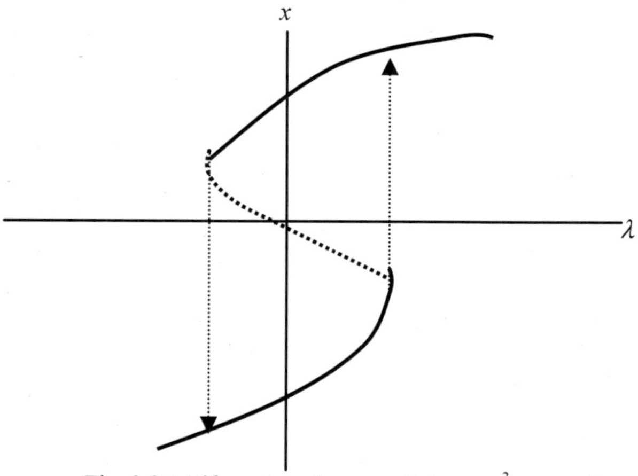

Fig. 3.3.5 Bifurcation diagram of $\dot{x} = -x^3 + x + \lambda$.

Example Pitchfork bifurcation.
Consider

$$\dot{x} = -x^3 + \lambda x = F(x, \lambda). \tag{3.3.6}$$

For $\lambda = 0$, the system is $\dot{x} = -x^3$. It is easy to confirm that for all positive λ, Eq. (3.3.6) has three equilibria and stable orbit structure. At $\lambda = 0$, the equilibria come together at the origin and the system is at a bifurcation point. For all $\lambda < 0$, the equation again has stable orbit structure. The bifurcation that Eq. (3.3.6) undergoes is called *pitchfork bifurcation*.

For this particular example, the pitchfork bifurcation is called *supercritical* because the additional equilibrium points which appear at the bifurcation value occur for the values of the parameter at which the equilibrium point is unstable. Figure 3.3.6a depicts this supercritical bifurcation diagram. When the additional equilibria occur for the values of the parameter at which the original equilibrium point is stable, the bifurcation is called *subcritical*. Figure 3.3.6b depicts a subcritical pitchfork bifurcation with $\dot{x} = x^3 + \lambda x$.

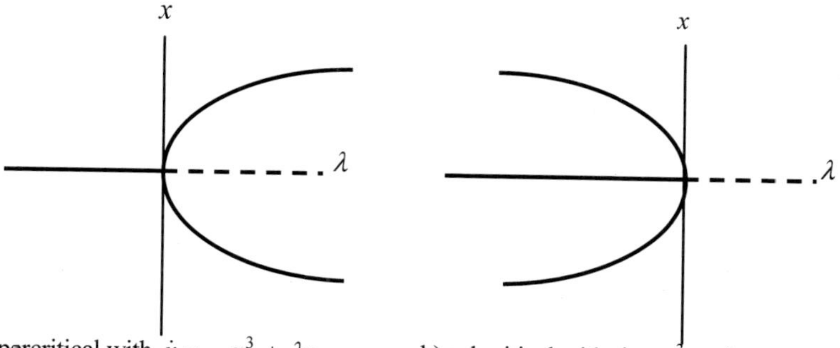

a) supercritical with $\dot{x} = -x^3 + \lambda x$ b) subcritical with $\dot{x} = x^3 + \lambda x$

Fig. 3.3.6 Pitchfork bifurcations.

Example Cusp bifurcation.
The differential equation is now

$$\dot{x} = -x^3 + dx + c = F(x, d, c), \tag{3.3.7}$$

where d and c are real parameters. The vector field (3.3.7) is the most general perturbation of the function $-x^3$ with lower order terms because any term involving x^2 can always be eliminated by an appropriate translation of the variable.[13] At bifurcation points, a differential equation must have a multiple equilibrium point, that is, $F(x, d, c) = 0$ and $F_x(x, d, c) = 0$. For the vector field (3.3.7), we have

$$-x^3 + dx + c = 0, \quad -3x^2 + d = 0.$$

Eliminating x from the above equations yields a *cusp*

[13] For $-x^3 + a_1x^2 + a_2x + a_3$, introduce $x \mapsto x + a_1/3$. The equation is transformed into the form (3.3.7).

$$4d^3 = 27c^2 . \tag{3.3.8}$$

Figure 3.3.7 depicts the graph of $4d^3 = 27c^2$ in the (c, d)-plane. In each appropriate region of that (c, d)-plane we sketch a graph for the function $F(x, c, d)$ and indicate the flow determined by Eq. (3.3.7).

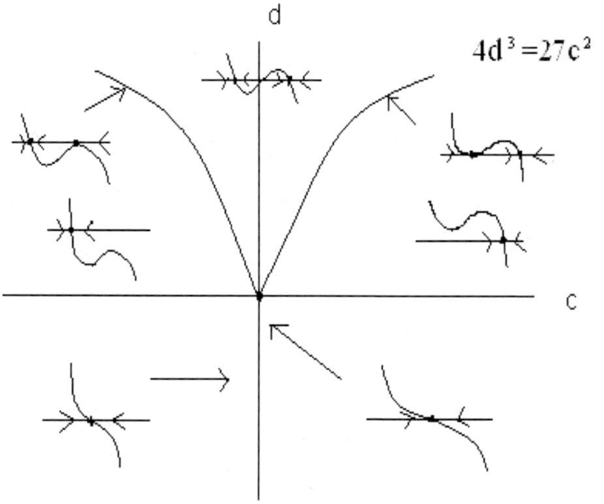

Fig. 3.3.7 Some representative phase portraits of Eq. (3.3.7).

The full bifurcation diagram of Eq. (3.3.7) in the three-dimensional (c, d, x)-space can be constructed from the equation

$$-x^3 + dx + c = 0,$$

see Fig. 3.3.8.

Lemma 3.3.3 Suppose that a one-dimensional system

$$\dot{x} = f(x, \alpha), \quad x \in R, \quad \alpha \in R^2,$$

with smooth f, has at $\alpha = 0$ the equilibrium $x = 0$, and let the cusp bifurcation conditions hold

$$\lambda = f_x(0, 0) = 0, \quad a = \frac{f_{xx}(0, 0)}{2} = 0 .$$

Assume that the following *genericity conditions* are satisfied

$$f_{xxx}(0,0) \neq 0 , \quad \left(f_{\alpha_1} f_{x\alpha_2} - f_{\alpha_2} f_{x\alpha_1}\right) \neq 0 ,$$

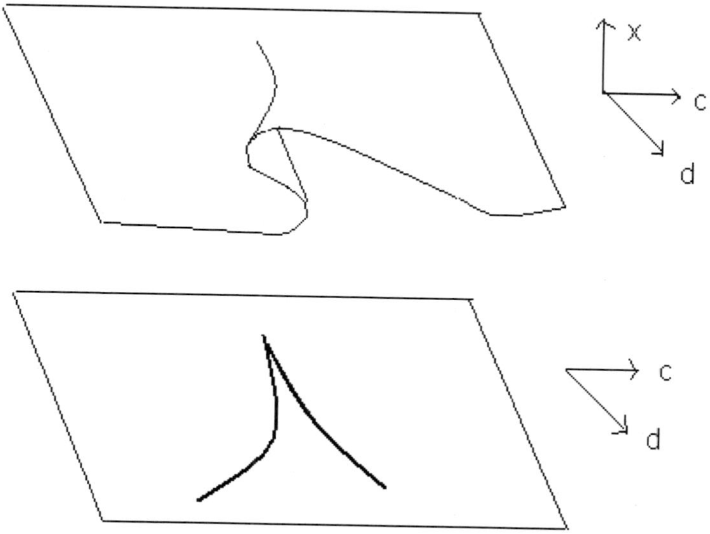

Fig. 3.3.8 The bifurcation diagram of Eq. (3.3.7).

at $(0, 0)$. Then, there are smooth invertible coordinate and parameter changes transforming the system into

$$\dot{\eta} = \beta_1 + \beta_2\eta \pm \eta^3 + O(\eta^4). \qquad (3.3.9)$$

This system with the $O(\eta^4)$ terms truncated is called the *approximate normal form* for the cusp bifurcation. It can be shown that higher-order terms do not actually change them. This justifies calling

$$\dot{\eta} = \beta_1 + \beta_2\eta \pm \eta^3, \qquad (3.3.10)$$

the *topological normal form* for the cusp bifurcation. It can be proved that Eq. (3.3.9) is locally topologically equivalent near the origin to Eq. (3.3.10).

Theorem 3.3.2 (Topological normal form for the cusp bifurcation) Any generic scalar two-parameter system $\dot{x} = f(x, \alpha)$ having an equilibrium $x = 0$ at $\alpha = 0$ exhibiting the cusp bifurcation is locally topologically equivalent near the origin to one of the normal forms (3.3.9).

When an n-dimensional system has a cusp bifurcation, the above theorem should be applied to the equation on the center manifold (see Chap. 9).

Exercise 3.3

1 Apply the implicit function theorem to show that there is a unique solution of the equation

$$x^3 + (1 - \lambda)x + \lambda = 0$$

near $(x, \lambda) = (1, -1)$.

2 Suppose the density of a population, $x(t)\ (\geq 0)$, changes according to the following differential equation

$$\dot{x}(t) = kx - cx^2 - h,$$

where all the coefficients, k, c, and h, are positive; k and c measure the intrinsic growth rate of the population and h is the rate of harvesting. For a positive initial population density, the population is exterminated if there is a finite value of t such that $\varphi(t, x_0) = 0$.

Without finding explicit solutions of the differential equation, show the following: (i) If $0 < h \leq k^2/4c$, then there is threshold value of the initial size of the population such that if the initial size is below the threshold value, then the population is exterminated. On the other hand, if the initial size is above the threshold value, then the population approaches an equilibrium point; (ii) If $h > k^2/4c$, the population is exterminated regardless of its initial size.

3.4 Periodic Solutions

This section provides some examples of existence of periodic solutions to differential equations. As the topic is complicated, this section only provides some simulation results.

Example The Van der Pol (VdP) equation.[14]
The VdP equation

$$\ddot{x} - \varepsilon\left(1 - x^2\right)\dot{x} + x = 0, \tag{3.4.1}$$

has played an important role in the development of nonlinear theory, because it displays limit cycles. The parameter ε is positive and small. Introduce $y = \dot{x}$ and the equation is reduced to a first order differential equation

$$\dot{x} = y,$$

$$\dot{y} = \varepsilon\left(1 - x^2\right)y - x. \tag{3.4.2}$$

As demonstrated in Fig. 3.4.1, the system exhibits a stable limit cycle when $\varepsilon = 1$.

As ε is increased, the circle becomes increasingly distorted. Figure 3.4.2 shows the limit cycles obtained numerically for $\varepsilon = 1$ and $\varepsilon = 3$.

The limit cycles for even larger values of ε are generated as in Fig. 3.4.3. As ε becomes larger, the VdP solution $x(t)$ displays so-called relaxation oscillations. As shown in Fig. 3.4.3, there are fast changes of $x(t)$ near certain values of t with relatively slowly varying regions in between. As ε is further increased, the slowly varying regions span longer and longer time intervals.

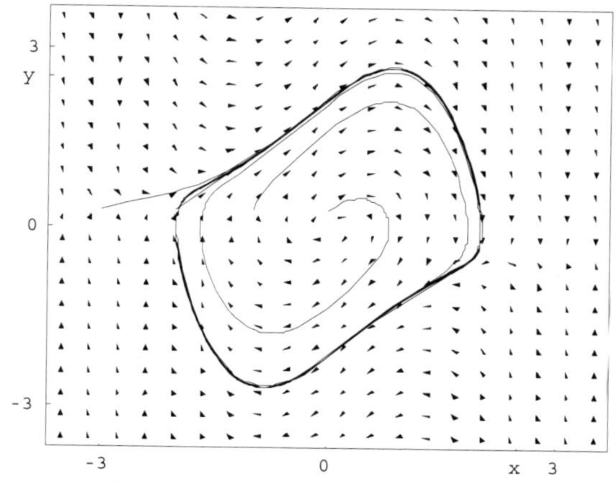

Fig. 3.4.1 The stable limit cycle in the VdP equation ($\varepsilon = 1$).

[14] The examination is based on Sec. 7.2 in Enns and McGuire (2001).

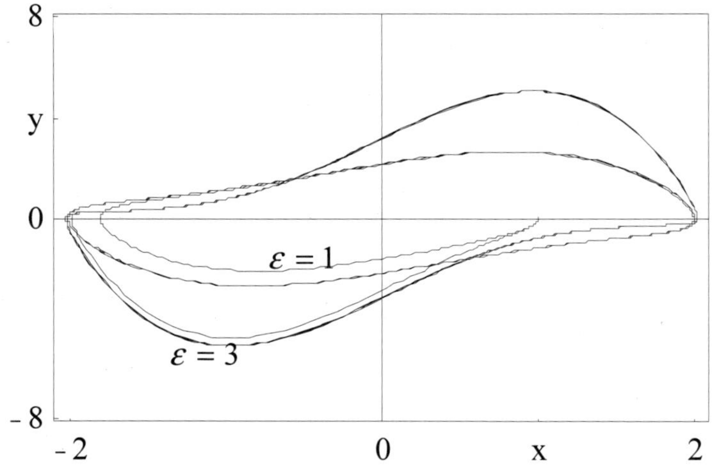

Fig. 3.4.2 Limit cycles of the VdP equation for $\varepsilon = 1$ and $\varepsilon = 3$.

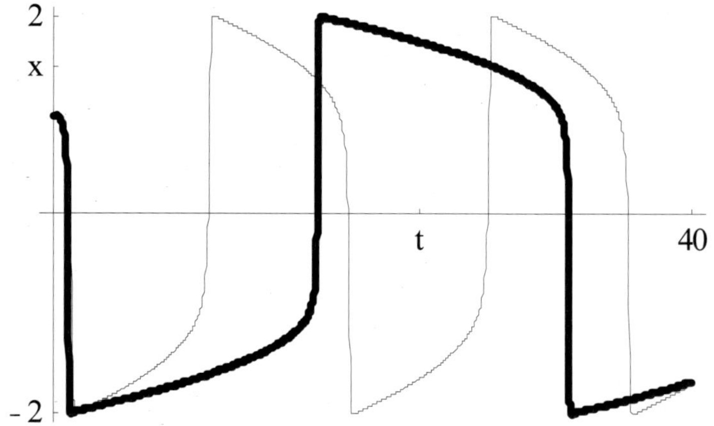

Fig. 3.4.3 Relaxation oscillations (thin for $\varepsilon = 9$ and thick for $\varepsilon = 18$).

To examine what happens for large ε, we rewrite the VdP equation in the form of

$$\dot{x} = \varepsilon(y - f(x)),$$

$$\dot{y} = -\frac{x}{\varepsilon},$$

where

$$f \equiv -x + \frac{x^3}{3}.^{15}$$

We rewrite the above equations in the form of

$$(y - f(x))\frac{dy}{dx} = -\frac{x}{\varepsilon^2}. \qquad (3.4.3)$$

For extremely large ε, the right-hand side of Eq. (3.4.3) is nearly zero. In this limit, either $y = f(x)$ or $dy/dx = 0$. We plot $y = f(x)$ in dashed line in Fig. 3.4.4 with $\varepsilon = 10$. The system slowly traverses the $f(x)$ curve in the sense of the arrows from A to B, jumps horizontally and quickly from B to C, again slowly moves along the $f(x)$ curve to D, and then jumps quickly back to A.

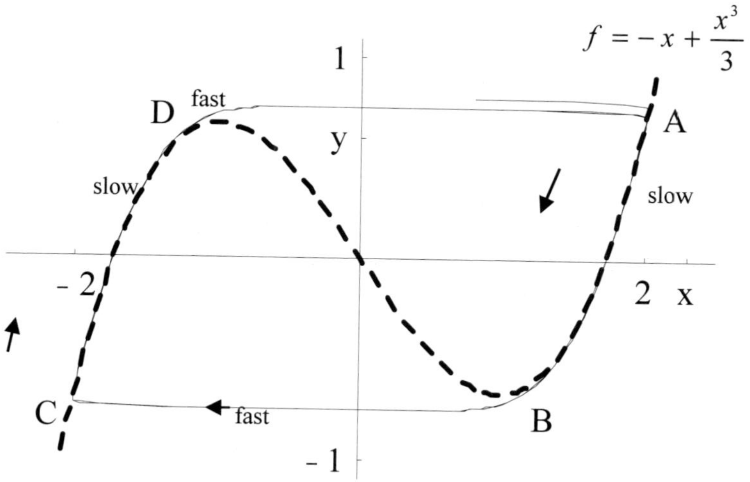

Fig. 3.4.4 Origin of fast and slow time scales for relaxation oscillations.

Example Duffing's equation.
Consider the forced spring equation, the so-called *Duffing equation*

$$\ddot{x} + 2\gamma\dot{x} + \alpha x + \beta x^3 = F\cos(\omega t), \qquad (3.4.4)$$

---

[15] The VdP equation (3.3.1) can be obtained by differentiating $\dot{x} = \varepsilon(y - f(x))$ and plugging $\dot{y} = -x/\varepsilon$ into the resulted equation.

where $x(t)$ is the displacement from equilibrium, γ is the damping coefficient, ω the driving frequency, and $F\ (\geq 0)$ is the force amplitude.

The Duffing equation is further classified according to the signs and values of the parameters of α and β. For $\alpha > 0$ and $\beta > 0$, it is known as the *hard spring Duffing equation.* A hard spring becomes harder to stretch for larger displacements from equilibrium. When $\alpha > 0$ and $\beta < 0$, the equation is referred to as the *soft spring Duffing equation.* The two other important categories are *nonharmonic* ($\alpha = 0$ and $\beta > 0$) and *inverted* ($\alpha < 0$ and $\beta > 0$) cases. Introducing

$$y = \dot{x}, \quad z = t,$$

we rewrite the Duffing equation as three coupled first-order ODEs with three state variables

$$\dot{x} = y,$$

$$\dot{y} = -2\gamma y - \alpha x - \beta x^3 + F\cos(\omega z),$$

$$\dot{z} = 1.$$

Figure 3.4.5 shows two plots of the behavior of the Duffing equation with

$$\alpha = -1,\ \beta = 1,\ \gamma = 0.25,\ x(0) = 0.09,\ \dot{x}(0) = 0,\ F = 0.34875.$$

The system exhibits a period-two solution as the pattern repeats every two oscillations when steady-state is achieved. The trajectory in the right-hand plot has wound onto a closed orbit which appears to cross itself. In three dimensions real trajectories do not cross. These crossings are an artifact resulting from the fact that we have projected a three dimensional phase trajectory onto a two dimensional plane. Figure 3.4.6 demonstrates chaotic behavior when F is increased to $F = 0.43$. The pattern has no irregular pattern even if a longer time range is chosen.

3.5 The Energy Balance Method and Periodic Solutions

We now introduce a method to solve an equation of the form

$$\ddot{x} + \varepsilon h(x, \dot{x}) + x = 0$$

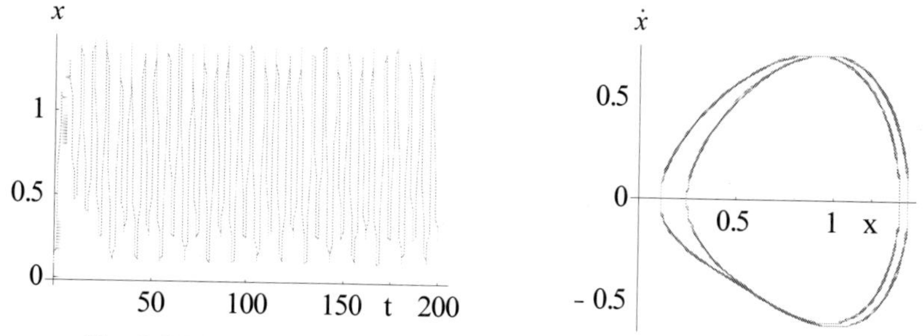

Fig. 3.4.5 Period-two behavior for the Duffing equation ($F = 0.34875$).

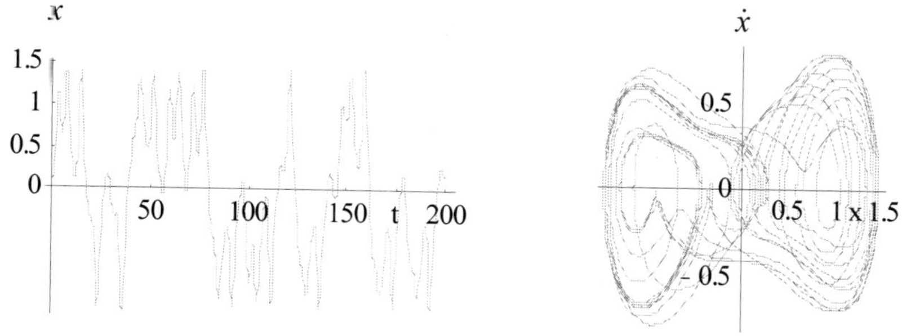

Fig. 3.4.6 Period-two behavior for the Duffing equation ($F = 0.43$).

where ε is small.[16] Such an equation is close to the equation

$$\ddot{x} + x = 0,$$

whose phase diagram consists of circles centered on the origin. We now use this fact to construct approximate solutions to the original equation.

Consider

$$\ddot{x} + \varepsilon h(x, \dot{x}) + x = 0. \tag{3.5.1}$$

The equation can be rewritten as

$$\dot{x} = y,$$

$$\dot{y} = -\varepsilon h(x, y) - x. \tag{3.5.2}$$

Assume $|\varepsilon| << 1$ and $h(0, 0) = 0$. The origin is an equilibrium point. When $\varepsilon = 0$, Eq. (3.5.1) becomes $\ddot{x} + x = 0$. Its general solution is

---

[16] This and the following sections are based on Secs. 4.1 and 4.2 in Jordan and Smith (1999).

$$x(t) = r\cos(t + \alpha),$$

where r and α are arbitrary constants. Without loss of generality, we require $a > 0$ and $\alpha = 0$. The family of phase paths for $\ddot{x} + x = 0$ is given by

$$x(t) = r\cos t, \quad y(t) = -r\sin t, \tag{3.5.3}$$

which is the family of circles,

$$x^2(t) + y^2(t) = r^2,$$

with period 2π. Introduce the total "energy" $E(t)$ as

$$E(t) = \frac{1}{2}x^2(t) + \frac{1}{2}y^2(t).$$

Taking derivatives of the total energy with regard to t and Eq. (3.5.2), we have

$$\dot{E}(t) = x\dot{x} + y\dot{y} = -y\varepsilon h(x, y).$$

Integrating the above equation from 0 to T yields

$$E(T) - E(0) = -\varepsilon\int_0^T yh(x, y)dt.$$

As we expect $x(t)$ and $y(t)$ to be periodic, E should return to its original value after one circuit. Hence, we should have

$$\int_0^T yh(x, y)dt = 0, \tag{3.5.4}$$

on the limit cycle. Now insert the approximation (3.5.3) into Eq. (3.5.4), we obtain the approximation equation

$$\int_0^{2\pi} h(r\cos t, -r\sin t)\sin t\, dt = 0. \tag{3.5.5}$$

The solution to Eq. (3.5.5) is denoted as r_0. Moreover, the stability of the cycle can also be determined. Define a function $g(r)$

$$g(r) = \varepsilon r\int_0^{2\pi} h(r\cos t, -r\sin t)\sin t\, dt.$$

On the limit cycle, $g(r_0) = 0$. The stability conditions are that if $g(r_0) = 0$, then the corresponding limit cycle is stable if $g'(r_0) < 0$ and unstable if $g'(r_0) > 0$.

Example The VdP equation.

The previous section illustrated periodic behavior of the VdP equation when ε is large. When ε is small, we may approximate periodic solutions. Comparing Eq. (3.4.1) and Eq. (3.5.1), we find

$$h(x, y) = -(1 - x^2)y.$$

Assuming $x(t) \approx r\cos t$, Eq. (3.5.5) becomes

$$\int_0^{2\pi} (r^2\cos^2 t - 1)\sin^2 t\, dt = 0.$$

This leads to $r^2/4 - 1 = 0$. Hence, the solution is $r = 2$.

We also have

$$g(r) = -\varepsilon r^2 \int_0^{2\pi} (r^2\cos^2 t - 1)\sin^2 t\, dt = -\varepsilon r^2 \pi\left(\frac{r^2}{4} - 1\right).$$

Therefore, $g'(r) = -\varepsilon r\pi(r^2 - 2)$. As $g'(2) = -2\varepsilon r\pi$, we conclude that the cycle is stable when $\varepsilon > 0$ and unstable when $\varepsilon < 0$.

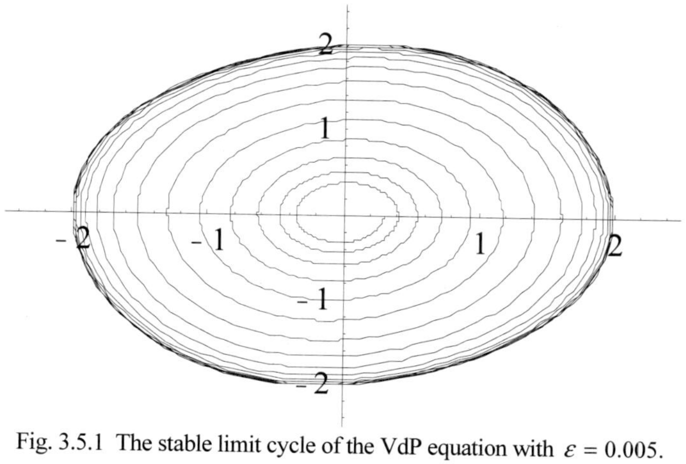

Fig. 3.5.1 The stable limit cycle of the VdP equation with $\varepsilon = 0.005$.

Exercise 3.5

1 Apply the energy balance method to find the amplitude and stability of any limit cycles in each of the following equations

(a) $\ddot{x} + \varepsilon\left(x^2 + \dot{x}^2 - 1\right)\dot{x} + x = 0$;

(b) $\ddot{x} + \varepsilon\left(\dfrac{\dot{x}^3}{3} - \dot{x}\right)\dot{x} + x = 0$;

(c) $\ddot{x} + \varepsilon\left(|x| - 1\right)\dot{x} + x = 0$.

3.6 Estimation of Amplitude and Frequency

We are still concerned with Eq. (3.5.1)

$$\ddot{x} + \varepsilon h(x, \dot{x}) + x = 0.$$

The equation is also written in the form of Eq. (3.5.2). Introduce polar coordinates

$$x(t) = r(t)\cos\theta(t), \quad \dot{y}(t) = r(t)\sin\theta(t).$$

Using

$$\cos^2\theta + \sin^2\theta = 1,$$

we obtain

$$r^2 = x^2 + y^2, \quad \tan\theta = \frac{y}{x}.$$

So we have

$$r\dot{r} = x\dot{x} + y\dot{y}, \quad r^2\dot{\theta} = x\dot{y} - \dot{x}y.$$

Substituting Eq. (3.5.2) into the above equations yields

$$\dot{r} = -\varepsilon h(r\cos\theta, r\sin\theta)\sin\theta,$$

$$\dot{\theta} = -1 - \frac{\varepsilon h(r\cos\theta, r\sin\theta)\cos\theta}{r}. \tag{3.6.1}$$

The differential equation for the phase paths is

$$\frac{dr}{d\theta} = \frac{\varepsilon h\sin\theta}{1 + \varepsilon h r^{-1}\cos\theta}. \tag{3.6.2}$$

Suppose that the system contains a limit cycle or one of the curves constituting a center. Let its time period be T. Then $r(t)$ and $\theta(t)$ all have time period T, meaning that $r(t_0 + T) = r(t_0)$ for every t_0. For very small $|\varepsilon|$, expanding the right-hand side of Eq. (3.6.2) in power of ε yields

$$\frac{dr}{d\theta} = \varepsilon h \sin\theta + O(\varepsilon^2) = O(\varepsilon). \tag{3.6.3}$$

Integrating this equation from 2π with respect to θ, we have

$$r(\theta) = r_0 + O(\varepsilon),$$

where $r_0 \equiv r(2\pi)$. Integrating $dr/d\theta = \varepsilon h \sin\theta + O(\varepsilon^2)$ from $\theta = 2\pi$ to $\theta = 0$, we get

$$0 = \varepsilon \int_{2\pi}^{0} h(r\cos\theta, r\sin\theta)\sin\theta\, d\theta + O(\varepsilon^2),$$

where we use $r(0) - r(2\pi) = 0$. Rearrange the above equation

$$\int_{0}^{2\pi} h(r\cos\theta, r\sin\theta)\sin\theta\, d\theta = O(\varepsilon).$$

Since the integral on the left does not depend on ε, a necessary condition for the phase path to be closed is

$$\int_{0}^{2\pi} h(r_0\cos\theta, r_0\sin\theta)\sin\theta\, d\theta = 0, \tag{3.6.4}$$

where we use $r(\theta) = r_0 + O(\varepsilon)$. This equation is used to approximate the amplitude, r_0. Integrating the expression for $\dot{\theta}$ with respect to t from 2π yields

$$\theta = 2\pi - t + O(\varepsilon).$$

We substitute this into Eq. (3.6.4) to obtain

$$\int_{0}^{2\pi} h(r_0\cos t, -r_0\sin t)\sin t\, dt = 0. \tag{3.6.5}$$

From Eq. (3.6.1), the period T is approximated by

$$T = \int_{0}^{T} dt = \int_{2\pi}^{0} \frac{d\theta}{\dot{\theta}} = \int_{0}^{2\pi} \frac{d\theta}{1 + \varepsilon r^{-1} h(r\cos\theta, r\sin\theta)\cos\theta}, \tag{3.6.6}$$

where we use the fact that the direction along the path for t increasing is clockwise and the positive direction for the polar coordinate θ is counterclockwise. From $r = r_0 + O(\varepsilon)$ and Eq. (3.6.6), we approximate T as

$$T = \int_0^{2\pi} \left[1 - \varepsilon r_0^{-1} h(r_0 \cos\theta, r_0 \sin\theta)\cos\theta + O(\varepsilon^2)\right] d\theta$$

$$\approx 2\pi - \frac{\varepsilon}{r_0} \int_0^{2\pi} h(r_0 \cos\theta, r_0 \sin\theta)\cos\theta \, d\theta .$$

The error is of order ε^2. The circular frequency of the periodic oscillation is

$$\omega = \frac{2\pi}{T} \approx 1 + \frac{\varepsilon}{2\pi r_0} \int_0^{2\pi} h(r_0 \cos\theta, r_0 \sin\theta)\cos\theta \, d\theta , \qquad (3.6.7)$$

where ε is small.[17] Such an equation is close to the equation

$$\ddot{x} + x = 0 ,$$

whose phase diagram consists of circles centered on the origin. We now use this fact to construct approximate solutions to the original equation.

Example The VdP equation.
Consider Eq. (3.4.1). Equation (3.6.5) is the same as Eq. (3.5.5). The amplitude is $r_0 = 2$ to order ε. By Eq. (3.6.7)

$$\omega = 1 + \frac{\varepsilon}{4\pi} \int_0^{2\pi} (4\cos^2\theta - 1)(2\sin\theta)\cos\theta \, d\theta = 1 .$$

Hence, the frequency is 1 with error $O(\varepsilon^2)$.

Example Analyze $\ddot{x} + \sin x = 0$. We note

$$\sin x \approx x - \frac{x^3}{6} .$$

The approximate equation is

---

[17] This and the following sections are based on Secs 4.1 and 4.2 in Jordan and Smith (1999).

$$\ddot{x} + x - \frac{x^3}{6} = 0.$$

Hence, $\varepsilon = -1/6$ and $h = x^3$. Equation (3.6.5) for the amplitude r_0 becomes

$$r_0^3 \int_0^{2\pi} \cos^3 \theta \sin t \, dt = 0.$$

The equation is satisfied for all r_0. Equation (3.6.7) becomes

$$\omega = 1 - \frac{1}{12\pi r_0} \int_0^{2\pi} r_0^3 \cos^4 \theta \, d\theta.$$

We have $\omega = 1 - r_0^2/16$.

Exercise 3.6

1 Obtain an approximation to the amplitude and frequency of the limit cycle for *Rayleigh's equation*

$$\ddot{x} + \varepsilon\left(\frac{\dot{x}^3}{3} - \dot{x}\right) + x = 0.$$

Chapter 4

Economic Dynamics with Scalar Differential Equations

This chapter applies concepts and theorems of the previous two chapters to analyze different models in economic model. Although the economic relations in these models tend to be complicated, we show that the dynamics of all these models are determined by motion of one-dimensional differential equations. Section 4.1 examines a one-sector growth model. As the economic mechanisms of this model will be applied to some other models in this book, we explain the economic structure in details. This section also applies the Liapunov theorem to guarantee global asymptotical stability of the equilibrium. Section 4.2 depicts the one-sector growth model proposed in Sec. 4.1 with simulation. Section 4.3 examines the one-sector-growth model for general utility functions. Section 4.4 examines a model of urban economic growth with housing production. In Sec. 4.5, we examine a dynamic model to see how leisure time and work hours change over time in association with economic growth. Section 4.6 examines dynamics of sexual division of labor and consumption in association of modern economic growth. We illustrate increases of women labor participation as a "consequence" of economic growth as well as changes of labor market conditions. Section 4.7 introduces the Uzawa two-sector model. In Sec. 4.8, we re-examine the Uzawa model with endogenous consumer behavior. The models of this chapter show the essence of economic principles in many fields of economics, such as equilibrium economics (as a stationary state of a dynamic economics), growth theory, urban economics, and gender economics. The basic ideas and conclusions of this chapter require some books to explain, if that is possible. This also proves power of differential equations theory.

4.1 The One-Sector Growth (OSG) Model

We are concerned with an economy of one production sector. The model proposed in this section is called the OSG model initially constructed by Zhang.[1] Most aspects of our model are similar to the Solow one-sector growth model. It is assumed that there is only one (durable) good in the economy under consideration. Households own assets of the economy and distribute their incomes to consume and save. Production sectors or firms use inputs such as labor with varied levels of human capital, different kinds of capital, knowledge and natural resources to produce material goods or services. Exchanges take place in perfectly competitive markets. Production sectors sell their product to households or to other sectors and households sell their labor and assets to production sectors. Factor markets work well; factors are inelastically supplied and the available factors are fully utilized at every moment.

Behavior of producers

Let $K(t)$ denote the capital existing at each time t and $N(t)$ the flow of labor services used at time t for production. The production process is described by some sufficiently smooth function

$$F(t) = F(K(t), N(t)). \tag{4.1.1}$$

We assume that $F(K(t), N(t))$ is neoclassical.[2] We assume that the production function exhibits constant returns to scale. It is straightforward to check that a linear homogeneous production $F(K, N)$ has the following properties:

(i) $F / N = F(k,1) \equiv f(k),\ k \equiv K / N$;

---

[1] Since some other models in this book are closely related to this model, we will explain it in details. See Zhang (1999) for further explanations.

[2] A production function $F(K, N)$ is called neoclassical if it satisfies the following conditions: (1) $F(K, N)$ is nonnegative if K and N are non-negative; (2) $F(0,0) = 0$; (3) F_K and F_N are nonnegative; (4) there exist second partial derivatives of F with respect to K and N; (5) the function is homogeneous of degree one, $F(\lambda K, \lambda N) = \lambda F(K, N)$, for all nonnegative λ; (6) the function is strictly quasi-concave.

(ii) $F_K = \partial F / \partial K = f'(k) > 0$;
(iii) $F_N = f(k) - kf'(k) > 0$; and
(iv) The *Euler theorem* holds

$$KF_K + NF_N = F.$$

We portray intensive form $f(k)$ of the aggregate production function in Fig. 4.1.1. As we move out to the right along the production function, output per worker increases as the capital/labor ratio $k(t)$ rises. The shape of $f(k)$ in the figure reflects the assumption that there are diminishing returns to increases in $k(t)$. The increment to output per worker declines as capital per worker rises. The slope of the production function becomes flatter from left to right. This means that although more capital always leads to more output, it does so at a decreasing rate.

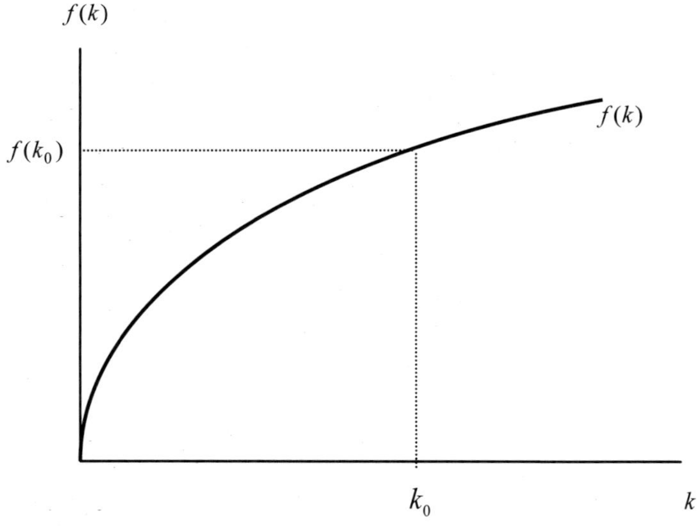

Fig.4.1.1 Intensive form of the aggregate production function.

We assume (identically numerous) one production sector. Its goal of economic production is to maximize its current profit

$$\pi(t) = p(t)F(t) - r(t)\,K(t) - w(t)N(t),$$

where $p(t)$ is the price of product, $r(t)$ is the rate of interest, and $w(t)$ is the wage rate.

We assume that the output good serves as a medium of exchange and is taken as numeraire. We thus set

$$p(t) = 1$$

and measure both wages and rental flows in units of the output good. Maximizing π with regard to K and N as decision variables yields

$$r = F_K = f'(k), \quad w = F_N = f(k) - kf'(k). \tag{4.1.2}$$

Since the production function is homogeneous of degree one, we have

$$KF_K + NF_N = F$$

or $rK + wN = F$. This result means that the total revenue is used up to pay all factors of the production. We thus conclude that if the production function is homogenous of degree one, the 'adding-up requirement' is satisfied.

Behavior of consumers

Consumers obtain income

$$Y = rK + wN = F,$$

from the interest payment rK and the wage payment wN. We call Y the *current income* in the sense that it comes from consumers' daily toils (payment for human capital) and consumers' current earnings from ownership of physical capital. The sum of money that consumers are using for consuming, saving, or transferring are not necessarily equal to the temporary income because consumers can sell wealth to pay, for instance, the current consumption if the temporary income is not sufficient for buying food and touring the country. Retired people may live not only on the interest payment but also have to spend some of their wealth. The total value of wealth that consumers can sell to purchase goods and to save is equal to $K(t)$. The *gross disposable income* is equal to

$$Y^* = Y + K.$$

The gross disposable income is used for saving and consumption and for paying the depreciation of the wealth.

We assume that consumers pay the depreciation of capital goods which they own. The total amount is equal to $\delta_k K(t)$ where

δ_k $(0 \le \delta_k < 1)$ is the depreciation rate of physical capital. At each point of time, consumers would distribute the total available budget among saving $S(t)$, consumption of goods $C(t)$, and payment for depreciation $\delta_k K(t)$. The budget constraint is given by

$$C(t) + \delta_k K(t) + S(t) = Y^*(t) = Y(t) + K(t).$$

Since the consumer has to pay the depreciation $\delta_k K(t)$, we call

$$Y(t) + K(t) - \delta_k K(t)$$

the *disposable income*, which equals the net income minus the depreciation loss

$$\hat{Y}(t) \equiv Y(t) + K(t) - \delta_k K(t)$$
$$= rK(t) + w(t)N(t) + \delta K(t), \tag{4.1.3}$$

in which

$$\delta \equiv 1 - \delta_k.$$

In our model, at each point of time, consumers have two variables to decide. A consumer decides how much to consume and to save. Consumption and saving exhaust the consumers' disposable personal income, i.e.

$$C(t) + S(t) = \hat{Y}(t). \tag{4.1.4}$$

We assume that utility level $U(t)$ that the consumers obtain is dependent on the consumption level $C(t)$ of commodity and the net saving $S(t)$. We use the Cobb-Douglas utility function to describe consumers' preferences

$$U(t) = C^{\xi}(t)S^{\lambda}(t), \quad \xi, \lambda > 0, \tag{4.1.5}$$

in which ξ and λ are respectively the propensities to consume goods and to own wealth. We assume

$$\xi + \lambda = 1$$

without loss of generality. Maximizing Eq. (4.1.5) subject to Eq. (4.1.3) yields

$$C^*(t) = \xi\hat{Y}(t), \quad S^*(t) = \lambda\hat{Y}(t). \tag{4.1.6}$$

The optimal choice is illustrated in Fig. 4.1.2.

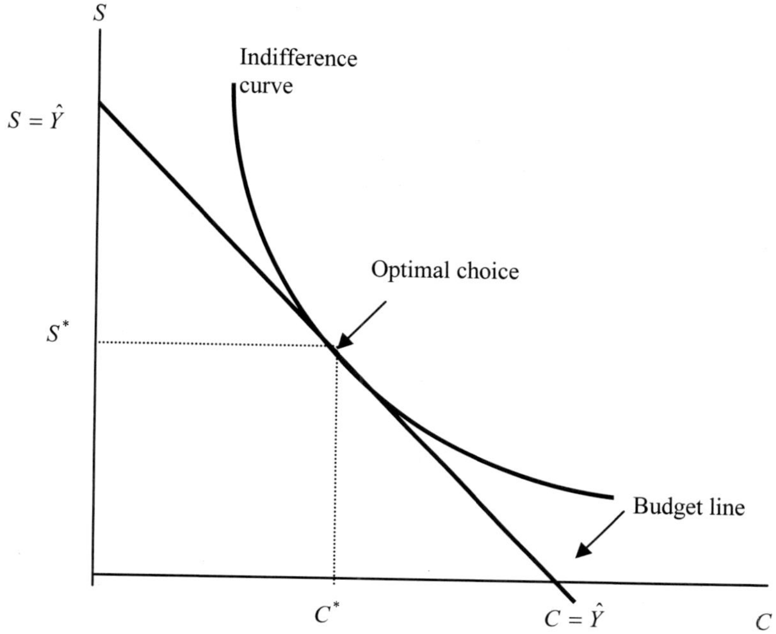

Fig. 4.1.2 Optimal choice at time t.

Dynamics in capital-labor ratio

It appears reasonable to consider population as independent of economic conditions, as a first approximation. Here, we assume that the population dynamics is exogenously determined in the following way

$$\dot{N}(t) = nN(t),$$

where n is a constant. The change in the households' wealth is equal to the net saving minus the wealth sold at time t, i.e.

$$\dot{K}(t) = S(t) - K(t). \tag{4.1.7}$$

The above equations determine all the variables, $K(t)$, $C(t)$, $S(t)$, $N(t)$, $F(t)$, $r(t)$, $w(t)$, $U(t)$, in the system. We call this dynamic system (with proper initial conditions) the one-sector growth (OSG) model. We now rewrite the dynamics in terms of per capita. From Eq. (4.1.6) and $Y(t) = F(t)$, we have

$$S(t) = \lambda \hat{Y}(t) = \lambda(F(t) + \delta K(t)).$$

Inserting the above equation into Eq. (4.1.7) yields

$$\dot{K}(t) = \lambda F(t) + \lambda \delta K(t) - K(t) = \lambda F(t) - \xi_k K(t), \tag{4.1.8}$$

where we use

$$\xi + \lambda = 1, \ \delta \equiv 1 - \delta_k, \ \xi_k \equiv \xi + \lambda \delta_k .$$

As $k(t) = K(t)/N(t)$, we have

$$\dot{k}(t) = \frac{\dot{K}(t)}{N(t)} - \frac{K(t)}{N(t)}\left(\frac{\dot{N}(t)}{N(t)}\right) = \frac{\dot{K}(t)}{N(t)} - nk(t). \tag{4.1.9}$$

Set Eq. (4.1.8) in Eq. (4.1.9)

$$\dot{k}(t) = \lambda f(k(t)) - (\xi_k + n)k(t). \tag{4.1.10}$$

The function $f(k)$ has the properties: $f(0) = 0$, $f'(k) > 0$ if k is nonnegative. It can be seen that once the capital per capita $k(t)$ is determined, all the variables in the system, such as K, F, Y, C, w, r and U can be calculated accordingly. We now examine some properties of Eq. (4.1.10).

The existence of a stable steady state

We now show that the economy will eventually arrives a steady state/equilibrium – a situation in which output per worker $y(t)\ (= Y(t)/N(t))$, consumption per worker $c(t)\ (= C(t)/N(t))$, and capital stock per worker $k(t)$ don't change over time.

Theorem 4.1.1 (The existence of a unique equilibrium and the stability) If δ and λ satisfy

$$0 < \frac{\xi_k + n}{\lambda} < f'(0), \tag{4.1.11}$$

then there exists a unique positive value k^* for Eq. (4.1.10) such that $\lambda f(k^*) = (\xi_k + n)k^*$. The equilibrium point k^* is asymptotically stable in the region $k > 0$.

Proof: We introduce function

$$\Phi(k) \equiv \lambda f(k) - \bar{\delta} k,$$

where $\bar{\delta} \equiv \xi + \lambda \delta_k + n$. For any $K > 0$, by continuity we have

$$0 = F(K, 0) = K\lim_{N \to 0}\left(\frac{N}{K}\right)F\left(\frac{K}{N}, 1\right).$$

Hence, we have

$$\lim_{k \to +\infty} \frac{f(k)}{k} = 0.$$

This guarantees that for any $\bar{\delta}/\lambda > 0$ there exists k^1 such that $f(k)/k < \bar{\delta}/\lambda$ for all $k > k^1$. On the other hand

$$\frac{\bar{\delta}}{\lambda} < f'(0) = \lim_{k \to 0}\left[\frac{f(k) - f(0)}{k}\right] = \lim_{k \to 0}\left[\frac{f(k)}{k}\right].$$

Hence, there exists $k^2 > 0$ such that $f(k)/k > \delta/\lambda$ for $0 < k < k^2$, which means $\Phi(k) > 0$ for $0 < k < k^2$. Since $\Phi(k)$ is continuous, the intermediate value theorem guarantees that there exists at least one point $k^* > 0$ such that $f(k^*)/k^* > \bar{\delta}/\lambda$, i.e., $\Phi(k^*) = 0$.

Let there be more than one positive solution and let $k^* > 0$ denote the one with minimum value. First, we note

$$\Phi' = \lambda f'(k) - \bar{\delta}, \quad f'(k) > 0, \quad \Phi''(k) = \lambda f'' < 0, \; \Phi(0) = 0,$$

$$\Phi'(0) > 0.$$

Since $\Phi(0) = 0$ and $\Phi(k^*) = 0$, there exists k^0, $0 < k^0 \le k^*$ such that $\Phi'(k^0) = 0$. It is trivial to check that

$$\Phi(k^*) = \Phi'(k^*) = 0.$$

This implies $(F_N =)\; f - kf' = 0$ which is impossible. Hence, we have $k^0 \le k^*$ and $\Phi'(k^*) < 0$. Because $\Phi''(k) < 0$ and $\Phi'(k^0) = 0$, we have $\Phi'(k) < 0$ for any $k^0 \le k^* \le k$. Since $\Phi(k^*) = 0$ and $\Phi'(k) < 0$ for any $k \ge k^*$, we conclude that it is impossible to find such a $k_1 > k^*$ at which $\Phi(k_1) = 0$. This means that the system has a unique equilibrium.

We now confirm that the equilibrium is stable. The equilibrium point of the system is asymptotically stable if for any admissible initial point k_0 the solution $\psi(t;t_0)$ to $\dot{k} = \lambda f - \bar{\delta}k$ satisfies

$$\lim_{t \to \infty} \psi(t;k_0) = k^*.$$

The asymptotical stability can be proved by applying the Liapunov theorem. Define the Liapunov function

$$V(x) \equiv |x|,$$

where $x \equiv k - k^*$ and k^* is the equilibrium value. We have $V(x) \geq 0$ and $V(x) = 0$ iff $x = 0$. Since

$$\dot{V} = \text{sgn}(x)\dot{k} = \text{sgn}(x)\{\lambda f(k) - \bar{\delta}k\}\begin{cases} < 0 & \text{if x} \neq 0, \\ = 0 & \text{if x} = 0. \end{cases}$$

Hence, the equilibrium $k = k^*$ is globally asymptotically stable.

The economic development can be described as follows. In the long run the economy will always converge smoothly to the unique equilibrium capital/labor ratio from any positive starting point. Moreover, along the balanced growth path, capital expands at the same rate as the population growth rate plus depreciation rate of capital. The importance of this model lies in the fact that it supplies a very simple consistent system to simultaneously determine all significant variables - labor and capital inputs to production, outputs, saving, consumption, investment - in economic development. Irrespective of its oversimplified assumptions about production function, saving and investment behavior, the role of monetary variables and so on, it is a powerful tool since it gives us a logical framework to analyze some aspects of economic development.

When the economy reaches the stationary capital intensity, capital per capita will remain the same as time passes, but the stock of capital $K(t)$ remains growing infinitely at the same predetermined rate as the labor force n. The sustainable growth rate of the model is exogenously given by n. This can be confirmed by

$$K(t) = kN_0 e^{nt}, \quad F(t) = f(k)N_0 e^{nt}, \quad C(t) = c(k)N_0 e^{nt}.$$

We now formally describe the properties of the dynamic system.

We have examined the dynamic properties of the OSG model. It has a long-run steady state at which growth rate of per capita consumption is zero. We now examine 'transitional dynamics' — a time-dependent process of how per capita income converges toward its long-run steady state. Dividing Eq. (4.1.10) by $k(t)$, we obtain growth rate $g_k(t)$ of per capita capital as

$$g_k(t) \equiv \frac{\dot{k}(t)}{k(t)} = \frac{\lambda f(k(t))}{k(t)} - (\xi_k + n).$$

Here, $g_k(t)$ stands for growth rate of per capita capital at time t. The above equation says that the growth rate of per capita capital equals the difference between two terms, $\lambda f / k$ and $\xi_k + n$, which we plot against k in Fig. 4.1.3. The first curve is a downward-sloping curve and the second term is a horizontal line. The vertical distance between the curve and the line equals the growth rate of per capita capital. As shown before, there is a unique equilibrium. The figure shows that to the left of the steady state, the curve lives above the line. Hence, the growth rate is positive and k increases over time. As k rises, the growth rate declines. Finally, k reaches k^* as the growth rate becomes zero. An analogous argument demonstrates that if the system starts from the right of the steady state, the growth rate is negative. As k declines, the growth rate rises and finally becomes zero.

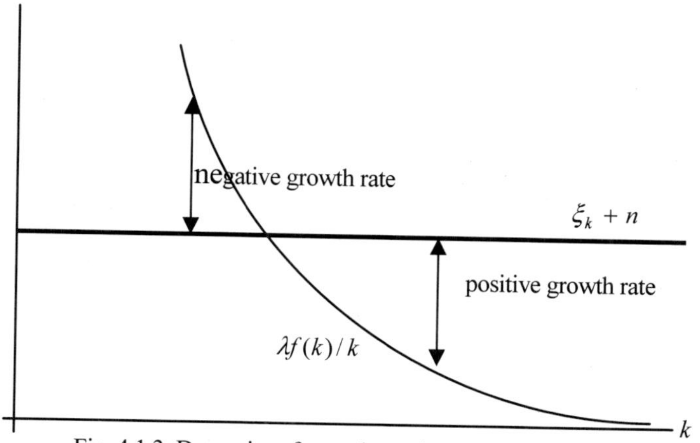

Fig. 4.1.3 Dynamics of growth rate in the OSG model.

Exercise 4.1

1 If the production function is taken on the Cobb-Douglas form

$$F(t) = Ae^{mt} K^{\alpha} N^{\beta},$$

where m is the rate of technical progress and A is a constant. Introduce ratio of capital and effective labor as follows

$$\hat{k}(t) \equiv \frac{K(t)}{A(t)N(t)}.$$

Show that the capital accumulation equation corresponding to Eq. (4.1.10) becomes

$$\dot{\hat{k}}(t) = \lambda \hat{k}(t)^{\alpha} - (\xi_k + n + m)\hat{k}(t).$$

Also find the equilibrium and stability conditions of the OSG model with exogenous technology.

4.2 The OSG Model with the Cobb-Douglas Production Function

This section solves the OSG model when the production function is taken on the Cobb-Douglas production function

$$F(t) = AK(t)^{\alpha} N(t)^{\beta}, \quad \alpha, \beta > 0, \ \alpha + \beta = 1, \tag{4.2.1}$$

where A is a number measuring overall productivity, and α and β are parameters. The parameter A is often referred to as *total factor productivity* or simply *productivity*. We summarize the OSG model in per capita terms under Eq. (4.2.1)

$$f = Ak^{\alpha}, \quad r = \alpha A k^{-\beta}, \quad w = \beta A k^{\alpha},$$

$$c = \xi\left(Ak^{\alpha} + \delta k\right), \quad s = \lambda\left(Ak^{\alpha} + \delta k\right),$$

$$\bar{s} = \lambda - \xi\delta\frac{k^{\beta}}{A}, \tag{4.2.2}$$

$$\dot{k} = \lambda A k^{\alpha} - (\xi_k + n)k. \tag{4.2.3}$$

We now simulate this model, specifying the parameters as

$$\alpha = 0.3, \quad n = 0.015, \quad \lambda = 0.55, \quad \delta_k = 0.015.$$

The population grows at annual growth rate of 1.5 percent and capital depreciates at rate of 1.5 percent. The propensity to own wealth is 0.55, which may be unreasonably low for a rich economy. We will discuss possible change of λ later on. We don't consider any technological change here and specify $A = 1$. The parameter α is set at 0.3. With initial conditions of $k(0) = 0.7$, we run the dynamics of the economy for 25 years. Figure 4.1.1a describes the dynamics of per-capita capital and per-capita income. The per-capita capital and per-capita income exhibit similar pattern of growth – in initial stages they grow very rapidly. The growth rates of these two variables are demonstrated in Fig. 4.2.1c. As the per-capita consumption and saving are positively proportionally related to the per-capita capital and per-capita income, they grow in the same pattern $k(t)$ and $y(t)$, as demonstrated in Fig. 4.2.1b. The wage rate grows not rapidly even during the initial stage of fast economic growth and it becomes stationary after a few years. Similarly but in the opposite direction, the rate of interest declines in the initial years but becomes stationary soon.

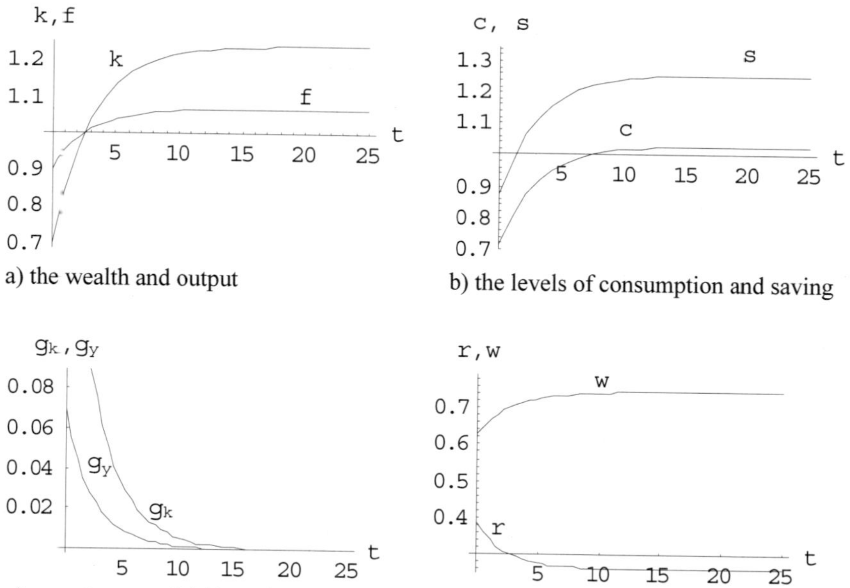

a) the wealth and output b) the levels of consumption and saving

c) growth rates of $k(t)$ and $y(t)$ d) the rate of interest and wage rate

Fig. 4.2.1 The dynamics of the OSG model with $\alpha = 0.3$.

We describe the dynamics of the model with the help of computer. In fact, we can analytically solve all the variables. Equaion (4.2.3) is a

Bernoulli equation in the variable $k(t)$. Inserting $z(t) = k^{\beta}(t)$ into Eq. (4.2.3) yields

$$\dot{z} + (1-\alpha)(\xi_k + n)z = (1-\alpha)\lambda A, \tag{4.2.4}$$

which is a standard first-order linear differential equation. The solution is given by

$$z(t) = \left(z(0) - \frac{\alpha A}{\xi_k + n} \right) e^{-\beta(\xi_k + n)} + \frac{\alpha A}{\xi_k + n}. \tag{4.2.5}$$

Substituting $z(t) = k^{\beta}(t)$ back to Eq. (4.2.5), we obtain

$$k^{\beta}(t) = \left(k^{\beta}(0) - \frac{\alpha A}{\xi_k + n} \right) e^{-\beta(\xi_k + n)t} + \frac{\alpha A}{\xi_k + n}, \tag{4.2.6}$$

where $k(0)$ is the initial value of the capital-labor ratio $k(t)$. This solution is what determines the time path of $k(t)$. Once we know $k(t)$, all the other points are explicitly determined at any point of time.

As $t \to +\infty$, the exponential expression will approach zero. Consequently, letting $t \to +\infty$ yields the unique steady-state capital ratio

$$k^* = \left(\frac{\lambda A}{\xi_k + n} \right)^{1/\beta}. \tag{4.2.7}$$

The capital-labor ratio will approach a constant as its equilibrium value. This steady state, as shown in the preceding section, varies directly with the propensity to save λ, the technology A, and inversely with the propensity to consume ξ, the population growth rate n, and the capital depreciation rate δ_k.

We mentioned that a rise in the propensity to own wealth may either increase or reduce consumption. We now simulate the model to demonstrate how the equilibrium values of y and c vary as the propensity to own wealth changes. We specify the parameters as follows

$$\alpha = 0.3, \quad n = 0.015, \quad \delta_k = 0.03, \quad A = 0.8.$$

Using

$$y = Ak^{\alpha}, \quad c = \xi\left(Ak^{\alpha} + \delta k\right), \quad s = \lambda\left(Ak^{\alpha} + \delta k\right), \quad \hat{y} = c + s,$$

we depict how c, s, y and $\hat{y}$ vary as λ changes for $\lambda \in [0.01, 0.99]$. The consumption per capita increases as λ rises until λ reaches 0.5628; after $\lambda = 0.5628$, the consumption per capita falls as λ rises. The simulation shows that from the long-run perspective, it is desirable to have a 'proper' propensity to own wealth. The saving per capita, the current income, and the disposable income rise as λ rises. A national economy may definitely become rich in this model by increasing the propensity to own wealth. If an economy 'over-saves', its income rises but consumption falls.

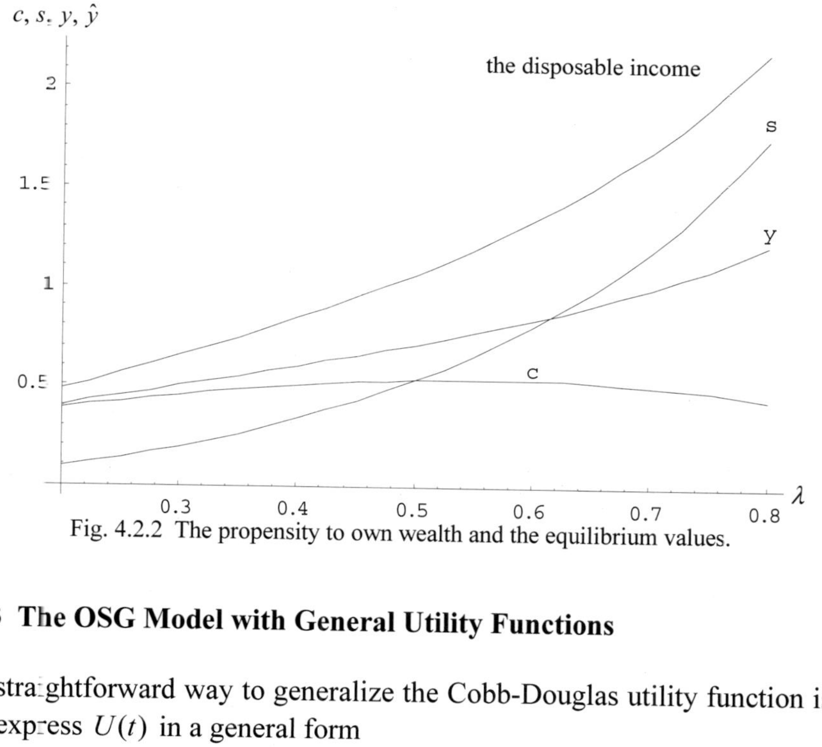

Fig. 4.2.2 The propensity to own wealth and the equilibrium values.

4.3 The OSG Model with General Utility Functions

A straightforward way to generalize the Cobb-Douglas utility function is to express $U(t)$ in a general form

$$U(t) = U(C(t), S(t); K(t), Y(t), t),$$

where $C(t)$ and $S(t)$ are variables that consumers decide and $K(t)$, $Y(t)$, and t are given parameters that may affect utility.

In the above formula, for instance, we may use t to express the age of the consumer. The age of the consumer is a key factor in affecting consumption in the life cycle theory. When we study a special individual or a certain age group, this parameter is significant for examining behavior pattern.

The consumer is to choose his most preferred bundle $(c(t), s(t))$ of consumption and saving under his budget constraint. The utility maximizing problem at any time is defined by

$$\underset{c,s\geq 0}{Max}\, U(c,s)\,,\ \text{s.t. } c(t)+s(t)\leq \hat{y}(t)\,. \tag{4.3.1}$$

The following theorem holds.

Theorem 4.3.1 Let $U(c,s): R_+^2 \to R^1$ be a C^1 function that satisfies the *monotonicity assumption*, which says that $\partial U/\partial c > 0$ and $\partial U/\partial s > 0$ for each (c, s) satisfying the constraint set in Eq. (4.3.1). Suppose that (c^*, s^*) maximizes U on the constraint set. Then, there is a scalar $\bar{\lambda}^* > 0$ such that

$$\frac{\partial U}{\partial c}(c^*, s^*) \leq \bar{\lambda}^*, \quad \frac{\partial U}{\partial s}(c^*, s^*) \leq \bar{\lambda}^*.$$

We have $\partial U/\partial c = \bar{\lambda}^*$ if $c^* \neq 0$ and $\partial U/\partial s = \bar{\lambda}^*$ if $s^* \neq 0$. If both $c^* > 0$ and $s^* > 0$, then

$$\frac{\partial U}{\partial c}(c^*, s^*) = \bar{\lambda}^*, \quad \frac{\partial U}{\partial s}(c^*, s^*) = \bar{\lambda}^*.$$

The budget constraint is binding,

$$c + s = \hat{y}.$$

Conversely, suppose that U is a C^1 function which satisfies the monotonicity assumption and that $(c^*, s^*) > 0$ satisfies the budget set and the first order conditions

$$\frac{\partial U}{\partial c} = \frac{\partial U}{\partial s} = \bar{\lambda}\,.$$

If U is C^2 and if

$$|\bar{H}| = \begin{vmatrix} 0 & 1 & 1 \\ 1 & U_{cc} & U_{cs} \\ 1 & U_{sc} & U_{ss} \end{vmatrix} = 2U_{cs} - U_{cc} - U_{ss} > 0\,,$$

then (c^*, s^*) is a strict local solution to the utility maximization problem. If U is quasiconcave and $\nabla U(c, s)$ for all $(c, s) \neq (c^*, s^*)$, then (c^*, s^*) is a global solution to the problem.

The proof of this proposition and other general properties of the problem can be found in advanced textbooks on microeconomics.[3]

We now specify some properties of U to obtain explicit conclusions. We require U to be a C^2 function, and satisfy $U_c > 0, U_s > 0$ for any $(c, s) > 0$. Construct the Lagrangian

$$L(c,s,\overline{\lambda}) = U(c,s) + \overline{\lambda}(\hat{y} - c - s).$$

The first-order condition for maximization is

$$U_c = U_s = \overline{\lambda}, \quad \hat{y} - c - s = 0. \tag{4.3.2}$$

The bordered Hessian for the problem is

$$|\overline{H}| = \begin{vmatrix} 0 & 1 & 1 \\ 1 & U_{cc} & U_{cs} \\ 1 & U_{sc} & U_{ss} \end{vmatrix} = 2U_{cs} - U_{cc} - U_{ss}.$$

The second-order condition tells that given a stationary value of the first-order condition, a positive $|\overline{H}|$ is sufficient to establish it as a relative maximum of U. It is known that the bordered Hessian is identical with the endogenous-variable Jacobian. Hence, if $|\overline{H}|$ is not equal to zero, we can directly apply the implicit function theorem to the problem. That is, the first-order condition has a solution as C^1 functions of the disposable income $\hat{y}$. Taking the derivatives of Eqs. (4.3.2) with respect to $\hat{y}$, we have

$$U_{cc}\frac{dc}{d\hat{y}} + U_{cs}\frac{ds}{d\hat{y}} = U_{sc}\frac{dc}{d\hat{y}} + U_{ss}\frac{ds}{d\hat{y}},$$

$$1 = \frac{dc}{d\hat{y}} + \frac{ds}{d\hat{y}}.$$

We solve these functions

$$\frac{ds}{d\hat{y}} = \frac{U_{sc} - U_{cc}}{2U_{cs} - U_{ss} - U_{cc}},$$

[3] see Mas-Colell, *et al.* (1995).

$$\frac{dc}{d\hat{y}} = \frac{U_{sc} - U_{ss}}{2U_{cs} - U_{ss} - U_{cc}}.$$

We see that $0 < ds/d\hat{y} < 1$ and $0 < dc/d\hat{y} < 1$ in the case of $U_{sc} \geq 0$ under the second-order condition of maximization. We denote an optimal solution as function of the disposable income

$$(c(t),\, s(t)) = (c(\hat{y}(t)),\, s(\hat{y}(t))).$$

The vector $(c(\hat{y}(t)),\, s(\hat{y}(t)))$ is known as the *Walrasian* (or *ordinary* or *market*) *demand function*, when it is single-valued for all positive disposable income.

The capital accumulation equation for general utility function is given by

$$\dot{K}(t) = s(\hat{y}(t))N(t) - K(t). \tag{4.3.3}$$

Inserting

$$\dot{k}(t) = \frac{\dot{K}(t)}{N(t)} - nk(t),$$

into Eq. (4.3.3) yields

$$\dot{k}(t) = s(\hat{y}(k)) - (1 + n)k(t), \tag{4.3.4}$$

where

$$\hat{y}(k(t)) = \frac{\hat{Y}(t)}{N(t)} = f(k(t)) + \delta k(t).$$

In a stationary state

$$s(\hat{y}(k)) = (1 + n)k. \tag{4.3.5}$$

We now show that this equation has a unique solution. Define

$$\Phi(k) \equiv \frac{s(\hat{y})}{(1 + n)k} - 1, \quad k \geq 0. \tag{4.3.6}$$

When k is approaching zero, $\hat{y}$ $(= f(k) + \delta k)$ is also approaching zero, and hence $s(\hat{y})$ is coming near zero. As $0 < s'(\hat{y}) < 1$ and $f'(k) \to \infty$ as $k \to 0$,

$$\lim_{k \to 0} \frac{s(\hat{y})}{(1 + n)k} = \frac{s'(0)(f'(0) + \delta)}{(1 + n)} > 1. \tag{4.3.7}$$

When k is approaching positive infinity, $\hat{y}$ tends to positive infinity. As $0 < s'(\hat{y}) < 1$ and $f'(k) \to 0$ as $k \to +\infty$, we have

$$\lim_{k \to +\infty} \frac{s(\hat{y})}{(1+n)k} = \frac{s'(+\infty)(f'(+\infty)+\delta)}{(1+n)} < 1.$$

Taking derivatives of Eq. (4.3.6) with respect to k yield

$$\frac{d\Phi}{dk} = \left[\frac{s'(\hat{y})k(f'(k)+\delta)}{s(\hat{y})} - 1\right]\frac{s(\hat{y})}{(1+n)k^2}.$$

We now show that $d\Phi / dk < 0$ for $k > 0$. To prove this, we use $ds / d\hat{y} < 1$ and the inequality $f'(k) < f(k)/k$ (which also guarantees $w > 0$). By Eq. (4.3.5) and the definition of $\hat{y}$, we have

$$\frac{s'(\hat{y})k(f'(k)+\delta)}{(1+n)s(\hat{y})} < \frac{s'(\hat{y})(f(k)+\delta k)}{(1+n)s(\hat{y})}$$

$$= \frac{s'(\hat{y})\hat{y}}{(1+n)s(\hat{y})} < 1, \quad (4.3.8)$$

where we use $s'(\hat{y}) < 1$ and $\hat{y}/s(\hat{y}) \le 1$ to guarantee the right inequality. We thus conclude $d\Phi / dk < 0$ for $k > 0$. The equation, $\Phi(k) = 0$ for $k > 0$ has a unique solution because of Eqs. (4.3.7) and (4.3.8), and $d\Phi / dk < 0$. We now demonstrate that the unique stationary state is stable.

For the steady state to be stable, the following conditions must prevail

$$\left.\frac{d[s(\hat{y}(k)) - (1+n)k]}{dk}\right|_{k=k^*}$$

$$= s'(\hat{y})(f'(k)+\delta) - (1+n) < 0. \quad (4.3.9)$$

From the equilibrium condition and inequalities (4.3.7), we have

$$\frac{s'(\hat{y})(f'(k)+\delta)}{1+n} = \frac{s'(\hat{y})k(f'(k)+\delta)}{s'(\hat{y})} < 1. \quad (4.3.10)$$

Accordingly, the inequality (4.3.8) is satisfied. We see that the conclusions for the Cobb-Douglas production and utility functions are also similarly held for more general production functions and utility functions. Summarizing the discussions, we obtain the following theorem.

Theorem 4.3.1 Given a production function that is 'neoclassical' and a utility function that is a C^2 function, and satisfies $U_c > 0, U_s > 0$ for any $(c(t), s(t)) > 0$. Let the bordered Hessian be positive for any nonnegative $(c(t), s(t))$. Then the capital-labor ratio converges monotonically to a unique positive steady state. The unique stationary state is stable.

The stability guaranteed above is local. We now show that if $s(\hat{y})$ is concave in $\hat{y}$, then the system is globally stable. Because of

$$\frac{d^2c}{dy^2} = -\frac{d^2s}{dy^2}$$

by equation

$$1 = \frac{dc}{d\hat{y}} + \frac{ds}{d\hat{y}},$$

concavity of s implies convexity of c. From the first-order conditions, it is straightforward to give that conditions under which s is concave, we omit the expression because we lack a clear economic interpretation.

Asymptotical stability can be proved by applying Liapunov's theorem. Define the Liapunov function

$$V(x(t)) \equiv x(t)^2,$$

where $x(t) \equiv k(t) - k^*$ and k^* is the equilibrium value. We have $V(x) \geq 0$ and $V(x) = 0$ iff $x = 0$. Differentiation of $V(x(t))$ with respect to t gives

$$\dot{V} = 2x\dot{k} = 2x\{s(\hat{y}(k)) - (1 + n)k\}$$
$$= 2x\{s(\hat{y}(k^* + x)) - (1 + n)(k^* + x)\},$$

where we use Eq. (4.3.3). By concavity of $f(x + k^*)$

$$f(k^* + x) \leq f(k^*) + xf'(k^*).$$

According to its definition, $\hat{y}(k)$ is also concave in k. Hence,

$$\hat{y}(k^* + x) \leq \hat{y}(k^*) + x\hat{y}'(k^*).$$

Since $ds / d\hat{y} > 0$, we have

$$\dot{V} \leq 2x\{s(\hat{y}(k^*) + x\hat{y}'(k^*)) - (1 + n)(k^* + x)\}.$$

Concavity of $s(\hat{y})$ estimates

$$\dot{V} \le 2x\{s(\hat{y}) + x\hat{y}'(k^*)s'(\hat{y}) - (1+n)(k^* + x)\}.$$

The equilibrium condition (4.3.4) rewrites the above inequality

$$\dot{V} \le 2x^2(1+n)\left\{\frac{s'(\hat{y})(f'(k^*) + \delta)}{(1+n)} - 1\right\}.$$

By the inequality (4.3.9), we conclude $dV/dt < 0$ if $x \ne 0$ and $dV/dt = 0$ at $x = 0$. Hence, the equilibrium $k = k^*$ is asymptotically stable.

4.4 Urban Growth with Housing Production[4]

We consider an isolated system. The population is homogenous. The households achieve the same utility level regardless of where they are located. All the markets are competitive. The system is geographically linear and consists of two parts - the CBD and the residential area. The isolated state consists of a finite strip of land extending from a fixed central business district (CBD) with constant unit width. All economic activities are concentrated in the CBD. The households occupy the residential area. The CBD is located at the left end of the linear territory, as illustrated in Fig. 4.4.1. As we can get similar conclusions if we locate the CBD at the center of the linear system, the special location will not essentially affect our discussion.

The system consists of two, industrial and housing, sectors. The industrial production is similar to that in the OSG model. The housing production is similar to that in the Muth model.[5] Housing is supplied with combination of capital and land. We select industrial good to serve as numeraire. We now describe the economic model of the isolated state. To describe the industrial sector, we introduce

[4] The urban model is based on Chap. 2 in Zhang (2002). The model may be extended with different urban elements (e.g., Arnott, 1979, Brueckner and Rabenau, 1981, Miyao, 1981, Wang, 1993). The literature on urban dynamics is referred to Zhang (2002).

[5] Muth (1973).

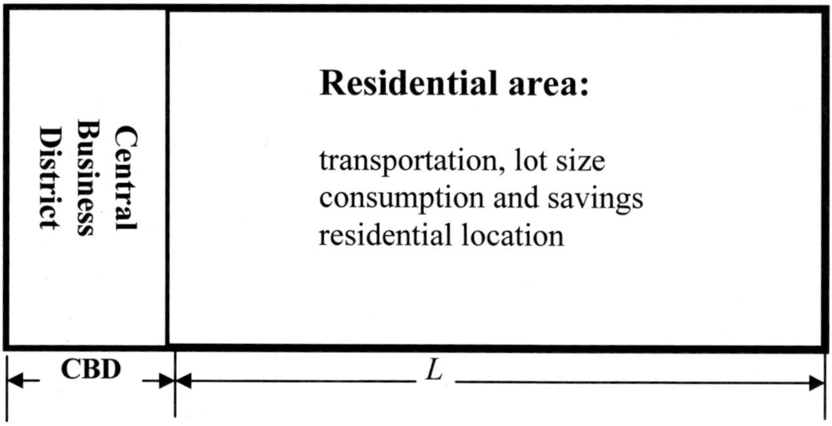

Fig. 4.4.1 Economic geography of the isolated state.

N = the fixed population;

$K_i(t)$ = the capital stocks employed by the industrial sector at time t;

$w(t)$ and $r(t)$ = the wage rate and the rate of interest,

$F(t)$ and $C(t)$ = the output of the industrial sector and the total consumption of the commodity;

L = the fixed (territory) length of the isolated state;

ω = the distance from the CBD to a point in the residential area;

$R(\omega, t)$ and $R_h(\omega, t)$ = land rent and housing rent per household at location ω;

$k(\omega, t)$ = capital stocks owned by the household at location ω;

$c(\omega, t)$ and $y(\omega, t)$ = the consumption and the net income of the household at location ω, respectively;

$n(\omega, t)$ and $L_h(\omega, t)$ = the residential density and the lot size of the household at location ω;

$K_h(t)$ = the capital stocks employed by the housing sector; and

$K(t)$ = the total capital stock of the economy.

We assume that industrial production is carried out by combination of capital and labor force in the form of

$$F(t) = K_i^{\alpha} N^{\beta}, \quad \alpha + \beta = 1, \quad \alpha, \beta > 0, \tag{4.4.1}$$

where α and β are parameters. The marginal conditions for profit maximization are given by

$$r = \frac{\alpha F}{K_i}, \quad w = \frac{\beta F}{N}. \tag{4.4.2}$$

Housing is produced with land and non-land inputs. Let us denote $c_h(\omega,t)$ housing service received by the household at location ω. We specify the housing service production function

$$c_h(\omega, t) = L_h^{\alpha_h}(\omega, t) k_h^{\beta_h}(\omega, t), \quad \alpha_h + \beta_h = 1, \quad \alpha_h, \beta_h > 0, \tag{4.4.3}$$

where $k_h(\omega, t)$ is the input level of capital per household at location ω. The marginal conditions are given by

$$r = \frac{\alpha_h c_h R_h}{k_h}, \quad R = \frac{\beta_h c_h R_h}{L_h}, \quad 0 \le \omega \le L. \tag{4.4.4}$$

According to the definitions of L_h and n, we have

$$n(\omega, t) = \frac{1}{L_h(\omega, t)}, \quad 0 \le \omega \le L. \tag{4.4.5}$$

The relationship between $k_h(\omega, t)$ and $K_h(t)$ is given by

$$K_h(t) = \int_0^L n(\omega, t) k_h(\omega, t) d\omega. \tag{4.4.6}$$

To define net income, we now specify land ownership. For simplicity, we assume the public ownership, which means that the revenue from land is equally shared among the population. The total land revenue is given by

$$\bar{R}(t) = \int_0^L R(\omega, t) d\omega.$$

The income from land per household is given by

$$\bar{r}(t) = \frac{\bar{R}(t)}{N}.$$

The current income $y(\omega, t)$ of the household at location ω consists of three parts: the wage income, the income from land ownership and the interest payment for the household's capital stocks. That is

$$y(\omega, t) = rk(\omega, t) + w(t) + \bar{r}(t). \tag{4.4.7}$$

The gross disposable income is

$$y^*(\omega, t) = y(\omega, t) + k(\omega, t).$$

Many previous models of residential location theory are developed with regard to rent theory since Alonso's seminal work.[6] In this approach residential location is modeled on the basis of the utility function. Location choice is closely related to the existence and quality of such physical environmental attributes as open space and noise pollution as well as social environmental quality. Basically following this approach, we assume that utility level U of the household at location ω is dependent on the temporary consumption level $c(\omega, t)$, housing conditions $c_h(\omega, t)$, the leisure time $T_h(\omega, t)$, the amenity $E(\omega, t)$, and the saving $S(\omega, t)$ as

$$U(\omega, t) = ET_h^{\sigma} c^{\xi} c_h^{\eta} S^{\lambda}, \quad \sigma, \xi, \eta, \lambda > 0, \tag{4.4.8}$$

where $E(\omega, t)$ and $T_h(\omega, t)$ are specified as

$$E(\omega) = \frac{\mu_1}{n(\omega)^{\mu}}, \quad T_h(\omega) = T_0 - \upsilon\omega, \quad \mu_1, \mu, \upsilon, T_0 > 0. \tag{4.4.9}$$

The function $E(\omega, t)$ implies that the amenity level at location ω is determined by the residential density at the location. The function $T_h(\omega, t)$ means that the leisure time is equal to the total available time T_0 minus the traveling time $\upsilon\omega$ from the CBD to the dwelling site. As the population is homogeneous, we will have

$$U(\omega_1, t) = U(\omega_2, t), \quad 0 \le \omega_1, \omega_2 \le L.$$

The budget constraint is given by

$$c(\omega, t) + R_h(\omega, t)c_h(\omega, t) + S(\omega, t) + \delta_k k(\omega, t) = y^*(\omega, t),$$

where δ_k is the rate of capital depreciation. Maximizing U subject to the budget constraint yields

---

[6] Alonso (1964).

$$R_h(\omega)c_h(\omega) = \eta\rho\hat{y}(\omega), \ \ c(\omega) = \xi\rho\hat{y}(\omega),$$
$$S(\omega) = \lambda\rho\hat{y}(\omega), \tag{4.4.10}$$

where

$$\hat{y}(\omega) \equiv y^*(\omega) - \delta_k k(\omega), \ \ \rho \equiv \frac{1}{\xi + \eta + \lambda}.$$

According to the definition of $S(\omega, t)$, the capital accumulation for the household at location ω is given by

$$\dot{k}(\omega) = S(\omega) - k(\omega), \ \ 0 \le \omega \le L.$$

Substituting $S(\omega, t)$ into Eqs. (4.4.10) into the above equation yields

$$\dot{k}(\omega) = sy(\omega) - \delta k(\omega), \ \ 0 \le \omega \le L, \tag{4.4.11}$$

where

$$s \equiv \lambda\rho, \ \ \delta \equiv \delta_k + (\xi + \eta)\rho_0, \ \ \rho_0 \equiv (1 - \delta_k)\rho.$$

As the state is isolated, the total population is distributed over the whole urban area. The population constraint is given by

$$\int_0^L n(\omega, t)d\omega = N. \tag{4.4.12}$$

Similarly, the consumption constraint is given by

$$\int_0^L n(\omega, t)c(\omega, t)d\omega = C(t). \tag{4.4.13}$$

We also have

$$\bar{S}(t) + C(t) = F(t), \tag{4.4.14}$$

where

$$\bar{S}(t) \equiv \int_0^L \{S(\omega) - k(\omega) + \delta_k k(\omega)\}n(\omega)d\omega.$$

The assumption that capital is always fully employed is given by:

$$K_i + K_h = K, \tag{4.4.15}$$

$$\int_0^L k(\omega)n(\omega)d\omega = K(t). \tag{4.4.16}$$

We have thus built the dynamic growth model with endogenous spatial distribution of wealth, consumption and population, capital accumulation and residential location. The system has 13 space-time-dependent variables, k, c, c_h, k_h, L_h, S, n, E, T_h, U, R_h, R, and y, and 10 time-dependent variables, F, K_i, K_h, K, C, w, r, $\bar{S}$, $\bar{R}$, and $\bar{r}$. The system contains 23 independent equations.

Before examining the dynamic properties of the system, we show that the dynamics can be described by the motion of a single variable $K(t)$. Multiplying Eq. (4.4.11) by $n(\omega, t)$ and then integrating the resulted equation from 0 to L with respect to ω yields:

$$\dot{K} = s\bar{Y} - \delta K, \tag{4.4.17}$$

where

$$\bar{Y} \equiv \int_0^L y(\omega)n(\omega)d\omega .$$

We now show that $\bar{Y}(t)$ can be expressed as a function of $K(t)$. Multiplying all the equations in Eqs. (4.4.10) by $n(\omega, t)$ and then integrating the resulted equations from 0 to L with respect to ω, we obtain

$$\frac{rK_h}{\alpha_h} = \eta\rho\bar{Y} + \eta\rho_0 K, \quad C = \xi\rho\bar{Y} + \xi\rho_0 K,$$

$$\bar{S} = \lambda\rho\bar{Y} + \lambda\rho_0 K, \tag{4.4.18}$$

where we use $c_h R_h = rk_h / \alpha_h$ in Eqs. (4.4.4). Substituting $\bar{S}$ and C from Eqs. (4.4.18) into Eq. (4.4.14) yields

$$(\xi + \lambda)\rho\bar{Y} - \eta\rho_0 K = F. \tag{4.4.19}$$

From $r = \alpha F / K_i$ and $rK_h / \alpha_h = \eta\rho\bar{Y} + \eta\rho_0 K$, we have

$$\frac{\alpha K_h F}{\alpha_h K_i} = \eta\rho\bar{Y} + \eta\rho_0 K .$$

Substituting

$$\bar{Y} = \frac{\eta\rho_0 K + F}{\rho(\xi + \lambda)}$$

obtained from Eq. (4.4.19) into the above equation yields

$$\Omega(K_i) \equiv \frac{\alpha K K_i^{-\beta}}{\alpha_h} - A K_i^{\alpha} - \frac{\eta \rho_0 K}{(\xi + \lambda)\rho N^{\beta}} = 0, \qquad (4.4.20)$$

where we use $K_h = K - K_i$ and

$$A \equiv \frac{\alpha}{\alpha_h} + \frac{\eta}{\xi + \lambda}.$$

We now show that for any given $K > 0$,

$$\Omega(K_i) = 0$$

has a unique solution for $0 < K_i < K$. As $\Omega(0) > 0$, $\Omega(K) < 0$ and $d\Omega / dK_i < 0$, we see that $\Omega(K_i) = 0$ has a unique solution as a function of K. Let us represent this unique relationship by: $K_i(t) = \Lambda(K(t))$. From Eq. (4.4.20), we have

$$\frac{d\Lambda}{dK} = \frac{A\Lambda}{(\beta K / \alpha_h \Lambda + A)\alpha K} > 0. \qquad (4.4.21)$$

That is, an increase in the total capital stock will always increase the capital stocks employed in the industrial production. From $K_i = \Lambda(K)$ and $K_h = K - K_i$, we see that the capital stocks, K_i and K_h, employed by the industrial and housing sectors are uniquely determined as functions of the total capital stocks K at any point of time. From $K_h = K - K_i$ and Eq. (4.4.21), we have the impact of changes in K on K_h as follows

$$\alpha K \left(\frac{\beta K}{\alpha_h \Lambda} + A \right) \frac{dK_h}{dK} = \frac{\alpha \beta K^2}{\alpha_h \Lambda} + \alpha A K - A\Lambda$$

$$= \left(\frac{\alpha K}{\alpha_h \Lambda} - A \right) \beta K + A(K - \Lambda) > 0,$$

where we use $K > \Lambda$ and

$$\frac{\alpha K}{\alpha_h \Lambda} - A = \frac{\eta \rho_0 K}{\rho N^{\beta} \Lambda^{\alpha} (\xi + \lambda)} > 0.$$

From $F = \Lambda^{\alpha} N^{\beta}$, F is a unique function of K. Substituting

$$\overline{Y} = \frac{\eta \rho_0 K + F(\Lambda)}{\rho(\xi + \lambda)}$$

into Eq. (4.4.17), we have

$$\dot{K} = s^* F\{\Lambda(K)\} - \delta^* K, \tag{4.4.22}$$

where

$$s^* \equiv \frac{s}{(\xi + \lambda)\rho} > 0, \;\; \delta^* = \delta - \eta \rho_0 s^* > 0 .$$

At any point of time, the dynamic equation (4.4.22) determines the value of the total capital stocks K. We can show that all the other variables are uniquely determined as a function of K and ω ($0 \le \omega \le L$) at any point of time.

Proposition 4.4.1 For any given (positive) level of the total capital stocks $K(t)$ at any point of time, all the other variables in the system are uniquely determined as functions of $K(t)$ and ω ($0 \le \omega \le L$). The dynamics of $K(t)$ is given by Eq. (4.4.22).

Proof: We already uniquely determined K_i, K_h and F as functions of K. The rate of the interest rate r and the wage rate w are uniquely determined by Eqs. (4.4.2). From Eqs. (4.4.4) and (4.4.5), we have $n(\omega)k_h(\omega) = \alpha_h R(\omega)/\beta_h r$. Substituting this into Eqs. (4.4.6), we have $\bar{R} = \beta_h r K_h / \alpha_h$ where r and K_h are functions of K. We directly have: $\bar{r} = \bar{R}/N$. From Eq. (4.4.7), we see that $y(\omega)$ is a known function of K and $k(\omega)$ (as r, w and $\bar{r}$ are functions of K). We determine $k(\omega)$ as a function of K. We can get $c(\omega)$, $S(\omega)$ and $c_h(\omega)R_h(\omega)$ directly from Eqs. (4.4.10). We obtain $\bar{S}$ and $k_h(\omega)$ from Eqs. (4.4.14) and $k_h = \alpha_h c_h R_h / r$ from Eqs. (4.4.4), respectively.

We now have five space-time dependent variables, L_h, n, c_h, R_h, and R, to determine. From Eqs. (4.4.10), we see that c, $c_h R_h$ and S are known functions of K. Substituting Eqs. (4.4.10) into $U(\omega)$ in Eq. (4.4.8), it is obvious to see that we may have $U(\omega)$ in the form of

$$U(\omega, K) = \frac{f(\omega, K)}{R_h^\eta(\omega, K)},$$

where $f(\omega, K)$ is a function of ω and K. On the other hand, substituting $c_h = L_h^{\alpha_h} k_h^{\beta_h}$ into $r = \alpha_h R_h c_h / k_h$, we have

$$R_h = \frac{rn^{\alpha_h} k_h^{\alpha_h}}{\alpha_h},$$

where r and k_h are given functions of ω and K. Substituting R_h into $U(\omega, K)$ and then using $U(\omega) = U(0)$, we have

$$n(\omega, K) = \left\{ \frac{f(\omega, K)}{f(0, K)} \right\}^{1/\alpha_h \eta} n(0, K).$$

Hence, if we can determine $n(0, K)$ as a function of K, then $n(\omega)$ is given. Substituting $n(\omega, K)$ in the above equation into the population constrain equation (4.4.12), we can explicitly get $n(0, K)$ as a function of K. From $L_h = 1/n$, $c_h = L_h^{\alpha_h} k_h^{\beta_h}$, $r = \alpha_h R_h c_h / k_h$ and $R = \beta_h R_h c_h / L_h$, we directly get L_h, c_h, R_h and R as functions of ω and K.

We can thus explicitly determine the motion of the system over time and space. It should be remarked that the result that the dynamics can be explicitly given in the simple form as Eq. (4.4.22) is important. This makes it possible to explicitly determine stability of the system.

We now examine problems of equilibrium and stability. From Eq. (4.4.22), equilibrium is determined as a solution of the following equation

$$s^* F\{\Lambda(K)\} = \delta^* K. \tag{4.4.23}$$

From Eq. (4.4.20), we have

$$K = \frac{A\Lambda^{\alpha}}{\alpha\Lambda^{-\beta} / \alpha_h - \eta\rho_0 / \rho N^{\beta}(\xi + \lambda)}. \tag{4.4.24}$$

Substituting Eq. (4.4.24) into Eq. (4.4.23) yields

$$\Lambda = \left[\frac{\alpha}{\alpha_h \{\delta^* A / s^* + \eta\rho_0 / \rho(\xi + \lambda)\}} \right]^{1/\beta} N, \tag{4.4.25}$$

where we use $F = \Lambda^{\alpha} N^{\beta}$. Substituting Eq. (4.4.25) into Eq. (4.4.24), we directly determine K as a function of the parameters in the system. That is, the dynamic system has a unique equilibrium. From

$$\frac{d(s^* F - \delta^* K)}{dK} = \frac{d(s^* F)}{dK} - \delta^* = -\frac{\beta \delta^* K}{\beta K + \alpha_h A \Lambda} < 0,$$

where we use $s^* F = \delta^* K$ and Eq. (4.4.21), we see that the unique equilibrium is stable.

We now determine equilibrium values of the other variables. The capital employed by the industrial sector is given by: $K_i = \Lambda$. From Eqs. (4.4.1), (4.4.2) and (4.4.15), we directly get F, w, r and K_h. From Eq. (4.4.17), we have $\bar{Y} = \delta K / s$. Substituting this into Eqs. (4.4.18), we have

$$C = \frac{\xi K}{\lambda}, \quad \bar{S} = \delta_k K .$$

We also have

$$\bar{R} = \frac{\beta_h r K_h}{\alpha_h}, \quad \bar{r} = \frac{\bar{R}}{N} .$$

From Eq. (4.4.11), we have $y = \delta k / s$. Substituting Eq. (4.4.7) into this equation yields

$$k = \frac{sW}{\delta - sr}, \tag{4.4.26}$$

where

$$W \equiv w + \bar{r} = \frac{\beta - \beta_h + \alpha \beta_h K / K_i}{\alpha_h N} F , \tag{4.4.27}$$

in which we use Eqs. (4.4.2). From $y = \delta k / s$ and Eqs. (4.4.10), we obtain

$$c_h R_h = \frac{\eta k}{\lambda}, \quad c = \frac{\xi k}{\lambda}, \quad S = \delta_k k . \tag{4.4.28}$$

From Eqs. (4.4.4), we have

$$k_h = \frac{\alpha_h c_h R_h}{r} .$$

Substituting $c_h R_h$ from Eqs. (4.4.28) into this equation yields

$$k_h = \frac{\alpha_h \eta R_h}{\lambda r} . \tag{4.4.29}$$

Substituting Eq. (4.4.29) into $U(\omega)$ and then using $U(0) = U(\omega)$, we have

$$\frac{R_h(\omega)}{R_h(0)} = \left\{\frac{E(\omega)T(\omega)^\sigma}{E(0)T(0)^\sigma}\right\}^{1/\eta}.$$

Substituting $c_h = L_h^{\alpha_h} k_h^{\beta_h}$ into $k_h = \alpha_h c_h R_h / r$, we have

$$R_h = \frac{r n^{\alpha_h} k_h^{\alpha_h}}{\alpha_h}.$$

Substituting this equation and Eq. (4.4.29) into $R_h(\omega)/R_h(0)$ in the above equation, we have

$$n(\omega) = n(0)\left(1 - \frac{\upsilon\omega}{T_0}\right)^B, \quad 0 \le \omega \le L, \tag{4.4.30}$$

where we use Eqs. (4.4.9) and $B \equiv \sigma/(\alpha_h \eta + \mu) > 0$. From the above equation, we can analyze how the amenity parameter, the available time, the housing technology parameter, the traveling speed parameter and the propensity to use leisure time affect the residential density distribution. It can be seen that $R_h(\omega)$ and $n(\omega)$ are decreasing functions of the distance ω. Substituting Eq. (4.4.30) into Eq. (4.4.12), we have

$$n(0) = \frac{\upsilon N(1 + B)}{T_0 - (1 - \upsilon L / T_0)^{1+B} T_0},$$

where we use Eq. (4.4.26). We thus obtained $n(\omega)$. Here, we require the following inequality:

$$\upsilon L < T_0.$$

That is, the household's available time is more than that needed from traveling from the CBD to the boundary of the economic system. We get L_h, R_h, R and c_h, respectively, by Eq. (4.4.5), $R_h = r n^{\alpha_h} k_h^{\alpha_h} / \alpha_h$, Eqs. (4.4.4) and (4.4.3). We have thus explicitly solved all the variables in the dynamic system with endogenous economic geography. Summarizing the above discussion, we have the following proposition.

Proposition 4.4.2 The dynamic system has a unique stable equilibrium.

On the basis of Eq. (4.4.22), we can illustrate the unique equilibrium and its stability as in Fig. 4.4.2. It can be seen that the macrodynamic aspects of the model are quite similar to the Solow model. In the remainder of this study, we examine effects of changes in some parameters on the economic growth and geography.

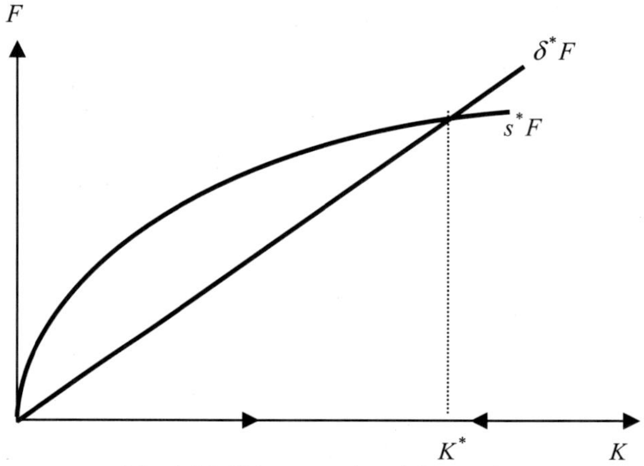

Fig. 4.4.2 Urban growth and the steady state.

4.5 Endogenous Time in the OSG Model

In most parts of the world the value of their time is very low. Work is hard, wage rate is low, life is harsh. In a few advanced economies, the value of the time is high. As recorded in Schultz, real wages measured in terms of the cost of food are changeable over history, as illustrated in Table 4.5.1.[7] Schultz explained that "The high price of human time is a clue to many puzzles. These puzzles include the shift in institutional support from the rights of property to that of human rights, the decline in fertility, the increasing dependence of economic growth on value added by labor relative to that added by materials, the increases in labor's share of national income, the decline in hours worked, and the high rate at which human capital increases.[8]"

[7] Source: Schultz (1993: 83).
[8] Schultz (1993: 84).

This section proposes a dynamic interdependence of economic growth, consumption and time allocation on the basis of the one-sector growth model developed in Sec. 4.1. This section is a simplified version of the dynamic trade model with time distribution proposed by Zhang in 1995.[9]

Table 4.5.1 An illustration of evolution of time values

	2 weeks of wages in bushels of wheat
Time of Ricardo (1817) England	1
Marshall's time (1890) United States	20
Eighty years later (1970) United States	300
India field laborer in 1970	2

The production aspects of the economic system are similar to the one-sector growth model. The parameters, N, δ_k, ξ, λ, variables, $K(t)$, $F(t)$, $S(t)$, $C(t)$, $r(t)$ and $w(t)$, are defined the same as in Sec. 4.1. We introduce two variables $T(t)$ and $T_h(t)$ to stand for the working time and leisure time of each worker. We assume that labor and capital are always fully employed. The total labor force $N^*(t)$ is given by

$$N^*(t) = T(t)N .$$

Here, we omit any other possible impact of working time on productivity. The production function of the economy is specified as $F = AK^{\alpha} N^{*\beta}$, $\alpha + \beta = 1$. Before simulation, let $A = 1$. The marginal conditions are given by

$$r = \frac{\alpha F}{K}, \quad w = \frac{\beta F}{N^*}. \tag{4.5.1}$$

The current income Y consists of the wage income and payment interest for its capital, i.e.

$$Y = rK + wTN .$$

By Eqs. (4.5.1), we have $Y(t) = F(t)$. The gross disposable income is $Y^*(t) = Y(t) + K(t)$. As in Sec. 4.1, the budget constraint is

$$C + \delta_k K + S = Y^* = rK + wTN + K .$$

---

[9] Zhang (1995).

The time constraint requires that the amounts of time allocated to each specific use add up to the time available

$$T(t) + T_h(t) = T_0 .$$

Substituting this equation into the above budget constrain yields

$$C + wT_h N + S = \hat{Y} ,$$

where

$$\hat{Y} = rK + wT_0 N + \delta K, \ \ \delta \equiv 1 - \delta_k .$$

We assume that at each point of time consumers' preferences over leisure time, consumption and saving can be represented by the following utility function

$$U(t) = T_h^{\sigma}(t) C^{\xi}(t) S^{\lambda}(t), \ \ \sigma, \xi, \lambda > 0 ,$$

where σ is called propensity to use leisure time, ξ, propensity to consume, and λ, propensity to own wealth. Consumers' problem is to choose current consumption, leisure, and saving in such a way that utility levels are maximized. Maximizing $U(t)$ subject to the budget constraints yields

$$wT_h N = \sigma\rho\hat{Y}, \ \ C = \xi\rho\hat{Y}, \ \ S = \lambda\rho\hat{Y} , \tag{4.5.2}$$

where

$$\rho \equiv \frac{1}{\sigma + \xi + \lambda} .$$

The change in the households' wealth is equal to the net saving minus the wealth sold at time t, i.e. $\dot{K} = S - K$. By Eqs. (4.5.2), $K(t)$ evolves according to

$$\dot{K} = \lambda\hat{Y} - K .$$

We have built the dynamic model. The dynamics consist of one-dimensional differential equation for $K(t)$.

In order to analyze the properties of the dynamic system, it is necessary to express the dynamics in terms of one variable at any point of time. From $\dot{K}(t) = \lambda\hat{Y}(t) - K(t)$ and the definition of $\hat{Y}(t)$, it is sufficient to express $T(t)$ or $T_h(t)$ as function of $K(t)$. To show this, first we substitute $r(t) = \alpha F(t)/K(t)$ and $w(t) = \beta F(t)/N^*(t)$ into $T_h(t) = \sigma \hat{Y}(t)/w(t)N$ in Eqs. (4.5.1). We get

$$\frac{\beta N T_h}{\sigma \rho K} - \frac{\beta T_0 N}{K} = \frac{\alpha N^*}{K} + \delta \frac{N^*}{F},$$

where the definition of $\hat{Y}(t)$ is used. From the definition of N^* and $T + T_a = T_0$, we have

$$N^*(t) = (T(t) - T_0)N.$$

Substituting $N^* = (T - T_0)N$ into the above equation yields

$$\Phi(T_h) \equiv (T_0 - T_h)^{\alpha} \frac{\delta N^{\alpha}}{K^{\alpha}} - \frac{\overline{\alpha} T_h N}{K} + \frac{T_0 N}{K} = 0, \qquad (4.5.3)$$

where we use $F = AK^{\alpha} N^{*\beta}$ and $\overline{\alpha} \equiv \beta / \sigma\rho + \alpha > 1$. It is straightforward to show that for any $K(t) > 0$, the equation $\Phi(T_h(t)) = 0$ has a unique solution for $0 < T_h(t) < T_0$. That is, we can consider $T_h(t)$ as a unique function of $K(t)$. We thus obtain the following lemma.

Lemma 4.5.1 The dynamics are given by the following differential equation

$$\dot{K} = \lambda \hat{Y} - K = \lambda Y - (1 - \delta\lambda)K,$$

where $\hat{Y}$ is a function only of K. The variables are uniquely determined as functions of $K(t)$ at any point of time by the following procedure: T_h by Eq. (4.5.3) $\rightarrow$ $T = T_0 - T_h$ $\rightarrow$ $N^* = TN$ $\rightarrow$ $F = AK^{\alpha} N^{*\beta}$ $\rightarrow$ $r = \alpha F / K$ and $w = \beta F / N^*$ $\rightarrow$ $\hat{Y} = Y + \delta K$ $\rightarrow$ C and S by Eqs. (4.5.2).

Proposition 4.5.1 The dynamic system has a unique stable equilibrium. The variable K is given as below. All the other variables are given by the procedure given by Lemma 4.5.1.

Proof: We now show that the system has a unique equilibrium. First, from $w = \beta F / TN$, $wNT_h = \sigma\rho\hat{Y}$, and $\lambda\hat{Y} = K$, we get

$$T_h = \frac{\sigma T K}{\beta \lambda F}. \qquad (4.5.4)$$

Substituting $\lambda\hat{Y} = K$ into C and S in Eqs. (4.5.1) yields $C = \xi K / \lambda$ and $S = \delta_k K$. Substituting these equations into

$$F = C + S - K + \delta_k K,$$

we get $F = \bar{\lambda} K$, where $\bar{\lambda} \equiv \xi_k / \lambda$. From Eq. (4.5.4) and $F = \bar{\lambda} K$

$$T_h = \frac{\sigma T}{\xi_k \beta}.$$

With this equation and $T + T_h = T_0$, we solve the equilibrium time distribution

$$T = \frac{\xi_k \beta T_0}{\sigma + \beta \xi_k}, \quad T_h = \frac{\sigma T_0}{\sigma + \beta \xi_k}.$$

By $F = \bar{\lambda} K$ and $F = K^{\alpha} N^{*\beta}$, we directly solve

$$K = \frac{TN}{\bar{\lambda}^{1/\beta}}.$$

By the procedure given in Lemma 4.5.1, we explicitly solve all the variables as functions of K. We thus found the unique equilibrium of the dynamic system.

To show that the unique equilibrium is stable, we calculate the derivative of the left-hand side of $\dot{K} = \lambda\rho\hat{Y} - K$ with respect to K. If the derivative evaluated at the equilibrium is negative (positive), then the equilibrium is stable (unstable). By $r = \alpha F / K$, $w = \beta F / N^*$ and the definition of $\hat{Y}(t)$, we get

$$\hat{Y} = \frac{\beta T_0 F}{T} + \alpha F + \delta K,$$

where T is a function of T_h. We have

$$\frac{d(\lambda\rho\hat{Y} - K)}{dK} = \lambda\rho \frac{\partial \hat{Y}}{\partial T}\frac{dT}{dK} + \lambda\rho \frac{\partial \hat{Y}}{\partial K} - 1.$$

This equation holds at any state. In equilibrium, we have

$$\lambda\rho \frac{\partial \hat{Y}}{\partial K} - 1 = -\left(\frac{\beta T_0}{T} + \alpha\right)\frac{\beta\rho\lambda F}{K} < 0,$$

where we use $\lambda\rho\hat{Y} = K$. We also have

$$\lambda \upsilon \frac{\partial \hat{Y}}{\partial T}\frac{dT}{dK} = -\frac{\alpha\beta\lambda\rho T_h F}{T^2}\frac{dT}{dK}.$$

By Eq. (4.5.3), we have

$$\frac{dT}{dK} = -\frac{\delta\beta T}{\{\delta\alpha K + (\alpha + \beta/\sigma\rho)F\}} < 0.$$

Using the above equations in equilibrium, we have

$$\frac{d(\lambda\rho\hat{Y} - K)}{dK} = \frac{\alpha\beta^2\delta\lambda\rho T_h F}{\{\delta\alpha K + (\alpha + \beta/\sigma\rho)F\}T} - \frac{\beta T_0/T + \alpha}{K}\beta\rho\lambda F$$

$$= -\frac{\{\delta\alpha + (\alpha + \beta/\sigma\rho)(1 + \beta T_h/T)F/K\}\beta\rho\lambda F}{\delta\alpha K + (\alpha + \beta/\sigma\rho)F} < 0,$$

where we use $\alpha + \beta = 1$ and $T + T_h = T_0$.

For simulation, we specify the following parameters

$A = 0.5,\ N = 1,\ T_0 = 24,\ \alpha = 0.3,\ \delta_k = 0.03,\ \lambda = 0.25,\ \sigma = 0.40$.

With the initial condition of $T(0) = 12$, we simulate the motion of $T(t)$, $K(t)$, $Y(t)$, $C(t)$, and $w(t)$ with period of 10 years. Figure 4.5.1 depicts the motion of these variables during the given period. Figure 4.5.1a shows that work time declines as time passes. In the initial state, half of the total time is spent on working. Workers gradually reduce work hours. Figure 4.5.1b shows that capital stock increases over time. The two plots show that capital accumulation moves in the same direction as leisure changes. As capital stock increases, the current income and the consumption rise. As capital stock increases, time value declines. Figure 4.5.1c demonstrates that the consumption and the current income rise rapidly in the initial period but the growth rates decline as the system approaches equilibrium. Figure 4.5.1d predicts that as work time declines, the wage rate declines.

We now examine impact of changes on dynamic processes of the system. First, we examine the case that all the parameters, except ρ, are the same as in Fig. 4.5.1. We reduce the propensity to enjoy time from 0.40 to 0.35. The simulation results are demonstrated in Fig. 4.5.2. The solid lines in Fig. 4.5.2 are the same as in Fig.6.3.1, representing the values of the corresponding variables when $\sigma = 0.40$; the dashing lines in Fig. 4.5.2 represent the new values of the variables when $\sigma = 0.35$. Figure 4.5.2a shows that as the propensity to enjoy leisure falls, work time increases from the initial time rather than declines as in the old situation with $\sigma = 0.40$. Figure 4.5.2b demonstrates that as work time rises, the wealth declines as time passes. Although the wealth declines as time passes in the case of $\sigma = 0.35$, the wealth in case $\sigma = 0.35$ is more than the wealth $\sigma = 0.40$ at each point of time. The reason is that as the propensity to enjoy time falls, the propensity to consume rises (as we fix the propensity to own wealth). As shown in Fig. 4.5.2c, both the consumption and the current income fall as time passes. In Fig. 4.5.2d, we see that the wage rate converges in the long term; but in the initial period, the wage rate in the new case declines as time passes.

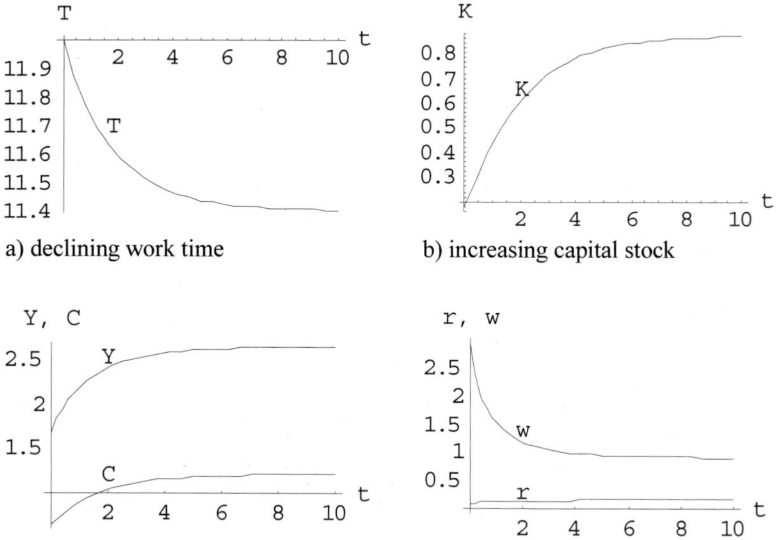

a) declining work time b) increasing capital stock

c) the income and consumption d) the wage rate and rate of interest

Fig. 4.5.1 Simulating the OSG model with endogenous time.

We examine the case that all the parameters, except A, are the same as in Fig. 4.5.1. We consider that the total productivity rises from 0.50 to 0.60. The simulation results are demonstrated in Fig. 4.5.3. The solid lines in Fig. 4.5.3 are the same as in Fig. 6.3.1, representing the values of the corresponding variables when $A = 0.50$; the dashing lines in Fig.

4.5.3 represent the new values of the variables when $A = 0.60$. We can similarly interpret the new dynamics as we interpreted Fig. 4.5.2.

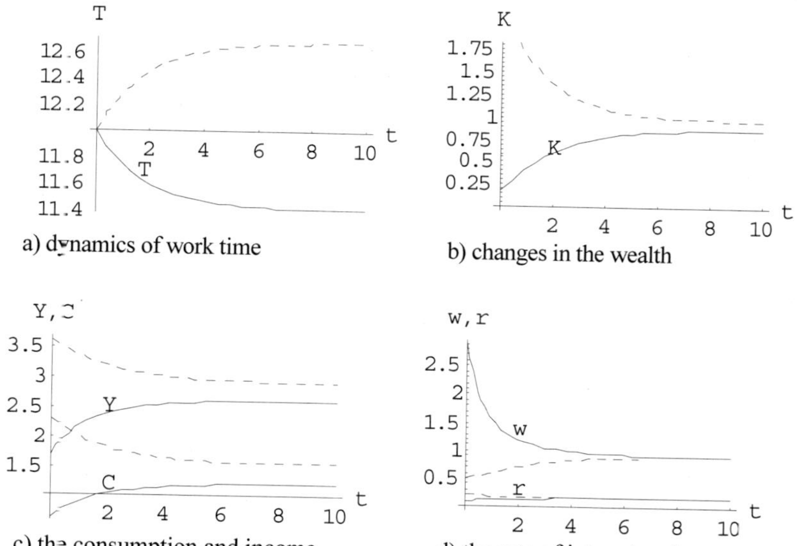

a) dynamics of work time

b) changes in the wealth

c) the consumption and income

d) the rate of interest and wage rate

Fig. 4.5.2 As the propensity to enjoy leisure σ declines from 0.40 to 0.35.

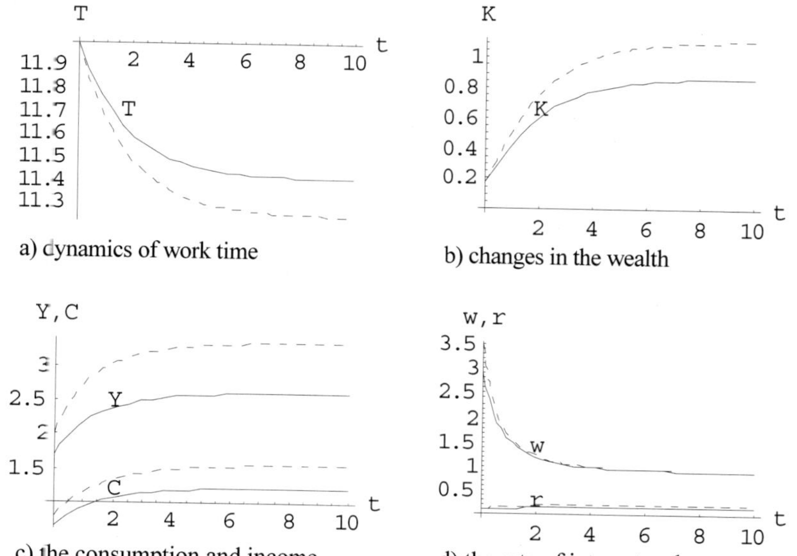

a) dynamics of work time

b) changes in the wealth

c) the consumption and income

d) the rate of interest and wage rate

Fig. 4.5.3 As the productivity A increases from 0.5 to 0.6.

It should be noted that another way to explore evolution of time distribution is to consider possibility of value change as an endogenous process of economic evolution. For instance, we may generally propose that the propensity to own capital and the propensity to enjoy time are related to the wealth and current income as

$$\lambda(t) = \phi(K, Y), \quad \sigma(t) = \varphi(K, Y),$$

where $\phi(K, Y)$ and $\varphi(K, Y)$ are proper functions of $K(t)$ and $Y(t)$. The two functions, $\phi(K, Y)$ and $\varphi(K, Y)$, may be taken on different forms because taste changes vary among individuals. Specifying taste change patterns, we can simulate the model.

4.6 The OSG Model with Sexual Division of Labor and Consumption

This section is concerned with another type of economic evolution with group differences. Different from the previous sections, we classify population into two groups based on gender. For simplicity, we are only concerned with an ideal — a very simple case — when each woman has only one husband and every adult must be married.

Dynamic interactions between economic growth and sexual division of labor and consumption have caused attention of economists. Yet there are only a few theoretical economic models which explicitly take account of these interactions within a compact framework. Over the years there have been a number of attempts to modify the neoclassical consumer theory to deal with economic issues about endogenous labor supply, family structure, working hours and the valuation of traveling time.[10] It has been argued that the increasing returns from human capital accumulation represent a powerful force creating a division of labor in the allocation of time between the male and female population.[11] There are studies on the relationship between economic growth and the family distribution of income.[12] There are studies of the female labor supply.

---

[10] See Becker (1976), Chiappori (1988, 1992), Folbre (1986), Mills and Hamilton (1985), Ashenfelter and Layard (1992).

[11] Refer to Becker (1985).

[12] The topic is referred to Fei, Ranis, and Kuo (1978).

Women choose levels of market time on the basis of wage rates and incomes. Lifetime variations in costs and opportunities - due to children, unemployment of the spouse, and general business cycle variations - influence the timing of female labor participation.[13] There are studies on the relationship between home production and non-home production and time distribution. Possible sexual discrimination in labor markets has attracted much attention from economists.[14] The gains from marriage may be reduced as people become rich and educated. The growth in the female population's earning power may raise the forgone value of their time spent at child care, education and other household activities, which may reduce the demand for children and encourage a substitution away from parental activities. Divorce rates, fertility, and labor participation rates may interact in much more complicated ways. Decision making about on family size is extremely complicated.[15] Irrespective of numerous studies on the complexity of the family as a subsystem of economic production, family economics - swept into a pile labeled economic demography or labor economics - is often relegated to a somewhat obscure corner of the mainstream studies of economic growth and development.

In Sec. 3.3, we extended the one-sector growth model to include time distribution. This section synthesizes these two growth models with home capital and time distribution into a single framework with the dynamic interdependence of sexual division of labor and consumption.[16] We consider an economic system similarly to the one-sector growth model proposed before. We assume the same family structure. Each family consists of four members - father, mother, son, and daughter. The total population is equal to $4N$. There is division of labor in the family. The children consume goods and accumulate knowledge through education. The parents have to do home work and find job for the family's living. The father and mother may either do home work or do business. The working time of the father and the mother may be different. We assume that working time of the two adults is determined

---

[13] See Mincer (1962), Smith (1977), and Heckman and Macurdy (1980).
[14] Lancaster (1966, 1971), Becker (1957), Cain (1986), and Lazear and Rosen (1990).
[15] See Becker (1976), and Weiss and Willis (1985).
[16] The model in this section is based on Zhang (1993c, 1999: Chap. 9).

by maximizing the family's utility function subject to the family and the available time constraints. We omit any possibility of divorce. We assume that the young people get educated before they get married and join labor market and the husband and the wife pass away at the same time. When the parents pass away, the son and the daughter respectively find their marriage partner and get married. The property left by the parents is shared equally by the two children. The children are educated so that they have the same level of human capital as their parents when they get married. When a new family is formed, the young couple join the labor market and have two children. As all the families are identical, the family structure is invariant over time under these assumptions.

We assume that labor markets are competitive. The total labor input $N^*(t)$ at time t is defined by

$$N^*(t) = N_1(t) + N_2(t), \quad N_j(t) = z_j T_j(t) N,$$

where $T_1(t)$ and $T_2(t)$ are respectively the husband's and the wife's working time and z_1 and z_2 are the levels of human capital at work of the husband and the wife, respectively. We specify production function of the economy

$$F(t) = K_i^{\alpha}(t)^{\alpha} N^{*\beta}(t), \quad \alpha + \beta = 1,$$

where $F(t)$ is the output level at time t, $K_i(t)$ is the level of capital input, and α and β are parameters. The marginal conditions are given by

$$r = \frac{\alpha F}{K_i}, \quad w_j = \frac{\beta z_j F}{N^*}, \tag{4.6.1}$$

where $r(t)$ is the rate of interest and $w_1(t)$ and $w_2(t)$ are respectively the wage rates per unity of working time of the husband and the wife. From Eqs. (4.6.1), the ratio of the wage rates per unity of time between the husband and the wife is given by $w_1(t)/w_2(t) = z_1/z_2$. The ratio is independent of capital stock and production scale and only dependent on the ratio of human capital. If $z_1/z_2 = 1$, the husband and the wife have the identical wage rate per unit of time. The current income $Y(t)$ of each family consists of the wage incomes and the interest payment for the family's capital. The current income at any point of time is given by

$$Y = rK + w_1 T_1 N + w_2 T_2 N .$$

Let us denote T_0 the husband's and the wife's total available time. The total available working time for any sex is distributed between leisure time and working time. The time constraint requires that the amounts of time allocated to each specific use add up to the time available

$$T_j(t) + T_{hj}(t) = T_0, \quad j = 1, 2 ,$$

where $T_{h1}(t)$ and $T_{h2}(t)$ are the husband's and the wife's leisure time, respectively. We assume that the family's utility level is dependent on the husband's leisure time, $T_{h1}(t)$, the wife's leisure time, $T_{h2}(t)$, the level of consumption, $C(t)$, home capital, $K_h(t)$, and the family's net wealth. We specify a typical family's utility function as follows

$$U(t) = T_{h1}^{\sigma_1} T_{h2}^{\sigma_2} C^{\xi} K_h^{\eta} S^{\lambda}, \quad \sigma_1, \sigma_2, \xi, \eta, \lambda > 0 ,$$

in which $\sigma_1, \sigma_2, \xi, \eta$ and λ are positive parameters. We call $\sigma_1, \sigma_2, \xi, \eta$, and λ, respectively, the family's propensities to use the husband's leisure time, to use the wife's leisure time, to consume goods, to utilize endurable goods, and to hold wealth. Each family makes decision on the 7 variables, $T_j(t)$, $T_{hj}(t)$, $(j = 1, 2)$, $K_h(t)$, $C(t)$, and $S(t)$ at any point of time.

Since a family consists of several members and each member has his/her own utility function, the family's behavior should be analyzed as the result of all members' rational decisions. The "collective utility function" should be analyzed within a framework which explicitly takes accounts of interactions within the family's members.[17] We might treat these issues, applying game theory. At this initial stage, we simplify the issues by assuming the existence of a family utility function.

The gross disposable income is

$$Y^*(t) \equiv Y(t) + K(t).$$

The financial budget constraint is given by

$$rK_h + \delta_k K + C + S = Y^* .$$

Substituting $T_j + T_{hj} = T_0$ into the above constraint, we get

[17] See Becker (1976), Heckman and Macurdy (1980), and Chiappori (1988).

$$rK_h + C + S + w_1 T_{h1} N + w_2 T_{h2} N = \hat{Y},$$

where

$$\hat{Y} = rK + w_1 T_0 N + w_2 T_0 N + \delta K(t), \quad \delta \equiv 1 - \delta_k .$$

Each family maximizes $U(t)$ subject to the above budget constraint. The optimal problem has the following unique solution

$$T_{hj} = \frac{\rho \sigma_j \hat{Y}}{w_j N}, \quad rK_h = \rho \eta \hat{Y}, \quad C = \rho \xi \hat{Y},$$

$$S = \rho \lambda \hat{Y}, \quad j = 1, 2, \tag{4.6.2}$$

where

$$\rho \equiv \frac{1}{\sigma_1 + \sigma_2 + \xi + \eta + \lambda}.$$

Substituting S in Eqs. (4.6.2) into the capital accumulation equation $\dot{K} = S - K$ yields

$$\dot{K} = \rho \lambda \hat{Y} - K. \tag{4.6.3}$$

The condition that the total capital stocks K is fully employed at each point of time is expressed by

$$K_i(t) + K_h(t) = K(t).$$

We have thus built the model. The system has 16 variables, K, K_i, K_h, N^*, F, Y, C, S, U, R, w_j, T_{hj}, and T_j $(j = 1, 2)$. It contains the same number of independent equations. We now examine properties of the dynamic system.

First, substituting $K_i + K_h = K$, $T_j + T_{hj} = T_0$, and the marginal conditions (4.6.1) into the definition of $\hat{Y}$, we get

$$\hat{Y} - rK_h - w_1 T_{1h} N - w_2 T_{2h} N = F + \delta K .$$

As

$$\hat{Y} - rK_h - w_1 T_{1h} N - w_2 T_{2h} N = C + S,$$

we have

$$C + S = F + \delta K .$$

Substituting $C = \rho \xi \hat{Y}$ and $S = \rho \lambda \hat{Y}$ from Eqs. (4.6.2) into the above equation yields

$$\hat{Y} = \frac{F + \delta K}{(\xi + \lambda)\rho}. \tag{4.6.4}$$

Substituting r and w_j from Eqs. (4.6.1) into

$$\hat{Y} - rK_h - w_1 T_{1h} N - w_2 T_{2h} N = F + \delta K,$$

we get

$$\hat{Y} = \left(\frac{\alpha K}{K_i} + \frac{\beta z T_0 N}{N^*} \right) F + \delta K,$$

where $z \equiv z_1 + z_2$. By this equation and Eqs. (4.6.4)

$$\left\{ \frac{\alpha K}{K_i} + \frac{\beta z T_0 N}{N^*} - \frac{1}{(\xi + \lambda)\rho} \right\} \frac{F}{K} = \sigma, \tag{4.6.5}$$

where

$$\sigma \equiv \frac{\delta(\sigma_1 + \sigma_2 + \eta)}{\xi + \lambda}.$$

By $K_h = \rho \eta \hat{Y} / r$ in Eqs. (4.6.2) and $r = \alpha F / K_i$ in Eqs. (4.6.1), we have

$$\hat{Y} = \frac{\alpha K_h F}{\rho \eta K_i}.$$

By this equation, Eq. (4.6.4) and $K_h = K - K_i$, we obtain

$$N^{*\beta} = \frac{\delta K}{(\lambda_1 K / K_i - \lambda_1 - 1) K_i^\alpha}, \tag{4.6.6}$$

where

$$\lambda_1 \equiv \frac{(\xi + \lambda)\alpha}{\eta}.$$

Substituting Eq. (4.6.6) into Eq. (4.6.5) yields

$$\Phi(K_i) \equiv \frac{\alpha K}{K_i} + \frac{\beta z T_0 N K_i^{\alpha/\beta} (\lambda_1 K / K_i - \lambda_1 - 1)^{1/\beta}}{\delta K} - \frac{1}{(\xi + \lambda)\rho}$$

$$-\frac{(\lambda_1 K / K_i - \lambda_1 - 1)\sigma}{\delta} = 0, \quad \frac{\lambda_1 K}{(1+\lambda_1)} > K_i > 0. \tag{4.6.7}$$

It is straightforward to check

$$\Phi\left[\frac{\lambda_1 K}{1+\lambda_1}\right] = +\infty, \quad \Phi(0) = -\infty,$$

$$\frac{d\Phi}{dK_i} = \frac{(\sigma_1 + \sigma_2)\alpha K}{\lambda K_i^2} +$$

$$\frac{(\alpha\lambda_1 + \alpha + \beta\lambda_1 K / K_i)\beta z T_0 N K_i^{\alpha/\beta}}{(\delta K)^{\alpha/\beta}(\lambda_1 K / K_i - \lambda_1 - 1)^{\alpha/\beta}} > 0. \tag{4.6.8}$$

The equation $\Phi(K_i) = 0$ has a unique solution in the interval of

$$\frac{\lambda_1 K}{1+\lambda_1} > K_i > 0.$$

This implies that for any given positive $K(t)$ at any point of time $K_i(t)$ is uniquely determined as a function of $K(t)$. Summarizing the above discussion, we have the following lemma.

Lemma 4.6.1 For any given positive $K(t)$ at any point of time, the other variables in the system are uniquely determined as functions of $K(t)$ by the following procedure: K_i by Eq. (4.6.7) $\to K_h = K - K_i \to$ N^* by Eq. (4.6.6) $\to F = K_i^\alpha N^{*\beta} \to r,\ w_j,\ j = 1, 2,$ by Eqs. (4.6.1) $\to$ $\hat{Y}$ by Eq. (4.6.4) $\to C,\ S$ and T_j by Eq. (4.6.2) $\to T_{hj} = T_0 - T_j$.

By the above lemma and Eq. (4.6.3), we conclude that the dynamics of the system are given by the following differential equation

$$\dot{K} = \rho\lambda\hat{Y}(K) - K, \tag{4.6.9}$$

in which $\hat{Y}(K)$ is a unique function of K. From Eq. (4.6.9), we determine $K(t)$ at each point of time. Then, by Lemma 4.6.1 we solve the values of the other variables in the system at any point of time.

In equilibrium, $\rho\lambda\hat{Y} = K$. From the equilibrium equation and $\hat{Y} = \alpha K_h F / \rho\eta K_i$, we solve

$$\frac{\alpha K_h K_i^{\alpha} N^{*\beta}}{\eta K_i} = \frac{K}{\lambda}.$$

Substituting Eq. (4.6.6) into this equation and using $K_h = K - K_i$, we obtain

$$K_i = \rho_0 K, \ \ K_h = \frac{\eta \rho_0 K}{\xi_k \alpha}, \tag{4.6.10}$$

where

$$\xi_k = \xi + \delta_k \lambda, \ \ \rho_0 \equiv \frac{\xi_k}{\xi_k + \eta / \alpha}.$$

Substituting K_i from Eqs. (4.6.10) into Eq. (4.6.7) yields

$$K = \frac{\beta \lambda^{1/\beta} T_0 z N}{(\xi_k + \eta / \alpha)^{\alpha/\beta} (\sigma_1 + \sigma_2 + \beta \xi_k)}. \tag{4.6.11}$$

Equation (4.6.11) gives a unique equilibrium value of K. By Lemma 3.1, we directly determine the unique equilibrium values of the other variables. To determine stability, we take derivatives of the left-side of Eq. (4.6.9) to get

$$\left(\frac{\xi}{\lambda} + 1\right) \frac{d(\rho \lambda \hat{Y} - K)}{dK} = \frac{\alpha F}{K_i} \frac{dK_i}{dK} + \frac{\beta F}{N^*} \frac{dN^*}{dK} - \frac{\xi}{\lambda} - \delta_k \tag{4.6.12}$$

in which we use Eq. (4.6.4), $F = K_i^{\alpha} N^{*\beta}$ and

$$\frac{\partial \Phi}{\partial K_i} \frac{dK_i}{dK} = - \frac{(1 + \lambda_1)(\lambda_1 K - \lambda_1 K_i - K_i)^{\alpha/\beta} z T_0 N}{\delta^{1/\beta} K^{\alpha/\beta}} + \frac{(\sigma_1 + \sigma_2)\alpha}{\lambda K_i},$$

$$\frac{\beta}{N^*} \frac{dN^*}{dK} = \frac{\{(K / K_i) dK_i / dK - 1\} \lambda_1}{\lambda_1 K - \lambda_1 K_i - K_i} + \frac{1}{K} - \frac{\alpha}{K_i} \frac{dK_i}{dK},$$

which $\partial \Phi / \partial K_i > 0$ is given by Eqs. (4.6.8), and Eqs. (4.6.7) and (4.6.6) are used. As we explicitly solved the equilibrium values of the variables, it is easy to calculate the left-hand side of Eq. (4.6.12). Summarizing the above discussion, we obtain the following proposition.

Proposition 4.6.1 The dynamic system has a unique equilibrium. The equilibrium is stable (unstable) if the left-hand side of Eq. (4.6.12) is negative (positive).

We now examine how the equilibrium values of variables are related to parameters. First, we examine how woman's human capital level affects the time distribution of the husband and wife. With z_2 to change, we specify the other parameters as follows

$$\alpha = 0.3, \ \ \delta_k = 0.03, \ \ \lambda = 0{,}5, \ \ \sigma_1 = 0.36, \ \ \sigma_2 = 0.43, \ \ \xi = 0.3,$$
$$\eta = 0.3, T_0 = 24, \ \ N = 1, \ \ z_1 = 0.7. \tag{4.6.13}$$

Figure 4.6.1 shows how man and woman's working hours vary as woman's human capital changes. For the fixed preferences, from Fig. 4.6.1a we observe that as woman's human capital rises for $z_2 \in [0.5, 0.9]$, man's working time declines and woman's working time rises. Their working hours become equal only when the wife accumulates more human capital than the husband because we have assumed that the wife makes more contributions to family life than the husband if they spend the same hours at home. Figure 4.6.1b shows that as woman's human capital rises, her wage rate increases.

We now examine the impact of woman's propensity to stay at home on the equilibrium. Except σ_2 and z_2, we specify the parameters as in (4.6.13). Let $z_2 = 0.6$ and σ_2 varies within $\sigma_2 \in [0.2, 0.6]$. As woman increasingly prefers staying at home, woman's work time declines but man's work time increases. As demonstrated in Fig. 4.6.2b, both man and woman's wage rates decline as woman's preference to stay at home becomes stronger.

Combining Figs. 4.6.1 and 4.6.2, we illustrate how man and woman's working hours vary with σ_2 and z_2. We observe that even when man's human capital and time preference remain invariant, man's working hours vary as woman's human capital and time preference change.

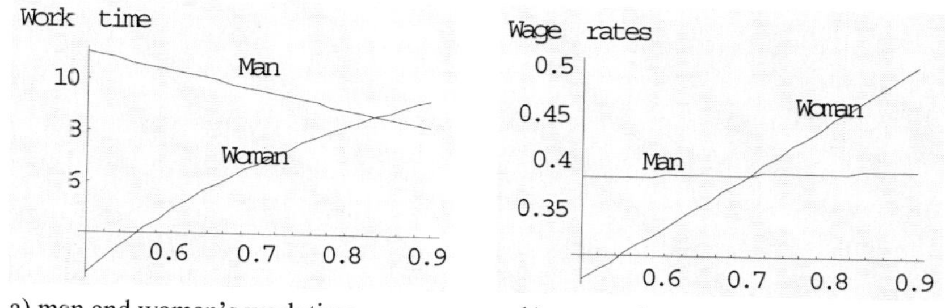

a) man and woman's work time b) man and woman's wage rates

Fig. 4.6.1 The impact of woman's human capital.

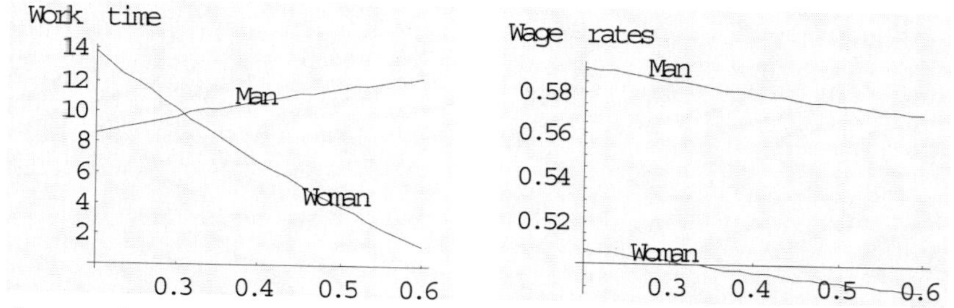

a) man and woman's work time b) man and woman's wage rates

Fig. 4.6.2 The impact of woman's propensity to stay at home.

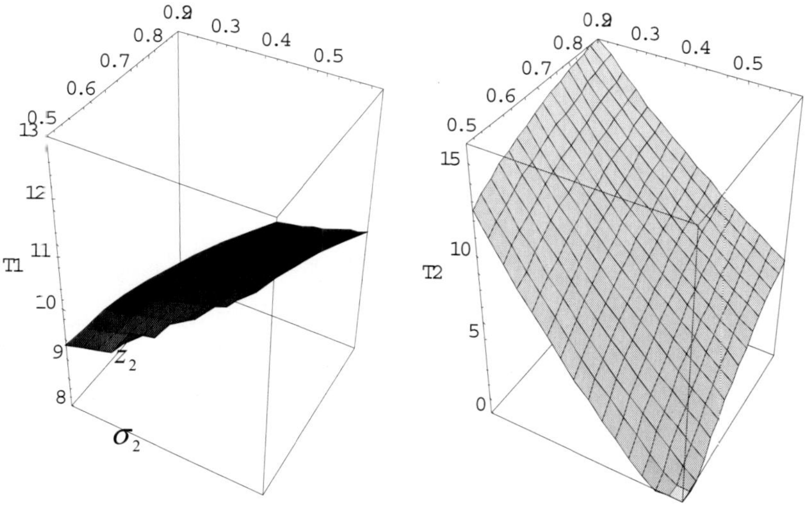

a) man's work time b) woman's work time

Fig. 4.6.3 The work hours and σ_2 and z_2.

4.7 The Uzawa Two-Sector Model

In Uzawa's two-sector growth model,[18] it is assumed that consumption and capital goods are different commodities, which are produced in two distinct sectors. Labor is homogeneous and labor grows at an exogenously given exponential rate n. There is only one malleable capital good, which can be used as an input in both sectors in the economy. Capital depreciates at a constant exponential rate δ_k, which is independent of the manner of use. The production functions are given by

$$Y_j = F_j(K_j, N_j), \quad j = 1, 2,$$

where Y_j are the output of sector j, K_j and N_j are respectively the capital and labor used in sector j, F_j the production functions, the subscripts 1 and 2 denote the capital good sector and the consumption good sector. Assume F_j to be neoclassical. We have $y_j = f_j(k_j)$ where $y_j \equiv Y_j / N_j$, $k_j \equiv K_j / N_j$, $f_j' > 0$, $f_j'' < 0$, $j = 1, 2$. It is assumed that the usual static efficiency conditions of pure competition hold at any time. This requirement means that the wages w_j, and the wage-rental ratio w_j^* in the two sectors are equal: $w = w_1 = w_2$, $w^* = w_1^* = w_2^*$. We have $w^* = f_j / f_j' - k_j$. As full employment of labor and capital is assumed, we have

$$K_1 + K_2 = K, \quad N_1 + N_2 = N,$$

which can be rewritten in the form of

$$n_1 k_1 + n_2 k_2 = k,$$

where $n_j \equiv N_j / N$, $j = 1, 2$. The gross saving propensities - both average and marginal - from wage incomes and profits are nonnegative constants denoted respectively by s_w and s_r. Thus, if the two propensities are equal (to s), the consumption is equal to a constant fraction $1 - s$ of the gross national product. If we denote the rental rate of the two sectors by r, then the total gross saving in the economy is

[18] The model was proposed by Uzawa (1961). Since its publication, the model has extended and generalized in different ways, see Hahn (1965), Takayama (1965), Zhang (1999). In particular, Solow (1961) made some important comments on the economic mechanism of the model; Sato (1965) provided some further analysis of stability properties of the model.

equal to $s_r rK + s_w wN$. As the investment in the economy comes from the production of new capital and saving is always equal to investment, we have

$$P_1 Y_1 = s_r rK + s_w wN ,$$

where $P_1(t)$ is the price of new capital. As $F_{1K}(t) = r(t)/P_1(t)$, the above equation becomes

$$n_1 f_1(k_1) = f_1'(s_r k + s_w w^*).$$

As

$$\dot{K} = Y_1 - \delta_k K ,$$

we have

$$\dot{k} = f_1'(s_r k + s_w w^*) - (n + \delta_k)k . \qquad (4.7.1)$$

Under certain conditions, the dynamic system is 'causal'. If we assume that the conditions are satisfied, then the right-hand side of Eq. (4.7.1) can be written as a function of k

$$\dot{k}(t) = H(k) = kh(k).$$

The functional form of h is referred to Burmeister and Dobell.[19] Let us denote the 'extended Jacobian' by J and define two numbers a and b as

$$a = \max\left\{\lim_{k_j \to 0}\left(\frac{f_j}{f_j'} - k_j\right), \ j = 1, 2\right\},$$

$$b = \min\left\{\lim_{k_j \to 0}\left(\frac{f_j}{f_j'} - k_j\right), \ j = 1, 2\right\}.$$

Then the following theorems hold.

Theorem 4.7.1 (Local stability) Let k^* be any root of $h = 0$. If $s_w/(n + \delta_k)$ is not larger than $k/f_1(k_1)$ or $s_r/(n + \delta_k)$ is not less than $k/f_1(k_1)$, then the equilibrium is locally stable.

---

[19] See Burmeister and Dobell (1970).

Theorem 4.7.2 (Uniqueness and Stability) If any of the following conditions are satisfied for all $a < w^* < b$, then it is proved that any equilibrium of the dynamic system is unique and stable: (i) s_r is not less than s_w, while k_1 is not larger than k_2; (ii) the wage elasticity of capital intensity $(w^* / k)dk / dw^*$ is not less than unity; (iii) The substitution elasticity of the consumption sector is not less than unity; (iv) $s_r = 1$ and $J > 0$; (v) $s_w = 0$ and $J > 0$, in which all functions and variables are evaluated at the equilibrium of the system.

It should be mentioned that even if all the conditions (i)-(v) are violated, it is still possible to find a unique and balanced growth path, and unstable balanced growth paths may exist.

4.8 Refitting the Uzawa Model within the OSG Framework[20]

Output of the capital sector goes entirely to investment and that of the consumption sector entirely to consumption. Labor is homogeneous and labor grows at an exogenously given exponential rate n. Capital depreciates at a constant exponential rate δ_k, which is independent of the manner of use. The production functions are given by

$$F_j = F_j(K_j, N_j), \quad j = i, s ,$$

where $F_j(t)$ are the output of sector j, $K_j(t)$ and $N_j(t)$ are respectively the capital and labor used in sector j, $F_j(t)$ the production functions, the subscripts i and s denote the capital good sector and the consumption good sector, respectively. Assume $F_j(t)$ to be neoclassical. We have

$$f_j = f_j(k_j),$$

$$y_j \equiv \frac{Y_j}{N_j}, \quad k_j(t) \equiv \frac{K_j}{N_j}, \quad f_j'(k_j) > 0, \quad f_j''(k_j) < 0, \quad j = i, s .$$

The marginal conditions are

[20] This section is based on Zhang (1996a, 1999).

$$r = f_i'(k_i) = pf_s'(k_s),$$

$$w = f_i(k_i) - k_i f_i'(k_i) = p(f_s(k_s) - k_s(t) f_s'(k_s)), \tag{4.8.1}$$

where $p(t)$ is the price of consumption good and price of new capital is always equal to 1.

As full employment of labor and capital is assumed, we have

$$K_i + K_s = K, \ \ N_i + N_s = N,$$

which can be rewritten in the form of

$$n_i k_i + n_s k_s = k, \ \ n_i + n_s = 1, \tag{4.8.2}$$

where

$$k = \frac{K}{N}, \ \ n_j = \frac{N_j}{N}, \ \ j = i, s.$$

The current income Y is given by

$$Y = rK + wN = F_i + pF_s.$$

The consumer problem is defined by

$$\text{Maximize } U = C^{\xi} S^{\lambda}, \ \ \text{s.t: } pC + S = \hat{Y} \equiv Y + \delta K,$$

where $\xi + \lambda = 1, \ \ \xi, \lambda > 0$. The optimal solution is

$$C(t) = \frac{\xi \hat{Y}(t)}{p(t)}, \ \ S(t) = \lambda \hat{Y}(t). \tag{4.8.3}$$

Capital accumulates according to

$$\dot{K} = \lambda \hat{Y} - K = \lambda Y - (1 - \delta \lambda) K. \tag{4.8.4}$$

As consumption good cannot be saved, we always have

$$C = F_s. \tag{4.8.5}$$

As the investment in the economy comes from the production of new capital and saving is always equal to investment, we have

$$S - K + \delta_k K = F_i.$$

It can be shown that this equation is redundant.

First, we have to describe the dynamics in terms of a single variable. Substituting $k(t) = K(t)/N(t)$ with $\dot{N}(t) = nN(t)$ into Eq. (4.8.4) yields

$$\dot{k} = \lambda\hat{y} - (1+n)k = \lambda y - (1+n-\delta\lambda)k, \tag{4.8.6}$$

where

$$\hat{y} = \frac{\hat{Y}}{N} = y + \delta k, \quad y \equiv \frac{Y}{N} = n_i f_i(k_i) + n_s f_s(k_s).$$

Equations (4.8.3) yield

$$c = \frac{\xi\hat{y}}{p}, \quad s = \lambda\hat{y},$$

where $c(t) \equiv C(t)/N(t)$ and $s(t) \equiv S(t)/N(t)$. From these equations and $S - \delta K = F_i$ and $C = F_s$, we have

$$\lambda\hat{y} - \delta k = n_i f_i(k_i), \quad \xi\hat{y} = pn_s f_s(k_s). \tag{4.8.7}$$

From $r(t) = f_i'(k_i) = pf_s'(k_s)$ in Eqs. (4.8.1), we get

$$p = \frac{f_i'(k_i)}{f_s'(k_s)}.$$

Substituting the above equation into

$$f_i(k_i) - k_i f_i'(k_i) = p(f_s(k_s) - k_s f_s'(k_s))$$

from Eqs. (4.8.1) yields

$$\Psi_i(k_i) \equiv \frac{w}{r} = \frac{f_i(k_i)}{f_i'(k_i)} - k_i = \frac{f_s(k_s)}{f_s'(k_s)} - k_s \equiv \Psi_s(k_s). \tag{4.8.8}$$

The function $\Psi_i(k_i)$ has the following properties:

$$\Psi_i(0) = 0, \quad \Psi_i(k_i) > 0 \text{ for } k_i > 0, \quad \Psi_i'(k_i) = -\frac{f_i(k_i)f_i''(k_i)}{f_i'^2(k_i)} \geq 0.$$

The function $\Psi_s(k_s)$ has the same properties in k_s. We see that for any given $k_i \geq 0$, Eqs. (4.8.8) determines $k_s \geq 0$ as a unique function of k_i, denoted by

$$k_s = \Omega(k_i).$$

We have $\Omega' > 0$.

By Eqs. (4.8.2) and $k_s = \Omega(k_i)$, we have

$$n_i(k_i - \Omega_i(k_i)) + \Omega_i(k_i) = k.$$

That is

$$n_i \equiv \Lambda_i(k, k_i) = \frac{k - \Omega_i(k_i)}{k_i - \Omega_i(k_i)},$$

$$n_s = 1 - n_i \equiv \Lambda_s(k, k_i) = \frac{k_i - k}{k_i - \Omega_i(k_i)}. \qquad (4.8.9)$$

We can thus solve n_i and n_s as functions of k and k_i. By Eqs. (4.8.7) and $p = f_i'(k_i)/f_s'(k_s)$, we solve

$$\frac{\Lambda_s(k, k_i) f_s(k_s)}{\Lambda_i(k, k_i) f_i(k_i) + \delta k} = \frac{\xi f_s'(\Omega(k_i))}{\lambda f_i'(k_i)}. \qquad (4.8.10)$$

Equation (4.8.10) contains only two variables, k and k_i. Assume that this equation determines k_i as a function of k, i.e., $k_i = \Lambda(k)$. From the procedure to obtain Eq. (4.8.10), we can express any variable in the dynamic system as a function only of k. In particular, since $y(t)$ is a function of $k(t)$, we can determine the dynamics by a one-dimensional differential equation

$$\dot{k} = \lambda y(k) - (1 + n - \delta\lambda)k. \qquad (4.8.11)$$

As further analysis would not provide results with easy interpretations, we examine behavior of the two-sector model with Cobb-Douglas production functions

$$F_j = A_j K_j^{\alpha_j} N_j^{\beta_j}, \ \alpha_j + \beta_j = 1, \ \alpha_j, \beta_j > 0.$$

We thus have

$$r = \frac{\alpha_i f_i}{k_i} = \frac{\alpha_s p f_s}{k_s}, \ w = \beta_i f_i = \beta_s p f_s, \ f_j = A_j k_j^{\alpha_j}. \qquad (4.8.12)$$

By Eqs. (4.8.8), we immediately get

$$k_s(t) = \alpha k_i(t),$$

where

$$\alpha \equiv \frac{\alpha_s \beta_i}{\alpha_i \beta_s}.$$

The capital-labor ratios are proportional. The parameter α is not related to the productivity parameters. Figure 4.8.1 shows how α varies as a function of α_s $(0.1 \le \alpha_s \le 0.9)$ and α_i $(0.1 \le \alpha_i \le 0.9)$.

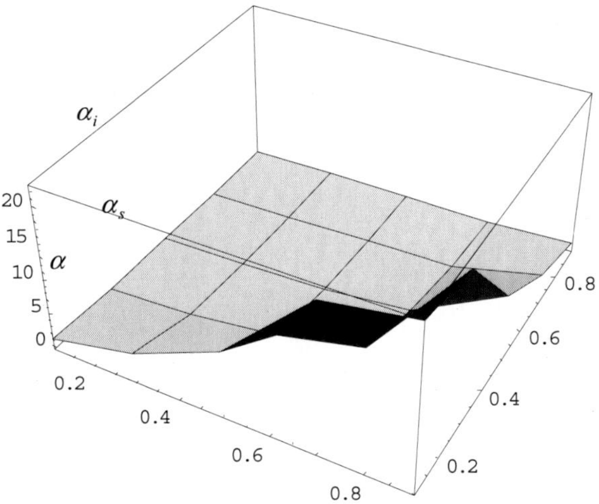

Fig. 4.8.1 α varies as a function of α_s and α_i.

By $k_s = \alpha k_i$ and $\beta_i f_i = \beta_s p f_s$, we solve

$$p = \frac{\beta_i A_i}{\beta_s \alpha^{\alpha_s} A_s} k_i^{\alpha_i - \alpha_s}. \tag{4.8.13}$$

If $\alpha_i = \alpha_s$, then the price is constant, $p = A_i / A_s$. The price of consumption good rises in productivity of the capital sector but falls in productivity of the consumption sector. In the remainder of this section, we require $\alpha_i \neq \alpha_s$. If the equality holds, then labor distribution is invariant in time.

Corresponding to Eqs. (4.8.9), we have

$$n_i = \frac{\alpha k_i - k}{(\alpha - 1)k_i}, \quad n_s = \frac{k - k_i}{(\alpha - 1)k_i}. \tag{4.8.14}$$

By Eq. (4.8.10), we solve

$$k = \frac{k_i}{A_0\left(1 + A k_i^{\beta_i}\right)}, \tag{4.8.15}$$

where we use Eqs. (4.8.13) and (4.8.3) and

$$A_0 \equiv \frac{1 + \beta_i \lambda / \beta_s \xi}{\alpha + \beta_i \lambda / \beta_s \xi}, \quad A \equiv \frac{(1 - \alpha)\delta / A_i}{1 + \beta_i \lambda / \beta_s \xi}.$$

By Eqs. (4.8.14) and (4.8.15) and according to the definitions of A and A_j, we find

$$n_i = \frac{\beta_i \lambda / \beta_s \xi - \alpha \delta k_i^{\beta_i} / A_i}{\alpha + \beta_i \lambda / \beta_s \xi} = 1 - \frac{\alpha + \alpha \delta k_i^{\beta_i} / A_i}{\alpha + \beta_i \lambda / \beta_s \xi}.$$

Hence, for $n_i(t)$ to satisfy $1 > n_i(t) > 0$, it is sufficient to have

$$k_i < \left(\frac{A_i \beta_i \lambda}{\alpha \delta \beta_s \xi} \right)^{1/\beta_i}.$$

By Eqs. (4.8.7),

$$\lambda \hat{y} = n_i f_i(k_i) + \delta k.$$

Substituting this equation into Eq. (4.8.6) yields

$$\dot{k} = n_i f_i(k_i) - \delta_0 k,$$

where

$$\delta_0 \equiv 1 + n - \delta > 0.$$

With $f_j = A_j k_j^{\alpha_j}$ and Eqs. (4.8.14) and (4.8.15), we find

$$\phi(k_i)\dot{k}_i = (\alpha A_0 - 1)\alpha_0 - (\delta_0 - \alpha \alpha_0 A_0 A) k_i^{\beta_i}, \tag{4.8.16}$$

where

$$\phi(k_i) \equiv \left[\frac{1 + \alpha_i A k_i^{\beta_i}}{1 + A k_i^{\beta_i}} \right] k_i^{\alpha_i} \geq 0, \quad \alpha_0 \equiv \frac{A_i}{(\alpha - 1)}.$$

Lemma 4.8.1 The dynamics of the two-sector model follows Eq. (4.8.16). For given value of $k_i(t)$ at any point of time, the other variables are determined by the following procedure: $k(t)$ by Eq. (4.8.15) $\rightarrow$ $k_s(t) = \alpha k_i(t)$ $\rightarrow$ $n_i(t)$ and $n_s(t)$ by Eqs. (4.8.14) $\rightarrow$ $r(t)$, $w(t)$, and $f_j(t)$, $j = i, s$ by Eqs. (4.8.12) $\rightarrow$ $p(t)$ by Eq. (4.8.13) $\rightarrow$ $N_j(t) = n_j(t)N$ $\rightarrow$ $K_j(t) = k_j(t)N_j(t)$ $\rightarrow$ $F_j(t) = f_j(t)N_j(t)$ $\rightarrow$ $Y(t) = F_i(t) + p(t)F_s(t)$ $\rightarrow$ $\hat{Y}(t) = Y(t) + \delta K(t)$ $\rightarrow$ $C(t) = \xi \hat{Y}(t)$ $\rightarrow$ $S(t) = \lambda \hat{Y}(t)$.

Once we determine $k_i(t)$, we solve all the variables. The unique equilibrium is

$$k_i^* = \left[\frac{(\alpha A_0 - 1)\alpha_0}{\delta_0 - \alpha_0 \alpha A A_0} \right]^{1/\beta_i}. \tag{4.8.17}$$

To demonstrate k_i to be a positive solution, we examine two terms $(\alpha A_0 - 1)\alpha_0$ and $(\delta_0 - \alpha\alpha_0 A_0 A)$. As shown in Fig. 4.8.2, α may be either larger or smaller than unity. In the case of $\alpha >$ (<) 1, according to their definitions, $\alpha_0 >$ (<) 0, $A <$ (>) 0, and $A_0 <$ (>) 1. We conclude

$$(\delta_0 - \alpha\alpha_0 A_0 A) > 0, \ (\alpha A_0 - 1)\alpha_0 = \frac{\beta_i \lambda A_i / \beta_s \xi}{\alpha + \beta_i \lambda / \beta_s \xi} > 0 .$$

This guarantees a unique positive solution as well as stability of the equilibrium. The system is stable because

$$\left. \frac{d\dot{k}_i}{dk_i} \right|_{k_i = k_i^*} = - \frac{(\delta_0 - \alpha\alpha_0 A_0 A)\beta_i}{k_i^{*\alpha_i} \phi(k_i^*)} < 0 .$$

We now show that for $k_i^* > 0$, $0 < n_i^* < 1$. By Eqs. (4.8.15) and (4.8.17), we have

$$n_i^* = \frac{\alpha k_i - k}{(\alpha - 1)k_i} = \frac{(\alpha A_0 - 1)}{(\alpha - 1)A_0} \frac{\delta_0}{(\delta_0 + \delta / (1 + \beta_i \lambda / \beta_s \xi))} .$$

As

$$0 < \frac{\delta_0}{(\delta_0 + \delta / (1 + \beta_i \lambda / \beta_s \xi))} < 1 ,$$

$$0 < \frac{(\alpha A_0 - 1)}{(\alpha - 1)A_0} = \frac{\beta_i \lambda / \beta_s \xi}{1 + \beta_i \lambda / \beta_s \xi} < 1 ,$$

we guaranteed $0 < n_i^* < 1$.

Theorem 4.8.1 The dynamic system has a unique stable equilibrium.

It is straightforward to examine the impact of changes in parameter values on the equilibrium structure. Finally, we specify the parameter values as follows

$$\alpha_i = 0.45, \ \alpha_s = 0.35, \ n = 0.01, \ \delta_k = 0.05,$$

$$\lambda = 0.65, \ A_i = 1.1, \ A_s = 0.9 .$$

We note that the productivity of the capital good sector is higher than that of the consumption good sector; the value of the capital good sector's α is also higher than that of the consumption good sector. We simulate the motion of the economic system over 15 years with $k_i(0) = 1.2$. The equilibrium value of k_i^* is 2.376. Figure 4.8.3a shows the growth of the per-worker input in the two sectors and per-capita wealth. These variables experience dramatic growth during the study period. Figure 4.8.3b shows the growth of the per-worker output levels of the two sectors. The growth rate of per-worker output the capital good sector is about 75 percent over 15 years; the growth rate of per-worker output of the consumption good sector is about 50 percent. Figure 4.8.3c describes the motion of the labor distribution and the output ratio between the two sectors. It can be seen that the ratio steadily falls over the study period. The labor participation rate in the capital good sector of the total labor force also declines steadily over the study period. The consumption good sector absorbs more and more labor force. Figure 4.8.3d shows how the price, the real wage rate, and the rate of interest change over time. The price of consumption good (in comparison to capital good) rises over the period. The real wage rate increases; the rate of interest falls during the study period. Figure 4.8.3e demonstrates the current income per capita and consumption per capita. Although the two variables increase over the study period, the consumption per capita grows faster than the current income per capita. Figure 3.3.3f depicts the dynamics of the shares of the two sectors in the GNP. The share of the capital good sector, denoted by $yi \equiv F_i / Y$ falls, and that of consumption good sector, denoted by $ys \equiv pF_s / Y$, rises.

It appears that the share of the capital good sector falls too rapidly. One reason is that we neglect capital goods, such as TVs, cars, videos, boats, houses, computers, washing machines, clothes, carpets, paintings, radios, used by households. It should be noted that it is possible to make the model more suitable for national data. One method is to introduce proper — either endogenously or exogenously — technological changes. The two sectors may experience different paths of technological changes. The other method is to make preference changeable over time.

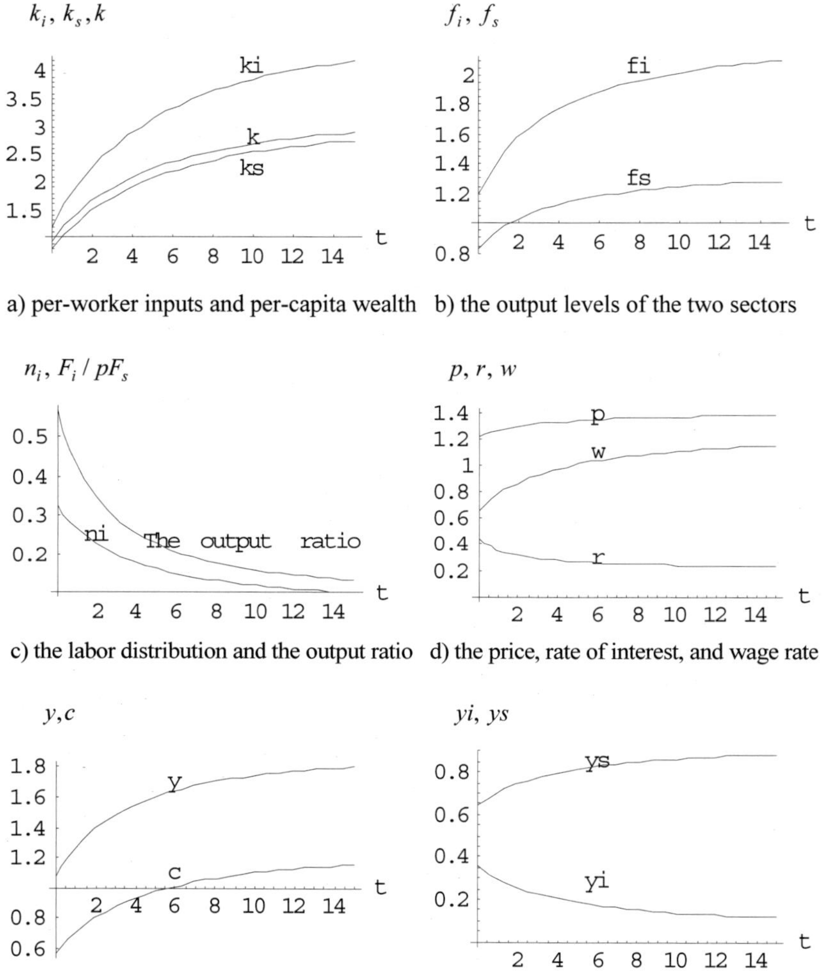

a) per-worker inputs and per-capita wealth b) the output levels of the two sectors

c) the labor distribution and the output ratio d) the price, rate of interest, and wage rate

e) the income and consumption per capita f) the share of the outputs in GNP

Fig. 4.8.3 Simulating the two-sector model.

We now examine impact of changes on dynamic processes of the system. First, we examine the case that all the parameters, except λ, are the same as in Fig. 4.8.3. We reduce the propensity to own wealth from 0.65 to 0.60. The simulation results are demonstrated in Fig. 3.3.4. The solid lines in Fig. 4.8.4 are the same as in Fig. 4.8.3, representing the values of the corresponding variables when $\lambda = 0.65$; the dashing lines in Fig. 4.8.4 represent the new values of the variables when $\lambda = 0.60$.

As the propensity to own wealth declines, the per-worker inputs of the two sectors and wealth per capita fall as shown in Fig. 4.8.4a. Figure 4.8.4b shows that the per-worker output levels of the two sectors fall. The labor participation ratio in the capital good sector and the output ratio of the capital good sector and the consumption good sector fall, as illustrated in Fig. 4.8.4c. Figure 4.8.4d shows that the price of consumption good falls, the wage rate declines, and the rate of interest rises. From Fig. 4.8.4e, we observe that both the current income and consumption rise as the propensity to save falls. Figure 4.8.4f demonstrates that the share of output of the capital good sector in the GNP falls, and that of the consumption good sector rises.

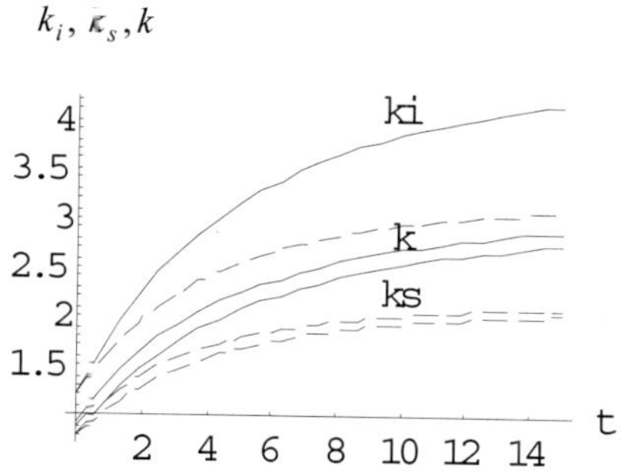

a) the capita intensities and wealth

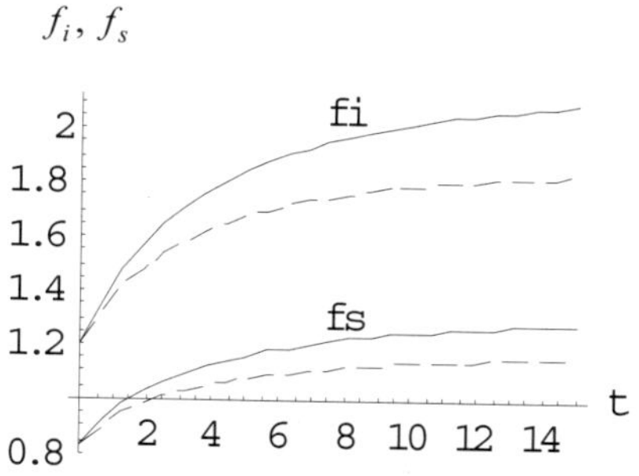

b) the output levels of the two sectors

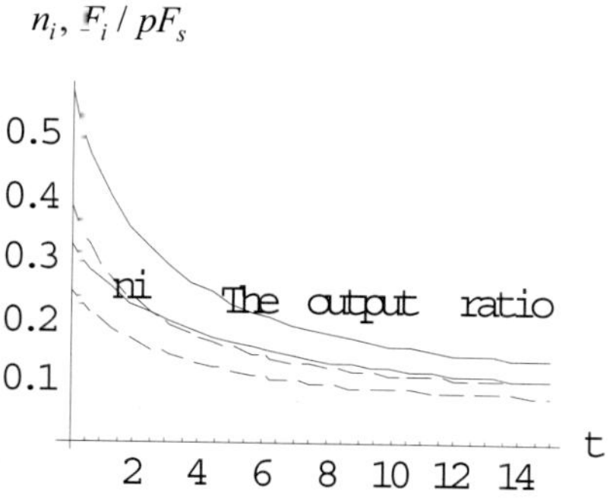

c) the labor distribution and the output ratio

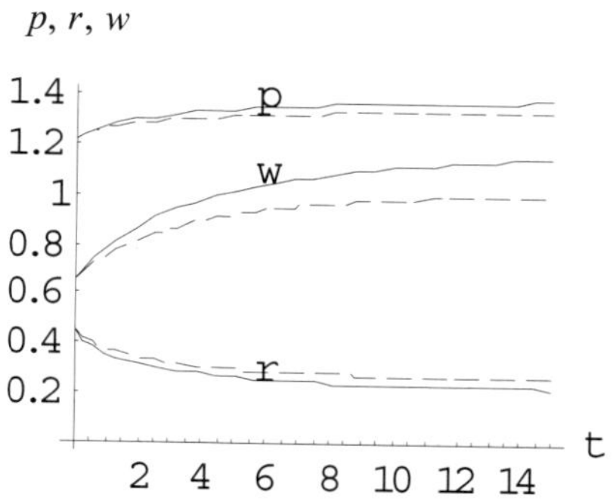

d) the price, interest rate, and wage rate

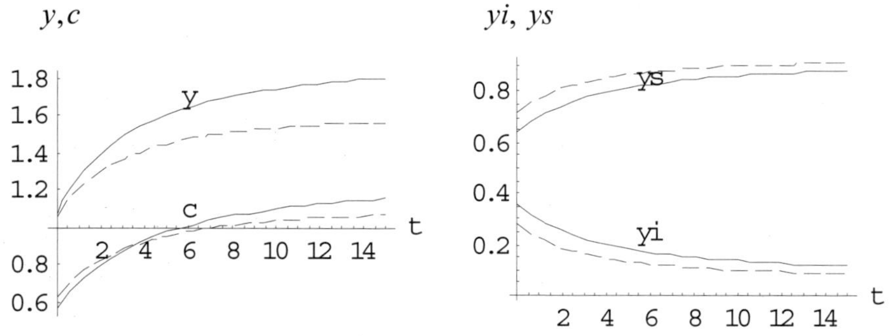

e) the income and consumption per capita f) the share of the outputs in GNP

Fig. 4.8.4 As λ declines from 0.65 (with the solid lines) to 0.60 (the dashing lines).

Figure 4.8.4e shows that as the propensity to save falls, consumption will increase until the system reaches the equilibrium. This situation may not occur under other circumstances. For instance, Fig. 4.8.5 portrays the case that when λ falls from 0.65 to 0.40 (with all the other parameters fixed), we find that after it rises two years (in comparison to the value before) consumption level begins to fall.

We now examine the case that all the parameters, except A_i, are the same as in Fig. 4.8.3. We increase the productivity of the capital good sector from 1.1 to 1.4. The simulation results are demonstrated in Fig. 3.3.6. The solid lines in Fig. 4.8.6 are the same as in Fig. 4.8.3, representing the values of the corresponding variables when $A_i = 1.1$; the dashing lines in Fig. 4.8.6 represent the new values of the variables when $A_i = 1.4$. As the productivity rises, the per-worker inputs of the two sectors and wealth per capita increase as shown in Fig. 4.8.6a. Figure 4.8.6b shows that the per-worker output levels of the two sectors rise. The labor participation ratio in the capital good sector and the output ratio of the capital good sector and the consumption good sector become higher, as illustrated in Fig. 4.8.6c. Figure 4.8.6d shows that the price of consumption good, the wage rate rise, and the rate of interest rise. From Fig. 4.8.6e, we observe that both the current income and consumption rise as the productivity rises. Figure 4.8.6f demonstrates that the share of output of the capital good sector in the GNP rises, and that of the consumption good sector falls.

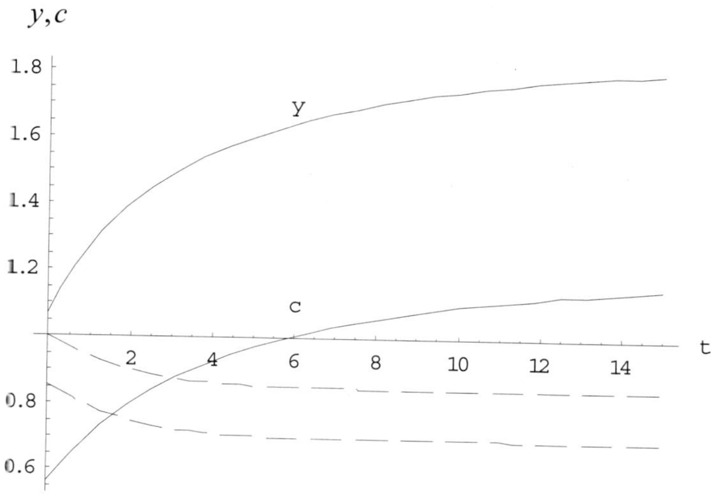

Fig. 4.8.5 As λ declines from 0.65 to 0.40, the consumption falls.

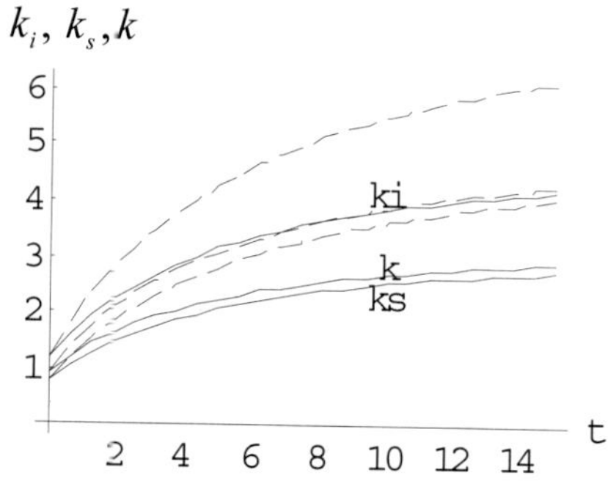

a) the capita intensities and wealth

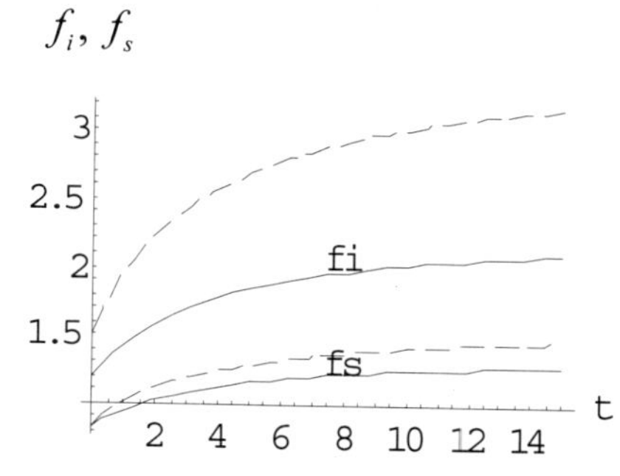

b) the output levels of the two sectors

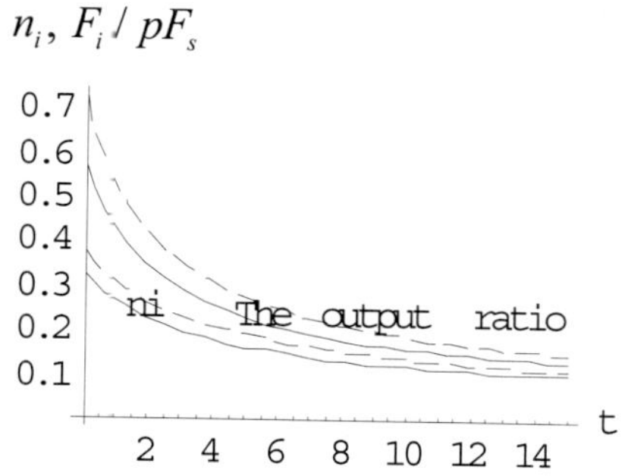

c) the labor distribution and the output ratio

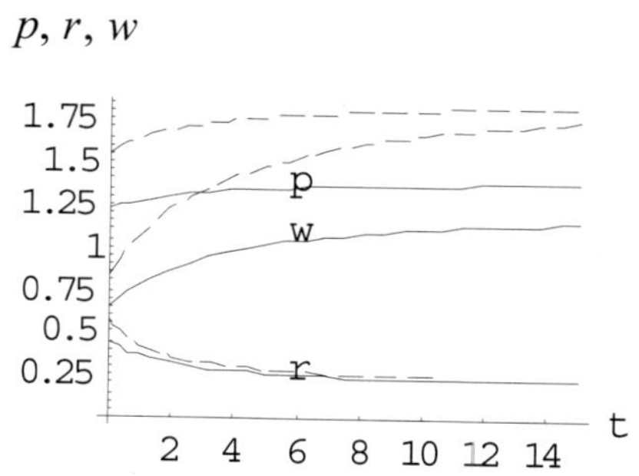

d) the price, interest rate, and wage rate

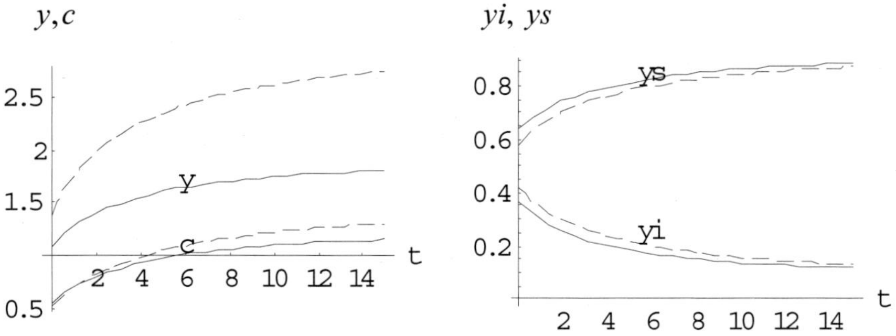

e) the income and consumption per capita f) the share of the outputs in GNP

Fig. 4.8.6 As A_i rises from 1.1 (with the solid lines) to 1.4 (the dashing lines).

Part II

Dimension Two

Chapter 5

Planar Linear Differential Equations

This chapter studies planar linear differential equations. Section 5.1 gives general solutions to planar linear first-order homogeneous differential equations. We also depict phase portraits of typical orbits of the planar systems. Section 5.2 introduces some concepts, such as positive orbit, negative orbit, orbit, limit set, and invariant set, for qualitative study. Section 5.3 shows how to calculate matrix exponentials and to reduce planar differential equations to the canonical forms. In Sec. 4, we introduce the concept of topological equivalence of planar linear systems and classify the planar linear homogeneous differential equations according to the concept. Section 5.5 studies planar linear first-order non-homogeneous differential equations. This section examines dynamic behavior of some typical economic models, such as the competitive equilibrium model, the Cournot duopoly model with constant marginal costs, the Cournot duopoly model with increasing marginal costs, the Cagan model with sluggish wages. Section 5.6 solves some types of constant-coefficient linear equations with time-dependent terms.

5.1 Planar Linear First-Order Homogeneous Differential Equations

We now consider a system of two linear first-order homogeneous differential equations

$$\dot{x}_i(t) = a_{i1}x_1(t) + a_{i2}x_2(t), \quad i = 1, 2,$$

where a_{ij} are parameters and $x_i(t)$ are variables. In vector notation, the equations are written as

$$\dot{x}(t) = Ax(t), \tag{5.1.1}$$

where $x(t)$ is the vector and A is the matrix [1]

$$x(t) = \begin{bmatrix} x_1(t) \\ x_2(t) \end{bmatrix}, \quad A = \begin{bmatrix} a_{11} & a_{12} \\ a_{21} & a_{22} \end{bmatrix}.$$

To solve the system, we attempt a solution of the form $x(t) = Ce^{\rho t}$ where C is a vector $C = [c_1 \quad c_2]^T$ and ρ is a scalar. Substituting $x(t) = Ce^{\rho t}$ into $\dot{x}(t) = Ax(t)$ yields

$$AC = \rho C .$$

We see that a nontrivial solution to this equation for a given ρ is an eigenvector and ρ is the corresponding eigenvalue. A necessary condition for C to be a nonzero to $AC = \rho C$ is $Det(A - \rho I) = 0$, that is

$$\rho^2 - \rho TrA + DetA = 0 , \tag{5.1.2}$$

which has either two real roots and a pair of complex conjugate roots. Here

$$TrA = a_{11} + a_{22}, \quad DetA = a_{11}a_{22} - a_{21}a_{12} .$$

The following lemma is held for Eq. (5.1.2).

Lemma 5.1.1 The eigenvalues ρ_1 and ρ_2 of A satisfy $\mathrm{Re}\,\sigma_j < 0$ if and only if

$$\rho_1\rho_2 = DetA > 0 , \ \rho_1 + \rho_2 = TrA < 0 .$$

They are pure imaginary if and only if the trace is zero. Moreover, $\rho_1 < 0 < \rho_2$ (or $\rho_2 < 0 < \rho_1$) if and only if $DetA < 0$.

We now consider the case that A is nonsingular and its eigenvalues ρ_1 and ρ_2 are distinct. In this case, the corresponding eigenvectors C_1 and C_2 are linearly independent vectors in R^2. We solve C_j by $AC_j = \rho_j C_j$. We thus obtain two special solutions $C_j e^{\rho_j t}$ to Eq. (5.1.1). It is straightforward to check that a linear combination of $C_j e^{\rho_j t}$ is also a

[1] App. A. 1 introduces some elementary concepts, such as eigenvalues and eigenvectors, and basic theorems in linear algebra and matrix theory. See also Gilbert and Gilbert (1995), Berman and Plemons (1979), and Peterson and Sochachi (2002).

solution to Eq. (5.1.1). We have the general solution to Eq. (5.1.1) as follows[2]

$$x(t) = \alpha_1 e^{\rho_1 t} C_1 + \alpha_2 e^{\rho_2 t} C_2 , \tag{5.1.3}$$

where α_1 and α_2 are scalars uniquely determined by the initial condition $x(0)$ by

$$\begin{bmatrix} c_{11} & c_{12} \\ c_{21} & c_{22} \end{bmatrix} \begin{bmatrix} \alpha_1 \\ \alpha_2 \end{bmatrix} = \begin{bmatrix} x_1(0) \\ x_2(0) \end{bmatrix}.$$

Example Consider a linear second-order equation

$$\ddot{y} + y = 0 .$$

Setting $x_1 = y$ and $x_2 = \dot{y}$, we see that the original equation is transformed to the system of linear first-order equations

$$\begin{bmatrix} \dot{x}_1 \\ \dot{x}_2 \end{bmatrix} = \begin{bmatrix} 0 & 1 \\ 1 & 0 \end{bmatrix} \begin{bmatrix} x_1 \\ x_2 \end{bmatrix}.$$

The two eigenvalues and eigenvectors are given by $\rho_{1,2} = \pm 1$, $C_1 = [1 \quad 1]^T$ and $C_2 = [1 \quad -1]^T$ respectively. Hence, the solution is

$$\begin{bmatrix} x_1 \\ x_2 \end{bmatrix} = \alpha_1 e^t \begin{bmatrix} 1 \\ 1 \end{bmatrix} + \alpha_2 e^{-t} \begin{bmatrix} 1 \\ -1 \end{bmatrix}.$$

We thus solve

$$y = x_1 = \alpha_1 e^t + \alpha_2 e^{-t} .$$

We now consider the case that the two eigenvalues are a pair of complex conjugate roots

$$\rho_{1,2} = \sigma \pm \phi i ,$$

where σ and ϕ ($\neq 0$ assumed) are real.[3]. For the solution $x(t)$ to be real, we have $x(t) = \bar{x}(t)$ where the bar indicates complex conjugation.

As

---

[2] Remark: every vector in R^2 can be written as a linear combination of two independent vectors.

[3] The condition $\theta \neq 0$ guarantees distinct roots of the characteristic equation.

$$e^{\rho_{1,2}t} = e^{\sigma t} e^{\pm i\phi t} = e^{\sigma t}\left(\cos\phi t \pm i\cos\phi t\right),$$

the condition $x(t) = \overline{x}(t)$ (where $x(t) = \alpha_1 e^{\rho_1 t} C_1 + \alpha_2 e^{\rho_2 t} C_2$) yields

$$2e^{2i\sigma t}\left(B_1 - \overline{B}_2\right) = \overline{B}_1 - B_2,$$

where $B_i \equiv \alpha_i C_i$. For the above equation to hold, we have $\overline{B}_1 = B_2$. Using these relations, we have

$$x(t) = e^{\rho_1 t} B_1 + e^{\rho_2 t} B_2 = e^{\sigma t}\left(b_1 \cos\phi t + b_2 \sin\phi t\right), \tag{5.1.4}$$

where $b_1 \equiv B_1 + B_2$ and $b_2 \equiv i(B_2 - B_1)$.

Example Consider

$$\begin{bmatrix} \dot{x}_1 \\ \dot{x}_2 \end{bmatrix} = \begin{bmatrix} 0 & 1 \\ -1 & 0 \end{bmatrix}\begin{bmatrix} x_1 \\ x_2 \end{bmatrix}.$$

The two eigenvalues and eigenvectors are given respectively by

$$\rho_{1,2} = \pm i, \; C_1 = [1 \quad i]^T, \; C_2 = [1 \quad -i]^T.$$

Hence,

$$B_1 = \alpha_1[1 \quad i]^T, \; B_2 = \alpha_2[1 \quad -i]^T.$$

For $\overline{B}_1 = B_2$ to hold, we have $\alpha = \alpha_1 = \alpha_2$. The solution is

$$\begin{bmatrix} x_1(t) \\ x_2(t) \end{bmatrix} = \begin{bmatrix} 2\alpha \\ 0 \end{bmatrix}\cos\phi t + \begin{bmatrix} 0 \\ 2\alpha \end{bmatrix}\sin\phi t.$$

When A is nonsingular, A has two distinct eigenvalues and the system $\dot{x}(t) = Ax$ has a unique equilibrium at the origin. We now discuss under what conditions the origin is stable or unstable. In the case that ρ_1 and ρ_2 are real and negative, from Eq. (5.1.3) we see that $x(t) \to 0$ as $t \to \infty$, irrespective of initial conditions. This means that the origin is globally asymptotically. Similarly, if ρ_1 and ρ_2 are real and positive, $x(t) \to \infty$ as $t \to \infty$, irrespective of initial conditions. The origin repels all orbits and is unstable.

When the eigenvalues are complex with $\rho_{1,2} = \sigma \pm \phi i$, its solution is given by

$$x(t) = e^{\sigma t}\left(b_1 \cos\phi t + b_2 \sin\phi t\right).$$

We see that if $\sigma > (<)\ 0$, the origin is unstable (stable). If $\sigma = 0$, the solution is bounded. The orbits

$$x(t) = b_1 \cos\phi t + b_2 \sin\phi t,$$

are actually closed. For each different $x(0)$, there is a corresponding closed orbits (called a *cycle*) passing through this point. None of cycles intersect each other. We are left with the last possibility for distinct eigenvalues, $\rho_1 < 0 < \rho_2$. The origin is unstable. The equilibrium in this case is called a *saddle*. Figure 5.1.1 illustrates dynamics of the four cases we have discussed. The intersection of the two lines is the equilibrium point. Here, we neglect other cases, such as one or two eigenvalues being equal to zero.[4] In the x_1-x_2-plane, time does not explicitly appear. To compensate for the loss of time parametrization in orbits, we insert arrows to indicate the direction in which the solution is changing as time passes. The flow of a differential equation is then drawn as the collection of all its orbits together with the direction arrows; the resulting picture is called the phase portrait of the differential equation.

For $\dot{x} = Ax$, after a long time, each individual trajectory exhibits one of only three types of behavior. As $t \to +\infty$, each trajectory either approaches infinity, or approaches the equilibrium point $x = 0$,[5] or repeatedly traverses a closed curve surrounding the equilibrium point. We observe that if the eigenvalues are real and negative or complex with negative real part, all trajectories approach the equilibrium point $x = 0$ as $t \to +\infty$. The original is either a nodal or a spiral sink. If the eigenvalues are pure imaginary, all trajectories remain bounded but do not approach the equilibrium point as $t \to +\infty$. The origin is a center. If at least one of the eigenvalues is positive or if the eigenvalues have positive real part, some trajectory, and possibly all trajectories except $x = 0$ tend to infinity as $t \to +\infty$. The origin is either a nodal source, a spiral source, or a saddle point. Table 5.1.1 summarizes the stability properties of the equilibrium.[6]

---

[4] See Britton (1986).

[5] Here, we require $\det A \neq 0$, which guarantees a unique equilibrium solution $x = 0$.

[6] See Boyce and Diprima (2001: 468).

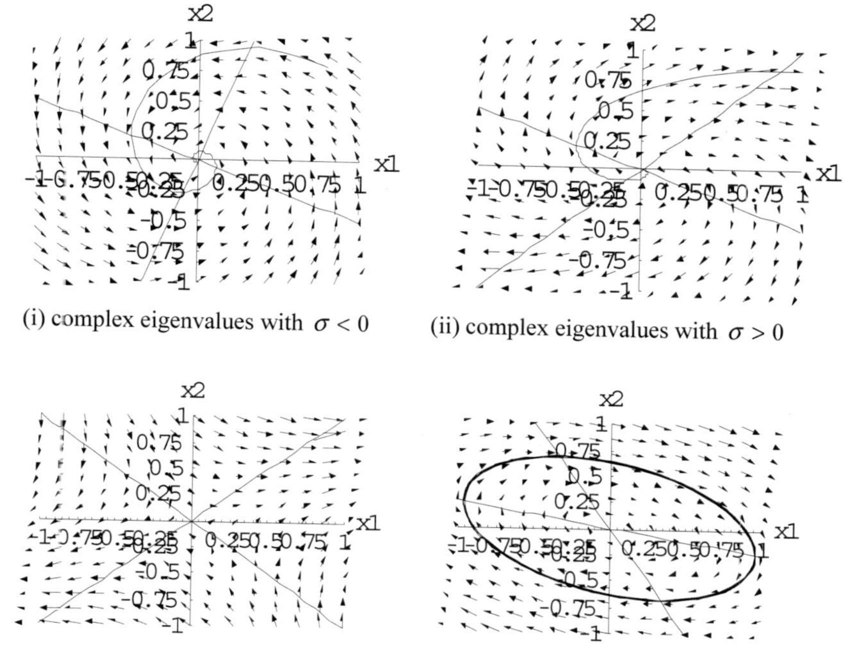

(i) complex eigenvalues with $\sigma < 0$ (ii) complex eigenvalues with $\sigma > 0$

(iii) real eigenvalues, opposite signs (iv) complex eigenvalues with $\sigma = 0$

Fig. 5.1.1 Phase portraits of typical orbits of $\dot{x}(t) = A_{2\times 2}x(t)$.[7]

Theorem 5.1.1 The equilibrium point $x = 0$ of the dynamic system

$$\dot{x} = A_{2\times 2}x\,, \quad \det A \neq 0\,,$$

is asymptotically stable if the eigenvalues ρ_1 and ρ_2 are real and negative or have negative real part; stable, but not asymptotically stable if ρ_1 and ρ_2 are pure imaginary; unstable if ρ_1 and ρ_2 are real and either is positive, or if they have positive real part.

Exercise 5.1

1 Prove Lemma 5.1.1.

---

[7] The matrices A for (i), (ii), (iii), and (iv) are respectively

$\begin{pmatrix} -1 & -2 \\ 3 & -1 \end{pmatrix}$, $\begin{pmatrix} 1 & 2 \\ -1 & 1 \end{pmatrix}$, $\begin{pmatrix} 1 & 1 \\ 1 & -1 \end{pmatrix}$, and $\begin{pmatrix} 1 & 4 \\ -2 & -1 \end{pmatrix}$.

Table 5.1.1 Stability properties of $\dot{x} = A_{2\times 2}x$ with $\det A \neq 0$.

Eigenvalues	Type of equilibrium	Stability
$\rho_1 > \rho_2 > 0$	Node	Unstable
$\rho_1 < \rho_2 < 0$	Node	Asymptotically stable
$\rho_1 < 0 < \rho_2$	Saddle point	Unstable
$\rho_1 = \rho_2 > 0$	Proper or improper node	Unstable
$\rho_1 = \rho_2 < 0$	Proper or improper node	Asymptotically stable
$\rho_{1,2} = \sigma \pm i\phi$	Spiral point	
$\sigma > 0$		Unstable
$\sigma < 0$		Asymptotically stable
$\rho_{1,2} = i\phi$	Center	Stable

5.2 Some Concepts for Qualitative Study

We have used some concepts, such as phase portrait without explanation. We now define a few concepts for qualitative study.

Consider an initial-value problem[8]

$$\dot{x}_j = f_j(x_1, x_2), \; j = 1, 2. \tag{5.2.1}$$

The corresponding initial-value problem is

$$\dot{x}_j = f_j(x_1, x_2), \; j = 1, 2, \; x(t_0) = x_0. \tag{5.2.2}$$

For convenience, choose $t_0 = 0$. If $f = (f_1, f_2)$ is a C^1 function, then for any $x_0 \in R^2$, there is an interval $I_{x0} \equiv (\alpha_{x0}, \beta_{x0})$ containing $t_0 = 0$ and a unique solution $\varphi(t, x_0)$ of the initial problem defined for all $t \in I_{x0}$, satisfying the initial condition $\varphi(0, x_0) = x_0$.

We now examine Eqs. (5.2.1) and its flow $\varphi(t, x_0)$ from a geometrical point of view. At each point of the $(t, x)-$space, the right-hand side of Eqs. (5.2.1) gives a value of the derivative dx/dt, which can be considered as the slope of a line segment at that point. The collection of all such line segments is called the *direction field* of the differential equations (5.2.1). Since the function f is independent of t, on any line parallel to the t-axis the segments of the direction field all

[8] This section is mostly referred to Sec. 7.2 in Hale and Koçak (1991).

have the same slope. It is thus natural to consider the projections of the direction field and the trajectories of Eqs. (5.2.1) onto the (x_1, x_2)-plane. We can assign to the point x the directed line segment from x to $x + f(x)$. The collection of all such vectors is called the *vector field* generated by Eqs. (5.2.1) or simply the vector field f. Projections of trajectories onto the (x_1, x_2)-plane are called orbit. We now define orbits.

Definition 5.2.1 The *positive orbit* $\gamma^+(x_0)$, *negative orbit* $\gamma^-(x_0)$, and *orbit* $\gamma(x_0)$ are defined, respectively, as the following subsets of

$$\gamma^+(x_0) = \bigcup_{t\in[0,\beta_{x0})} \varphi(t, x_0),$$

$$\gamma^-(x_0) = \bigcup_{t\in(\alpha_{x0},0]} \varphi(t, x_0),$$

$$\gamma(x_0) = \bigcup_{t\in(\alpha_{x0},\beta_{x0})} \varphi(t, x_0).$$

The simplest of orbits is an equilibrium point. In planar systems there can be another orbit of special interest, called a periodic orbit, which has no counterpart among the scalar autonomous differential equations.

Definition 5.2.2 A solution $\varphi(t, x_0)$ of $\dot{x} = f(x)$ is called a *periodic solution of period* T, with $T > 0$, if $\varphi(t + T, x_0) = \varphi(t, x_0)$ for all $t \in R$. The *minimum period* T is that period with the property that $\varphi(t, x_0) \neq x_0$ for $0 < t < T$. The orbit

$$\gamma(x_0) = \{\varphi(t, x_0), t \in R\},$$

of a periodic solution $\varphi(t, x_0)$ with period T is said to be a *periodic orbit* (also *closed orbit*) of period T.

To compensate for the loss of time parametrization in orbits, on the orbit $\gamma(x_0)$ we insert arrows to indicate the direction in which $\varphi(t, x_0)$ is changing as t increases. The flow of a differential equation is drawn as the collection of all its orbits together with the direction arrows; the resulting picture is called the *phase portrait* of the differential equation.

To examine asymptotic behavior of the system, we define limit point of orbits.

Definition 5.2.3 A point y is an $\omega-$*limit point* of the orbit $\gamma(x_0)$ if there exists a sequence t_j with $t_j \to \beta_{x_0}$ as $j \to \infty$ such that $\varphi(t_j, x_0) \to y$ as $j \to \infty$. That is, y is an $\omega-$limit point of the orbit $\gamma(x_0)$ if, for any $\varepsilon > 0$, there is a $t(\varepsilon)$ such that $\|y - \varphi(t(\varepsilon), x_0)\| < \varepsilon$. The set of all $\omega-$limit points of the orbit $\gamma(x_0)$ is called the $\omega-$*limit set* of $\gamma(x_0)$ and is denoted by $\omega(x_0)$.

An equivalent definition of $\omega(x_0)$ is

$$\omega(x_0) = \bigcap_{\tau \geq 0} \overline{\gamma^+(\varphi(\tau, x_0))}.$$

The concept of the α*-limit set* of $\gamma(x_0)$ can be defined similarly by reversing the direction of time. A point y is an ω*-limit point* of the orbit $\gamma(x_0)$ if there exists a sequence t_j with $t_j \to \alpha_{x_0}$ as $j \to \infty$ such that $\varphi(t_j, x_0) \to y$ as $j \to \infty$. The geometric definition of $\alpha(x_0)$ is

$$\alpha(x_0) = \bigcap_{\tau \leq 0} \overline{\gamma^-(\varphi(\tau, x_0))}.$$

Definition 5.2.4 An *invariant set* of a dynamic system

$$\dot{x} = f(x),\ (x, t) \in R^n \times R,$$

is a subset of $S \subset R^n$ such that $x_0 \in S$ implies $\phi(t, x_0) \in S$.

Clearly, an invariant set S consists of orbits of the dynamical system.

5.3 Matrix Exponentials and Reduction to Canonical Forms

The flow of the scalar linear differential equation $\dot{x} = ax$, is given by the exponential function

$$\varphi(t, x_0) = e^{at} x_0.$$

To obtain an analogous formula for the flow of linear multi-dimensional systems, we need the concept of the matrix exponential.

Definition 5.3.1 Let A be an $n \times n$ matrix. Then for $t \in R$

$$e^{At} = \sum_{k=0}^{\infty} \frac{A^k t^k}{k!}.$$

For $t = 0$, $e^{A0} = I_{n\times n}$.

Lemma 5.3.1 Let A be an $n\times n$ matrix. Then for $t \in R$

$$\frac{d}{dt}e^{At} = Ae^{At}.$$

For an $n\times n$ matrix A, e^{At} is an $n\times n$ matrix which can be computed in terms of the eigenvalues and eigenvectors of A.

Lemma 5.3.2 If P and T are linear transformations on R^n and $S = PTP^{-1}$, then $e^S = Pe^T P^{-1}$. If $PT = TP$, then $e^{S+T} = e^S e^T$.

Applying this lemma, we obtain that if $P^{-1}AP = diag[\rho_j]$ then $e^{At} = Pdiag[\rho_j]P^{-1}$. If A is invertible, then $(e^A)^{-1} = e^{-A}$. If

$$A = \begin{bmatrix} a & -b \\ b & a \end{bmatrix},$$

then

$$e^A = e^a \begin{bmatrix} \cos b & -\sin b \\ \sin b & \cos b \end{bmatrix}.$$

If

$$A = \begin{bmatrix} a & b \\ 0 & a \end{bmatrix},$$

then

$$e^A = e^a \begin{bmatrix} 1 & b \\ 0 & 1 \end{bmatrix}.$$

We can now compute the matrix e^{At} for any 2×2 matrix A.

Theorem 5.3.1 Let A be a 2×2 matrix with real entries. Then, there exists a real invertible 2×2 matrix such that the matrix $B = P^{-1}AP$ has one of the following forms in *Jordan normal form*:

$$B = \begin{bmatrix} \rho_1 & 0 \\ 0 & \rho_2 \end{bmatrix}, \quad B = \begin{bmatrix} \rho & 0 \\ 0 & \rho \end{bmatrix}, \quad B = \begin{bmatrix} a & -b \\ b & a \end{bmatrix}. \tag{5.3.1}$$

From the above discussions, we calculate these matrix exponentials as

$$e^{Bt} = \begin{bmatrix} e^{\rho_1 t} & 0 \\ 0 & e^{\rho_2 t} \end{bmatrix}, \quad e^{Bt} = e^{\rho t}\begin{bmatrix} 1 & t \\ 0 & 1 \end{bmatrix},$$

$$e^{Bt} = e^{at}\begin{bmatrix} \cos bt & -\sin bt \\ \sin bt & \cos bt \end{bmatrix}. \tag{5.3.2}$$

It is known that the solution to

$$\dot{x} = Bx \text{ with } x(0) = x_0,$$

is

$$x(t) = e^{Bt} x_0 .$$

From Eqs. (5.3.2), we can immediately solve $\dot{x} = Bx$ for B to take any matrix form of Eqs. (5.3.2).

For $\dot{x} = Ax$ which A is not in the Jordan normal form, we introduce $x = Py$. We thus have

$$\dot{y} = P^{-1}APy = By ,$$

where B is in Jordan normal form. We can explicitly solve y. Hence, we solve $\dot{x} = Ax$ by $x = Py$.

Exercise 5.3

1 Compute the exponentials of the following matrices

(i) $\begin{bmatrix} 1 & 0 \\ 0 & 3 \end{bmatrix}$;

(ii) $\begin{bmatrix} 1 & 2 \\ 0 & 3 \end{bmatrix}$;

(iii) $\begin{bmatrix} 1 & 0 \\ 5 & 3 \end{bmatrix}$;

(iv) $\begin{bmatrix} 2 & -1 \\ 1 & 2 \end{bmatrix}$.

2 Compute the exponentials of the following matrices

(i) $\begin{bmatrix} 1 & 0 & 0 \\ 0 & 2 & 0 \\ 0 & 0 & 3 \end{bmatrix}$;

(ii) $\begin{bmatrix} 1 & 0 & 0 \\ 0 & 2 & 1 \\ 0 & 0 & 3 \end{bmatrix}$;

(iii) $\begin{bmatrix} 1 & 0 & 0 \\ 1 & 2 & 0 \\ 0 & 1 & 3 \end{bmatrix}$.

3 Find 2×2 matrices A and B such that $e^{A+B} = e^A e^B$.

4 Solve $\dot{x} = Bx$ for

$$B = \begin{bmatrix} \rho_1 & 0 \\ 0 & \rho_2 \end{bmatrix}, \quad B = \begin{bmatrix} \rho & 0 \\ 0 & \rho \end{bmatrix}, \quad B = \begin{bmatrix} a & -b \\ b & a \end{bmatrix}.$$

5.4 Topological Equivalence in Planar Linear Systems

Two flows can be considered qualitatively equivalent if they have the same orbit structure, that is, if they have equal number of orbits and the directions of flows on the corresponding orbits are the same. We begin examining qualitative equivalence in linear systems with a precise definition of the equivalence.

Definition 5.4.1 Two planar linear systems $\dot{x} = Ax$ and $\dot{x} = Bx$ are said to be *topologically equivalent* if there is a *homeomorphism*

$$h:\ R^2 \to R^2$$

of the plane, that is, h is continuous with continuous inverse, that maps the orbits of $\dot{x} = Ax$ onto the orbits of $\dot{x} = Bx$ and preserves the sense of direction of time.

Since we are now concerned with the flows of planar linear systems, it is convenient to recast this definition in a somehow more quantitative form by mapping one flow to the other, that is

$$h\left(e^{At}x\right) = e^{Bt}h(x), \tag{5.4.1}$$

for every $t \in R$ and $x \in R^2$. A homeomorphism h satisfying Eq. (5.4.1) is a bit more special than the one required in the definition.

Theorem 5.4.1 Suppose that the eigenvalues of two 2×2 matrices A and B have nonzero parts. Then the two linear systems $\dot{x} = Ax$ and $\dot{x} = Bx$ are topologically equivalent if and only if A and B have the same number of eigenvalues with negative (and hence positive) real parts. Consequently, up to topological equivalence, there are three equivalence classes of hyperbolic planar linear systems with, for example, the following representatives

(i) $\begin{pmatrix} -1 & 0 \\ 0 & -1 \end{pmatrix}$: two negative eigenvalues;

(ii) $\begin{pmatrix} 1 & 0 \\ 0 & 1 \end{pmatrix}$: two positive eigenvalues;

(iii) $\begin{pmatrix} 1 & 0 \\ 0 & -1 \end{pmatrix}$: one negative and one positive eigenvalues.

From the topological viewpoint, there are only three cases of planar hyperbolic linear systems and they are determined solely by the signs of the real parts of the eigenvalues. It should be noted that a stable spiral and a stable node are topologically equivalent. When we study the qualitative features of planar linear systems, it is proper to use terms such as a *hyperbolic source*, a *hyperbolic sink*, and a *hyperbolic saddle*.

Theorem 5.4.2 Suppose that the eigenvalues of 2×2 matrix A have at least one eigenvalue with zero real part, then the planar linear system

$$\dot{x} = Ax$$

is topologically equivalent to precisely one of the following five linear systems

(i) $\begin{pmatrix} 0 & 0 \\ 0 & 0 \end{pmatrix}$: the zero matrix;

(ii) $\begin{pmatrix} -1 & 0 \\ 0 & 0 \end{pmatrix}$: one negative and one zero eigenvalues;

(iii) $\begin{pmatrix} 1 & 0 \\ 0 & 0 \end{pmatrix}$: one positive and one zero eigenvalues;

(vi) $\begin{pmatrix} 0 & 1 \\ 0 & 0 \end{pmatrix}$: two zero eigenvalues but one eigenvector;

(v) $\begin{pmatrix} 0 & 1 \\ -1 & 0 \end{pmatrix}$: two purely imaginary eigenvalues.

Figure 5.4.1 depicts phase portraits of the representatives of three hyperbolic and five nonhyperbolic linear systems. Formal proofs of the above two theorems are intricate.[9]

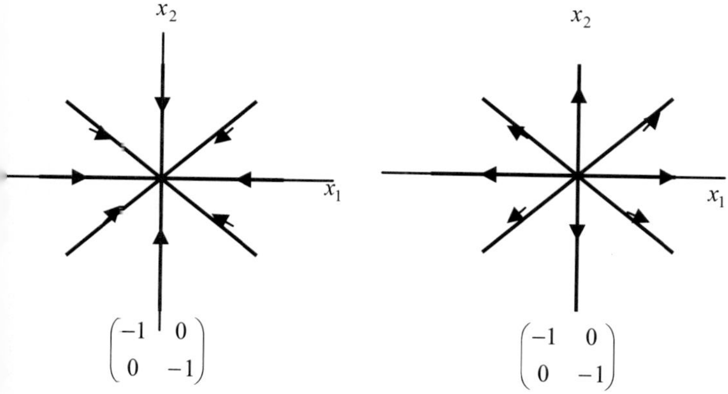

[9] See Chap. 8 in Hale and Koçak (1991).

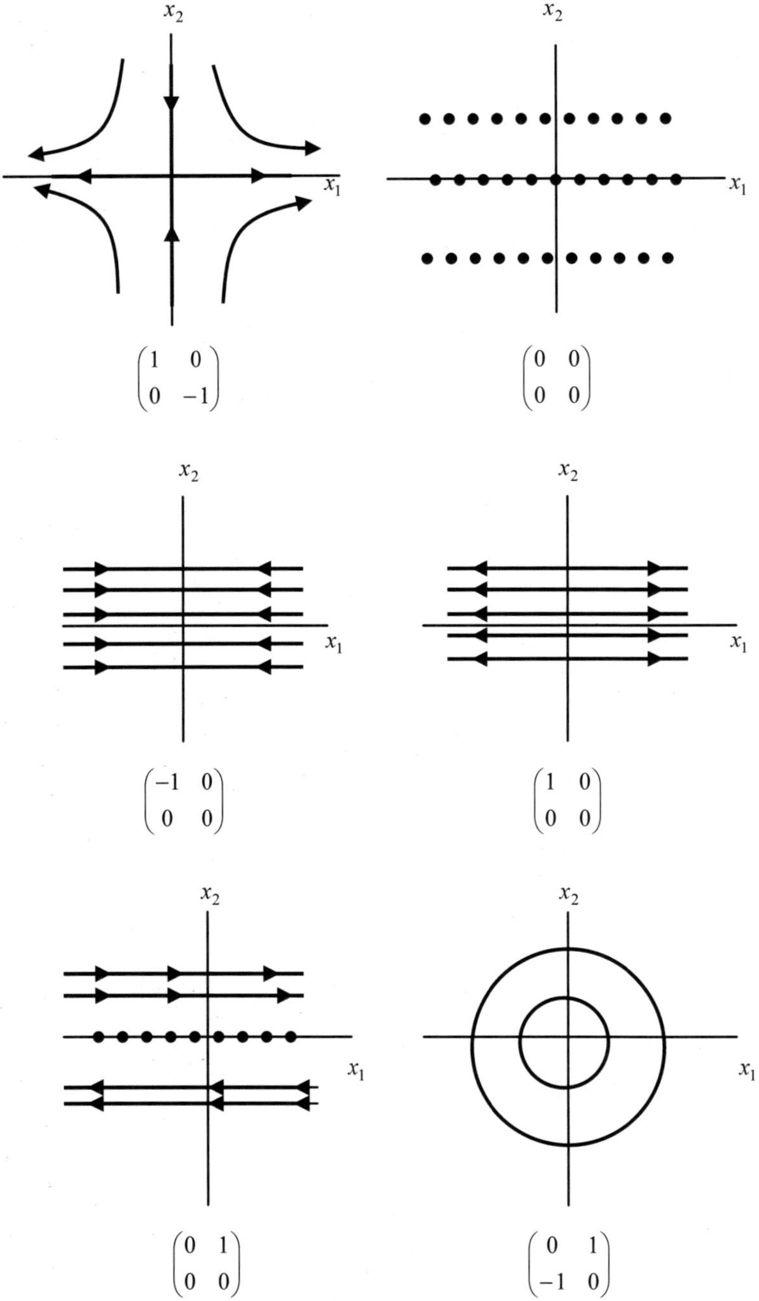

Fig. 5.4.1 Phase portrait of topological equivalence classes of planar linear systems.

Exercise 5.4

1 Show that the following two linear systems $\dot{x} = Ax$ are topologically equivalent

(i) $A = \begin{bmatrix} -2 & 0 \\ 0 & 2 \end{bmatrix}$;

(ii) $A = \begin{bmatrix} -2 & 1 \\ 0 & 2 \end{bmatrix}$.

5.5 Planar Linear First-Order Nonhomogeneous Differential Equations

We now concerned with

$$\dot{x}(t) = A_{2\times 2}x(t) + b_{2\times 1}, \tag{5.5.1}$$

where $b \neq 0$. Assume A is nonsingular and b is a constant vector. define $y(t)$ by

$$y(t) = x(t) + A^{-1}b.$$

Then Eq. (5.5.1) can be written as

$$\dot{y}(t) = A_{2\times 2}y(t),$$

which has exactly the same form as Eq. (5.1.1). Since the behavior of the solution for Eq. (5.5.1) can be completely described by that of $\dot{y}(t) = A_{2\times 2}y(t)$ via

$$x(t) = y(t) - A^{-1}b,$$

it is sufficient to study Eq. (5.1.1).

Example The competitive equilibrium and its stability.
Consider a competitive market composed of three commodities X_0, X_1, and X_2 with prices P_0, P_1, and P_2, respectively. Demand and supply functions of each commodity are related to the prices of the three commodities. Excess demand function of each commodity is denoted by $E_i(P_0, P_1, P_2)$. We assume E_i is homogeneous of degree zero and the market satisfies Walras's law. The homogeneity allows us to choose a commodity as numeraire

$$E_i = E_i(P_0, P_1, P_2) = E_i(1, p_1, p_2), \quad i = 0, 1, 2,$$

where $p_i \equiv P_i / P_0$, $i = 1, 2$. Walras's law implies

$$\sum_{i=0}^{2} P_i E_i ,$$

or

$$E_0 + p_1 E_1 + p_2 E_2 = 0. \tag{5.5.2}$$

This condition implies that we discuss market equilibrium and its stability solely in terms of commodities 1 and 2, neglecting commodity 0.

Suppose that prices change in proportion to their excess demand. That is

$$\dot{p}_i = k_i E_i(p_1, p_2), \quad k_i > 0, i = 1, 2. \tag{5.5.3}$$

We specify

$$k_1 = 2, \ k_2 = 3, \ E_1 = 3 - 16 + 3p_2, \ E_2 = 16 + 4p_1 - 8p_2 .$$

The price dynamics are

$$\dot{p}_1 = 2(3 - 6p_1 + 3p_2),$$

$$\dot{p}_2 = 3(16 + 4p_1 - 8p_2).$$

The system has a unique stable equilibrium $(p_1^*, p_2^*) = (2, 3)$.
We simulate the model with $(p_1(0), p_2(0)) = (3.8, 1)..$

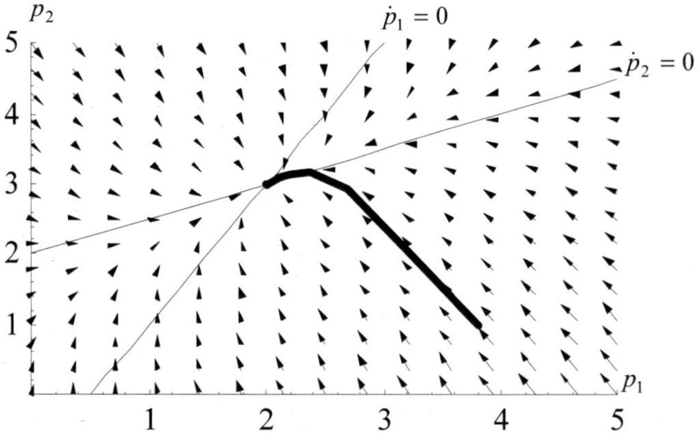

Fig. 5.5.1 The competitive equilibrium and its stability.

Example The Cournot duopoly model with constant marginal costs. Consider a Cournot duopoly, in which the two firms produce the same product and face constant marginal costs c_1 and c_2. The market price $P(t)$ is a function of the total quantity of output produced $Q(t)$, that is

$$P(t) = a_0 - a_1 Q(t), \quad a_0, a_1 > 0 .$$

Each firm wishes to maximize its profits but must adjust towards the profit maximizing output. At t, firm i's profit $\pi_i(t)$ is given by

$$\pi_i(t) = Q_i(t)P(t) - c_i Q_i(t), \quad c_i > 0 ,$$

where Q_i is firm i's output. Note that

$$Q = Q_1 + Q_2 .$$

In making its production plans, each firm assumes the other will hold its output level unchanged. It is straightforward to show that the following solution $\overline{Q}_i$ maximizes $\pi_i(t)$

$$\overline{Q}_i(t) = A_i - \frac{Q_j(t)}{2}, \quad i, j = 1, 2, i \neq j , \tag{5.5.4}$$

where

$$A_i \equiv \frac{a_0 - c_i}{2a_1} .$$

Assume that each firm adjusts its output $Q_i(t)$ towards $\overline{Q}_i(t)$ in the following way

$$\dot{Q}_i = \beta_i \left(\overline{Q}_i - Q_i \right), \quad \beta_i > 0 .$$

Substituting Eqs. (5.5.4) into the above equations yields

$$\begin{bmatrix} \dot{Q}_1 \\ \dot{Q}_2 \end{bmatrix} = A \begin{bmatrix} Q_1 \\ Q_2 \end{bmatrix} + \begin{bmatrix} \beta_1 A_1 \\ \beta_2 A_2 \end{bmatrix}, \quad A \equiv \begin{bmatrix} -\beta_1 & -\beta_1 / 2 \\ -\beta_2 / 2 & -\beta_2 \end{bmatrix} . \tag{5.5.5}$$

The system has a unique positive equilibrium

$$Q_1^* = \frac{2(2A_1 - A_2)}{3},$$

$$Q_2^* = \frac{2(2A_2 - A_1)}{3},$$

if $2A_1 > A_2 > A_1/2$. As

$$\mathrm{Tr}(\mathrm{A}) = -\beta_1 - \beta_2 < 0, \quad \mathrm{Det}(A) = 3\beta_1\beta_2/4,$$

$$\Delta = \mathrm{Tr}(\mathrm{A})^2 - \mathrm{Det}(A) = (\beta_1 - \beta_2)^2 + \beta_1\beta_2 > 0,$$

we see that the equilibrium is stable and the approach to it will not be cyclical (since the positive discriminant means that the roots are real). The phase diagram is given as in Fig. 5.5.2.

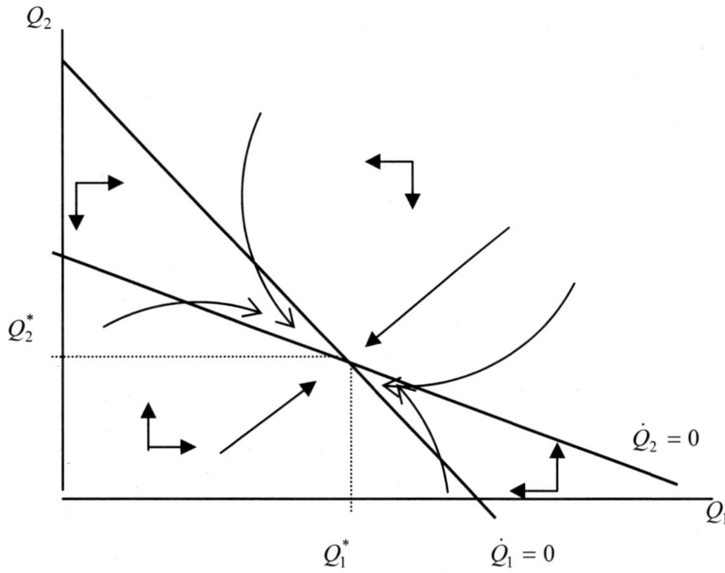

Fig. 5.5.2 The Cournot duopoly with constant marginal costs.

For illustration, take $\beta_1 = \beta_2 = \beta$, $a_0 = 9$, $a_1 = 1$, and $c_1 = c_2 = 3$. The system becomes

$$\begin{bmatrix} \dot{Q}_1 \\ \dot{Q}_2 \end{bmatrix} = \begin{bmatrix} -\beta & -\beta/2 \\ -\beta/2 & -\beta \end{bmatrix} \begin{bmatrix} Q_1 \\ Q_2 \end{bmatrix} + \begin{bmatrix} 3\beta \\ 3\beta \end{bmatrix}.$$

The solution is

$$Q_1(t) = 2 + \left(\frac{Q_{10} + Q_{20}}{2} - 2 \right) e^{-3\beta t/2} + (Q_{10} - Q_{20}) \frac{e^{-\beta t/2}}{2},$$

$$Q_2(t) = 2 + \left(\frac{Q_{10} + Q_{20}}{2} - 2\right)e^{-3\beta t/2} - (Q_{10} - Q_{20})\frac{e^{-\beta t/2}}{2}.$$

As $t \to \infty$, the solution approaches the equilibrium point $(2, 2)$.

Example The Cournot duopoly model with increasing marginal costs. Different from the previous example, we now consider that the total cost functions are specified as $3Q_i^2(t)$, where Q_i is firm i's output. The marginal costs, $6Q_i$, are not constant. The market price $P(t)$ is a function of the total quantity of output produced $Q(t)$ is specified as

$$P(t) = 9 - Q(t),$$

where

$$Q = Q_1 + Q_2.$$

Each firm wishes to maximize its profits but must adjust towards the profit maximizing output. At t, firm i's profit $\pi_i(t)$ is given by

$$\pi_i(t) = Q_i(t)P(t) - 3Q_i^2(t).$$

In making its production plans, each firm assumes the other will hold its output level unchanged. It is straightforward to show that the following solution $\overline{Q}_i$ maximizes $\pi_i(t)$

$$\overline{Q}_i(t) = \frac{9}{8} - \frac{Q_j(t)}{8}, \quad i, j = 1, 2, i \neq j. \tag{5.5.6}$$

Assume that each firm adjusts its output $Q_i(t)$ towards $\overline{Q}_i(t)$ in the following way

$$\dot{Q}_i = \beta_i(\overline{Q}_i - Q_i), \quad \beta_i > 0.$$

Substituting Eqs. (5.5.6) into the above equations yields

$$\begin{bmatrix} \dot{Q}_1 \\ \dot{Q}_2 \end{bmatrix} = \begin{bmatrix} -\beta_1 & -\beta_1/8 \\ -\beta_2/8 & -\beta_2 \end{bmatrix} \begin{bmatrix} Q_1 \\ Q_2 \end{bmatrix} + \begin{bmatrix} 9\beta_1/8 \\ 9\beta_2/8 \end{bmatrix}.$$

The system has a unique positive equilibrium $(1, 1)$. We calculate

$$\mathrm{Tr}(\mathrm{A}) = -\beta_1 - \beta_2 < 0, \quad \mathrm{Det}(A) = \frac{63}{64}\beta_1\beta_2 < 0,$$

$$\Delta = (\beta_1 - \beta_2)^2 + \frac{\beta_1\beta_2}{16} > 0.$$

The equilibrium is stable. For illustration, take $\beta_1 = \beta_2 = \beta$. The solution is

$$Q_1(t) = 1 + \left(\frac{Q_{10} + Q_{20}}{2} - 1\right)e^{-9\beta t/8} + (Q_{10} - Q_{20})\frac{e^{-7\beta t/8}}{2},$$

$$Q_2(t) = 1 + \left(\frac{Q_{10} + Q_{20}}{2} - 1\right)e^{-9\beta t/8} - (Q_{10} - Q_{20})\frac{e^{-7\beta t/8}}{2}.$$

As demonstrated in Fig. 5.5.3, the solution approaches the equilibrium point $(1, 1)$.

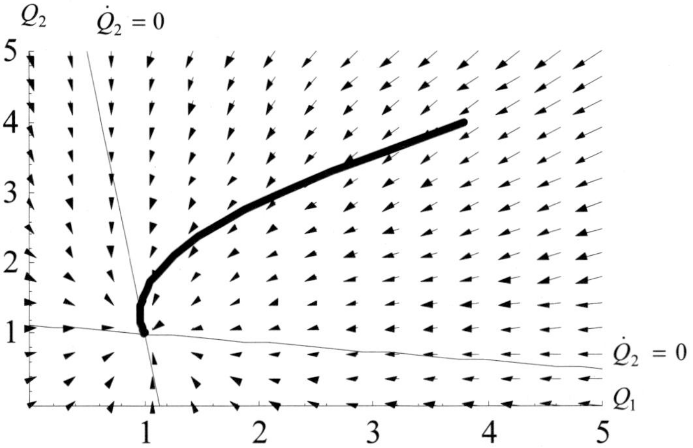

Fig. 5.5.3 The Cournot model with increasing marginal costs.

Example The Cagan model with sluggish wages.[10]
The Cagan model with sluggish wages is

$$M - P(t) = \alpha_1 Y(t) - \alpha_2 \dot{P}, \ \ \alpha_1, \alpha_2 > 0,$$

$$Y = c + (1 - \theta)N(t), \ \ 0 < \theta < 1,$$

$$W(t) - P = a - \theta N, \ \ \dot{W} = \gamma(N - \overline{N}), \ \ \gamma > 0,$$

where

[10] The model is based on Sect. 7.2 in Turnovsky (2000).

Y = output;
N = employment;
$\overline{N}$ = the full employment (exogenous);
P = price level;
M = the (fixed) nominal money stock; and
W = wage rate.

The first equation describes money market equilibrium under the perfect foresight assumption that the anticipated rate of inflation is equal to the actual rate of inflation. The second equation is a production function. The third equation describes the demand for labor as the corresponding marginal production condition. The last equation says that money wages evolves in accordance with the Phillips curve. The system can be reduced to the following pair of differential equations in W and P

$$\begin{bmatrix} \dot{W} \\ \dot{P} \end{bmatrix} = \begin{bmatrix} -\dfrac{\gamma}{\theta} & \dfrac{\gamma}{\theta} \\ a_0 & \dfrac{1}{\alpha_2} - a_0 \end{bmatrix} \begin{bmatrix} W \\ P \end{bmatrix} + \begin{bmatrix} \gamma\left(\dfrac{a}{\theta} - \overline{N}\right) \\ \dfrac{\alpha_1 c}{\alpha_2} - aa_0 - \dfrac{M}{\alpha_2} \end{bmatrix},$$

where

$$a_0 \equiv \frac{\alpha_1}{\alpha_2}\left(\frac{\theta - 1}{\theta}\right) < 0.$$

It can be shown that the determinant of the matrix is equal to $-\gamma/\theta\alpha_2 < 0$, which implies that the two eigenvalues are of opposite signs. We thus conclude that the equilibrium point is a saddle point.

Exercise 5.5

1 Find equilibrium of the following dynamic version of the IS-LM model and discuss its stability:

$$\dot{Y}(t) = \gamma(D(t) - Y(t)), \quad \gamma > 0,$$

$$\dot{r}(t) = \eta\left(L(t) - \overline{M}\right), \quad \eta > 0,$$

in which γ and η are parameters, $Y(t)$ is income, $r(t)$ is the interest rate, $L(t)$ is the demand for money, $\overline{M}$ is the fixed level of money supply. Here, the aggregate demand $D(t)$ is given by $D = C + I + G$. In the model, consumption $C(t)$, investment $I(t)$, and the level of government spending $G(t)$, the demand for money, are respectively given by: (1) the consumption function,

$$C(t) = C_0 + cY(t), 1 > c > 0;$$

(2) the investment function,

$$I(t) = I_0 + vY - \delta r,$$

where $1 > v > 0$ and $\delta > 0$; (3) the exogenous government expending, $G(t) = \overline{G}$; and (4) the money demand function,

$$L = L_0 - \beta r + \alpha Y,$$

where α, $\beta > 0$, where C_0, c, v, δ, α, and $\overline{G}$ are constant.

2 Re-examine the model of interaction of inflation and unemployment in the previous chapter. The expectations-augmented version of the Phillips relation is: $p = a - bU + h\pi$ ($0 < h \leq 1$), where a, b, and h are parameters, $p(t)$ is the rate of inflation, $U(t)$ the unemployment rate, and $\pi(t)$ is the expected rate of inflation. We also have

$$\dot{\pi} = j(p - \pi), \quad 0 < j \leq 1,$$

$$\dot{U} = -k(m - p), \quad k > 0,$$

where M is the nominal money and $m = \dot{M} / M$ its rate of growth.

3 Solve the Cournot model with constant marginal costs $c_1 = c_2 = 4$ and

$$P(t) = 20 - 3Q(t),$$

$$\dot{Q}_i = \beta_i(\overline{Q}_i - Q_i), \quad \beta_i > 0, i = 1, 2.$$

5.6 Constant-Coefficients Nonhomogeneous Linear Equations with Time-Dependent Terms

This section deals with the nonhomogeneous differential equations

$$\begin{bmatrix} \dot{x}_1 \\ \dot{x}_2 \end{bmatrix} = \begin{bmatrix} a_{11} & a_{12} \\ a_{21} & a_{22} \end{bmatrix} \begin{bmatrix} x_1 \\ x_2 \end{bmatrix} + \begin{bmatrix} h_1(t) \\ h_2(t) \end{bmatrix}. \qquad (5.6.1)$$

That is,

$$\dot{x} = Ax + h(t).$$

We explained how to solve the equation when $h(t)$ is constant. We now solve the problem when $h(t)$ takes on different forms. We now solve the question of finding a particular solution $x_p(t)$ to the nonhomogeneous equation.

First, we apply the *method of undetermined coefficients*. This method is appropriate when the entries of $h(t)$ are linear combinations of functions of the form $t^k e^{at}$. These include polynomials, plain exponentials, and functions such as $t^k \sin \omega t$ and $t^k \cos \omega t$.[11] When $h(t)$ is of the form $p(t)e^{at}$, where $p(t)$ is a vector of polynomial functions. In that case, try setting $x(t) = q(t)e^{at}$ where $q(t)$ is an unknown vector of polynomial functions of the same degree as $p(t)$.

Example Solve the differential equation

$$\begin{bmatrix} \dot{x}_1 \\ \dot{x}_2 \end{bmatrix} = \begin{bmatrix} -2 & -1 \\ 5 & 4 \end{bmatrix} \begin{bmatrix} x_1 \\ x_2 \end{bmatrix} + \begin{bmatrix} 0 \\ 1 \end{bmatrix} e^{2t}.$$

The solution to $\dot{x} = Ax$ is

$$\begin{bmatrix} x_1 \\ x_2 \end{bmatrix} = C_1 e^{3t} \begin{bmatrix} 1 \\ -5 \end{bmatrix} + C_2 e^{-t} \begin{bmatrix} 1 \\ -1 \end{bmatrix}.$$

Try substituting

$$\begin{bmatrix} x_1 \\ x_2 \end{bmatrix} = \begin{bmatrix} a_1 \\ a_2 \end{bmatrix} e^{2t},$$

into the nonhomogeneous equation. We find $a_1 = 1/3$ and $a_2 = -4/3$. The final solution is

$$\begin{bmatrix} x_1 \\ x_2 \end{bmatrix} = C_1 e^{3t} \begin{bmatrix} 1 \\ -5 \end{bmatrix} + C_2 e^{-t} \begin{bmatrix} 1 \\ -1 \end{bmatrix} + e^{2t} \begin{bmatrix} 1/3 \\ -4/3 \end{bmatrix}.$$

---

[11] Note: $\sin \omega t = \left(e^{i\omega t} - e^{-i\omega t}\right)/2i$ and $\cos \omega t = \left(e^{i\omega t} + e^{-i\omega t}\right)/2$.

The method of undetermined coefficients does not always work. To see why, assume the driving term $h(t)$ is of the form Ge^{at}, where G is a constant vector. Then, if we substitute $x = Qe^{at}$ into Eqs. (5.6.1), we find the following equation

$$aQe^{at} = AQe^{at} + Ge^{at},$$

which leads to $(aI - A)Q = G$. This linear equation is solvable only if the matrix $(aI - A)$ is invertible. The inverse only exists if a is not an eigenvalue of A. The phenomenon in which the driving term includes e^{at} for some eigenvalue a of the matrix A is called *resonance*. If a is an eigenvalue of A, instead of trying a solution of the form Qe^{at}, we try a solution of the form $P(t)e^{at}$, with $P(t)$ a vector of polynomials of degree equal to the multiplicity of a as a root of the characteristic polynomial of A.

Example Solve the differential equation

$$\begin{bmatrix} \dot{x}_1 \\ \dot{x}_2 \end{bmatrix} = \begin{bmatrix} -2 & -1 \\ 5 & 4 \end{bmatrix} \begin{bmatrix} x_1 \\ x_2 \end{bmatrix} + \begin{bmatrix} 0 \\ 1 \end{bmatrix} e^{-t}.$$

The solution to $\dot{x} = Ax$ is

$$\begin{bmatrix} x_1 \\ x_2 \end{bmatrix} = C_1 e^{3t} \begin{bmatrix} 1 \\ -5 \end{bmatrix} + C_2 e^{-t} \begin{bmatrix} 1 \\ -1 \end{bmatrix}.$$

As -1 is an eigenvalue of A, we substitute

$$\begin{bmatrix} x_1 \\ x_2 \end{bmatrix} = \begin{bmatrix} a_1 + b_1 t \\ a_2 + b_2 t \end{bmatrix} e^{-t},$$

into the nonhomogeneous equation. We find the final solution

$$\begin{bmatrix} x_1 \\ x_2 \end{bmatrix} = C_1 e^{3t} \begin{bmatrix} 1 \\ -5 \end{bmatrix} + C_2 e^{-t} \begin{bmatrix} 1 \\ -1 \end{bmatrix} + e^{-t} \begin{bmatrix} -c - 1/4 + t/4 \\ c - t/4 \end{bmatrix}.$$

It looks as if this solution depends on three arbitrary constants, but c and C_2 play the same role.

There is an explicit formula for solving nonhomogeneous equations. The formula is called *variation of parameters*.

Theorem 5.6.1 The solution to

$$\dot{x} = Ax + h(t)$$

with $x(t_0) = x_0$ is

$$x(t) = e^{(t-t_0)A}x_0 + \int_{t_0}^{t} e^{(t-s)A}h(s)ds. \tag{5.6.2}$$

To prove Theorem 5.6.1, introduce

$$y(t) = e^{-At}x(t).$$

Differentiate $y(t) = e^{-At}x(t)$ with respect to t

$$\dot{y} = e^{-At}\dot{x} - Ay.$$

Substituting $\dot{x} = Ax + h$ into the above equation, we obtain

$$\dot{y} = e^{-At}h.$$

We integrate the above equation

$$y(t) = C + \int_{t_0}^{t} e^{-As}h(s)ds. \tag{5.6.3}$$

As $y(t_0) = e^{-At_0}x_0$, we determine

$$C = e^{-At_0}x_0.$$

Inserting $C = e^{-At_0}x_0$ and $y = e^{-At}x$ into Eq. (5.6.3) yields Eq. (5.6.2).

Exercise 5.6

1 Solve the following differential equation

$$\begin{bmatrix} \dot{x}_1 \\ \dot{x}_2 \end{bmatrix} = \begin{bmatrix} -2 & -1 \\ 5 & 4 \end{bmatrix} \begin{bmatrix} x_1 \\ x_2 \end{bmatrix} + \begin{bmatrix} 0 \\ e^t + te^t \end{bmatrix}.$$

2 Suppose that A is a constant square matrix. The linear system $\dot{y} = -yA$, where y is a row vector, is called the *adjoint equation* for $\dot{x} = Ax$. Show that (i) the flow $\varphi(t, y_0)$ of the adjoint equation is given by $\varphi(t, y_0) = y_0 e^{-At}$; (ii) $\varphi(t, y_0)\phi(t, x_0) = y_0 x_0$ for all $t \in R$, where $\phi(t, x_0)$ is the flow of $\dot{x} = Ax$.

Chapter 6

Planar Nonlinear Differential Equations

The previous chapter dealt with planar linear differential equations and their applications to economic analysis. This chapter deals with nonlinear planar differential equations. Section 6.1 carries out local analysis and provides conditions for validity of linearization. We also provide relations between linear systems and almost linear systems with regard to dynamic qualitative properties. This section examines dynamic properties of some frequently-applied economic models, such as the competitive equilibrium model, the Walrasian-Marshallian adjustment process, the Tobin-Blanchard model, and the Ramsey model. Section 6.2 introduces the Liapunov methods for stability analysis. In Sec. 6.3, we study some typical types of bifurcations of planar differential equations. Section 6.4 demonstrates motion of periodic solutions of some nonlinear planar systems. Section 6.5 introduces the Poincaré-Bendixon Theorem and applies the theorem to the Kaldor model to identify the existence of business cycles. Section 6.6 states Lienard's Theorem, which provides conditions for the existence and uniqueness of limit cycle in the Lienard system. Section 6.7 studies one of most frequently applied theorems in nonlinear economics, the Andronov-Hopf bifurcation theorem and its applications in the study of business cycles.

6.1 Local Stability and Linearization

Consider

$$\dot{x}_1 = f_1(x_1, x_2) = -4x_1^2 + x_1 x_2,$$

$$\dot{x}_2 = f_2(x_1, x_2) = 2x_2 - x_1 x_2.$$

The system has two equilibria $(0, 0)$ and $(2, 8)$. The linearized system at $(2, 8)$ is

$$\dot{u}_1 = -8u_1 + 2u_2,$$

$$\dot{u}_2 = -8u_1,$$

where we use

$$u_1 = x_1 - 2, \quad u_2 = x_2 - 8.$$

The solution to this linear system is

$$u_1 = \left((c_2 - 4c_1) - 4c_2 t\right)e^{-4t},$$

$$u_2 = -8(c_1 + c_2 t)e^{-4t}.$$

As $t \to +\infty$, we have $u_1 \to 0$ and $u_2 \to 0$. This implies that if the initial state is sufficiently close to $(2, 8)$, the solution to the initial problem will approach $(2, 8)$ as $t \to +\infty$.

Similarly, the linearized system at $(0, 0)$ is

$$\dot{u}_1 = 0,$$

$$\dot{u}_2 = 2u_2.$$

The solution to this linear system is $u_1 = c_1$ and $u_2 = c_2 e^{2t}$. We see that if an initial state is close to the origin, then the solution will be infinite as $t \to +\infty$.

We now consider a general autonomous system of the form

$$\dot{x}_j(t) = f_j(x_1, x_2), \quad j = 1, 2, \tag{6.1.1}$$

where $f_j(x_1, x_2)$ are smooth functions. Denote

$$x = (x_1, x_2)^T, \quad f = (f_1, f_2)^T.$$

For this system, a point x is called an *equilibrium point* (also *critical point*, *steady state solution*, etc.) if $f(x) = 0$. Suppose x^* is a unique equilibrium of Eqs. (6.1.1). Introduce

$$X(t) = x(t) - x^*.$$

From the Taylor theorem for functions of two variables,[1] we know

---

[1] The Taylor series is referred to as App. A.4.

$$f_j(x) = f_j(x^*) + \frac{\partial f_j}{\partial x_1}(x^*)X_1 + \frac{\partial f_j}{\partial x_2}(x^*)X_2 + g_j(X),$$

$$j = 1, 2, \tag{6.1.2}$$

where $g_j(X)$ are higher order terms and

$$\frac{g_j(X)}{\|X\|} \to 0 \text{ as } \|X\| \to 0,$$

where

$$\|X\| \equiv \sqrt{X_1^2 + X_2^2}.$$

Using $\dot{x} = \dot{X}$ and $f(x^*) = 0$, Eqs. (6.1.1) can be expressed in vector form as

$$\dot{X} = AX + g(X), \tag{6.1.3}$$

where the matrix A ($= (\partial f_i / \partial x_j)_{2\times 2}$) is the *Jacobian matrix* of f at x^*. The linear system

$$\dot{X} = AX, \tag{6.1.4}$$

is called the *linearized system* of Eqs. (6.1.1). When

$$\frac{g_j(X)}{\|X\|} \to 0 \text{ as } t \to +\infty,$$

we call the system Eqs. (6.1.3) an *almost linear system.*

Since the nonlinear term $g(X)$ is small compared to the linear term AX when X is small, it is reasonable to hope that the trajectories of the linearized system are good approximations to those of the nonlinear system, at least near the origin. The following theorem summarizes the relations between Eqs. (6.1.1) and its linearized system $\dot{x} = Ax$.

Theorem 6.1.1 Let ρ_1 and ρ_2 be the eigenvalues of the linear system $\dot{x} = A_{2\times 2}x$ corresponding to the almost linear system (6.1.3). Then the type and stability of the critical point $(0, 0)$ of the linear system and Eq. (6.1.3) are shown in Table 6.1.1.[2]

---

[2] See Boyce and Diprima (2001: 484).

Table 6.1.1 Relations between linear and almost linear systems.

Eigenvalues	Linear system Type of CP	Linear system Stability	Almost linear system Type of CP	Almost linear system Stability
$\rho_1 > \rho_2 > 0$	Node	Unstable	Node	Unstable
$\rho_1 < \rho_2 < 0$	Node	AS	N	AS
$\rho_1 < 0 < \rho_2$	SP	Unstable	SP	Unstable
$\rho_1 = \rho_2 > 0$	PN	Unstable	Node or SP	Unstable
$\rho_1 = \rho_2 < 0$	PN	AS	Node or SP	AS
$\rho_{1,2} = \sigma \pm i\phi$	SP		SP	
$\sigma > 0$		Unstable		Unstable
$\sigma < 0$		AS		AS
$\rho_{1,2} = i\phi$	Center	Stable	Center or SP	Indeterminate

Note: AS, asymptotically stable; CP, critical point;
PN, proper or improper node; SP, spiral point/focus.

Moreover, a stable node or focus is called a *sink* of the linear system and an unstable node or focus is called a *source*. We see that the trajectories of the linear system are good approximations to those of the nonlinear system in most cases. In two sensitive cases, $\rho_{1,2} = i\phi$ and $\rho_1 = \rho_2$, types of critical points of the two systems may differ. The theorem tells that in many cases, the type and stability of the critical point of the nonlinear system can be determined from a study of the much simpler linear system. It should be noted that even if the critical point is of the same type as that of the linear system, the trajectories of the almost linear system may be considerably different in appearance from those of the corresponding system, except very near the critical point.[3] If one is concerned with actual paths of economic evolution, this implies that trajectories of the linearized system have to be interpreted with great caution as the reality following a nonlinear system may be far from equilibrium.

[3] It can be shown that the slopes at which trajectories "enter" or "leave" the critical point are given correctly by the linear system.

Example Consider the following model for competing species

$$\dot{x}_1 = x_1(1 - x_1 - x_2),$$

$$\dot{x}_2 = x_2(0.75 - x_2 - 0.5x_1). \tag{6.1.5}$$

The system (6.1.5) has four equilibrium points, $(0, 0)$, $(0, 0.75)$, $(1, 0)$, and $(0.5, 0.5)$. At any of the first three points, one or both species are extinct; only the last corresponds to the long-term survival of both species.

At $(0, 0)$, the corresponding linear system is

$$\begin{bmatrix} \dot{X}_1 \\ \dot{X}_2 \end{bmatrix} = \begin{bmatrix} 1 & 0 \\ 0 & 0.75 \end{bmatrix} \begin{bmatrix} X_1 \\ X_2 \end{bmatrix}.$$

The two eigenvalues and eigenvectors are respectively given by $\rho_{1,2} = 1, 0.75$, $C_1 = [1 \quad 0]^T$ and $C_2 = [0 \quad 1]^T$. Hence, the solution is

$$\begin{bmatrix} X_1 \\ X_2 \end{bmatrix} = \alpha_1 e^t \begin{bmatrix} 1 \\ 0 \end{bmatrix} + \alpha_2 e^{0.75t} \begin{bmatrix} 0 \\ 1 \end{bmatrix}.$$

Thus the origin is an unstable node of both the linear system and the nonlinear system (6.1.5). In similar way, we can show that $(0, 0.75)$ is a saddle point and is an unstable critical point of the linear system and of the nonlinear system; $(1, 0)$ is a saddle point and is an unstable critical point of the linear system and of the nonlinear system; and $(0.5, 0.5)$ is an asymptotically stable node of the linear system and of the nonlinear system. Please check these conclusions.

Example Consider

$$\begin{pmatrix} \dot{x}_1 \\ \dot{x}_2 \end{pmatrix} = \begin{pmatrix} 1 & 0 \\ 0 & 0.5 \end{pmatrix} \begin{pmatrix} x_1 \\ x_2 \end{pmatrix} + \begin{pmatrix} -x_1^2 - x_1 x_2 \\ -0.75 x_1 x_2 - 0.25 x_2^2 \end{pmatrix}.$$

The system has four equilibrium points

$$(0, 0), \ (0, 2), \ (1, 0), \ (0.5, 0.5).$$

The origin is an isolated equilibrium. To show that near the origin the system is an almost linear system, introduce

$$x_1 = r\cos\theta, \ \ x_2 = r\sin\theta .$$

Then $\|x\| = r$. We have

$$\frac{g_1}{\|x\|} = \frac{-x_1^2 - x_1x_2}{r} = -r\left(\cos^2\theta + \cos\theta\sin\theta\right) \to 0, \quad as\ r \to 0,$$

$$\frac{g_2}{\|x\|} = \frac{-0.75x_1x_2 - 0.25x_2^2}{r}$$

$$= -r\left(0.75\cos\theta\sin\theta + \sin^2\theta\right) \to 0, \quad as\ r \to 0.$$

We conclude that the system is almost linear near the origin.

Example Consider

$$\dot{x}_1 = x_2,$$

$$\dot{x}_2 = -x_2 - 2\sin x_1.$$

At any equilibrium, we have

$$A = \begin{bmatrix} 0 & 1 \\ -2\cos x_1 & -1 \end{bmatrix}_{x=x^*}.$$

The origin is an equilibrium. At $(0, 0)$, We have: $DetA = 2 > 0$ and $TrA = -1 < 0$. Hence, the origin is locally asymptotically stable for the linearized system. At another equilibrium $(0, \pi)$, $DetA = -2 < 0$ and $TrA = -1 < 0$, we conclude that the equilibrium $(0, \pi)$ is an unstable saddle equilibrium for the linearized system.

Example Rapoport's model for the arms race.
We now consider a model of arms race between two nations. We measure the competition with money expenditures, denoted by $x_1(t)$ and $x_2(t)$, by nations in their defence budgets.

In his survey of the defence budgets of France, Germany, Russia and the Austria-Hungaria empire for the pre-World War I years (1909–1913), Richardson found that the defence budgets of Group 1 (France and Russia) and Group 2 (Germany and Austria-Hungary) could be mathematically described by the relations that the rate at which the arms budget $x_1(t)$ of Group 1 increased was proportional to that $x_2(t)$ of Group 2 and vice versa,

$$\dot{x}_1 = a_1 x_2, \quad \dot{x}_2 = a_2 x_1,$$

where a_1 and a_2 are positive parameters.[4] This simple model did not fit the reality as time passes. Rapoport introduced the following model[5]

$$\dot{x}_1 = -m_1 x_1 + a_1 x_2 + b_1 x_2^2,$$

$$\dot{x}_2 = -m_2 x_2 + a_2 x_1 + b_2 x_1^2,$$

where all the parameters are positive and the decay terms, $-m_i x_i$, reflect the pressure within nations to reduce the defence budget and spent the money on non-defence items. Figure 6.1.1 shows the behavior of $x_1(t)$ and $x_2(t)$ for a particular choice of parameters and initial values

$$m_1 = 0.5, a_1 = 1, b_1 = 0.02, m_2 = 0.4, a_2 = 0.1, b_2 = 0.05,$$

$$x_0 = 0.8, y_0 = 0.9.$$

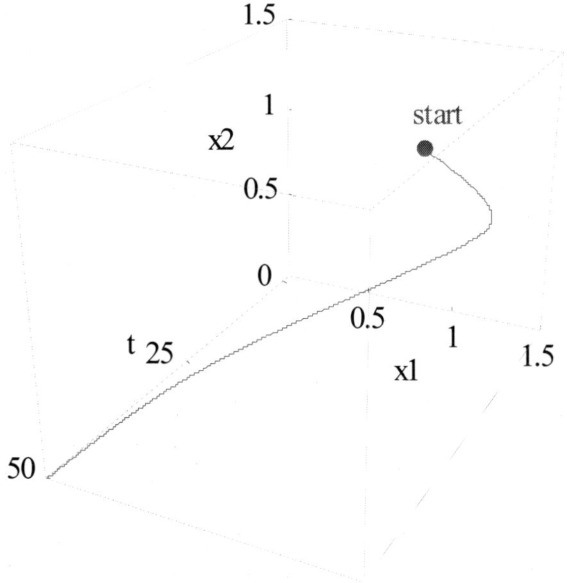

Fig. 6.1.1 The armies of two warring countries.

---

[4] See Richardson (1960). In fact, Richardson obtained a reasonable fit to the total budget expenditure of both groups by taking $a_1 = a_2 = a$, under which $x_1 + x_2 = ce^{at}$. See also Enns and McGuire (2001: 66–68).

[5] See Rapoport (1960).

The two countries in war build their armies as follows

$$\dot{x}_1 = \alpha x_1 - \beta x_1 x_2,$$

$$\dot{x}_2 = (1 + \alpha)x_2 - \gamma\beta x_1 x_2,$$

in which $x_j(t)$ is country j's number of individuals in the army, the coefficients α, β and γ are positive. Equilibrium points are given by

$$\alpha x_1 - \beta x_1 x_2 = 0,$$

$$(1 + \alpha)x_2 - \gamma\beta x_1 x_2 = 0.$$

The system has two equilibrium points

$$P_1 : \ x_1 = x_2 = 0,$$

$$P_2 : \ x_1 = \frac{(1 + \alpha)}{\gamma\beta}, \quad x_2 = \frac{\alpha}{\beta}.$$

It is straightforward to check that the equilibrium point P_1 is an unstable nodal point; and P_2 is a saddle point. We specify the coefficient values as

$$\alpha = 6, \ \gamma = 1.2, \ \beta = \frac{1}{3000}.$$

Figure 6.1.2 shows the phase portrait of the equations. The trajectories show that country 1 wins out in one case and country 2 in the other.

Example The competitive equilibrium and its stability re-examined. We consider the competitive equilibrium problem as the example in Sec. 5.2. As in Eqs. (5.2.3), the price dynamics are given by

$$\dot{p}_i = k_i E_i(p_1, p_2), \ k_i > 0, i = 1, 2. \tag{6.1.6}$$

We examined the system with linear excess demand functions and specified values of the parameters. We now examine the properties of the system when $E_i(p_1, p_2)$ are not specified. Suppose the system has a unique equilibrium and denote the equilibrium by (p_1^*, p_2^*). The Jacobian at the equilibrium is

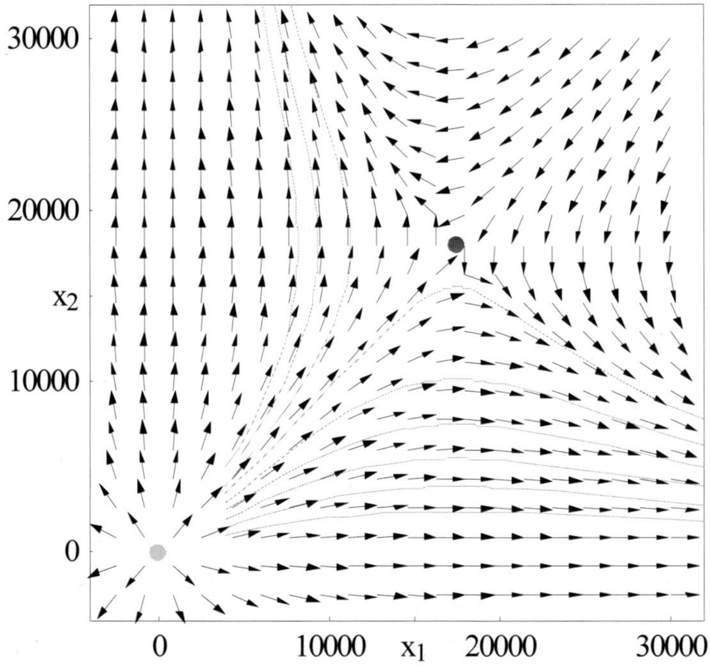

Fig. 6.1.2 Phase portrait plot of the armies of two warring countries.

$$J = \begin{bmatrix} k_1 E_{11} & k_1 E_{12} \\ k_2 E_{21} & k_2 E_{22} \end{bmatrix},$$

where $E_{ij} \equiv \partial E_i / \partial p_j$. Hence

$$TrJ = k_1 E_{11} + k_2 E_{22}, \quad DetA = k_1 k_2 \left(E_{11} E_{22} - E_{21} E_{12} \right).$$

For the real parts of the two eigenvalues to be negative, we should have $DetA > 0$ and $TrA < 0$. These conditions are satisfied if $E_{ii} < 0$ and $E_{11} E_{22} > E_{21} E_{12}$. We now show that the gross substitutability guarantees stability. Assume that commodities 1 and 2 are gross substitutes. This assumption implies

$$E_{ii} < 0, \quad E_{ij} > 0, \quad i \neq j. \tag{6.1.7}$$

We further require that commodities 0 and 1, and commodities 0 and 2 are gross substitutes. This means $E_{01} > 0$ and $E_{02} > 0$.

Remember that we required $E_i(P_0, P_1, P_2)$ to be homogeneous of degree zero. According to the definition, we have

$$E_i(xP_0, xP_1, xP_2) = E_i(P_0, P_1, P_2).$$

Take derivatives of

$$E_i(xP_0, xP_1, xP_2) = E_i(P_0, P_1, P_2)$$

with regard to x for $i = 1, 2,$ and then set $x = 1$

$$E_{i0} + p_1 E_{i1} + p_2 E_{i2} = 0, \quad i = 1, 2. \tag{6.1.8}$$

As $E_{i0} > 0$, Eqs. (6.1.8) yield

$$\frac{E_{11}}{E_{12}} < -\frac{p_2}{p_1}, \quad \frac{E_{21}}{E_{22}} > -\frac{p_2}{p_1},$$

where we use Eqs. (6.1.7). We thus have

$$(DetJ =)\ E_{11}E_{22} - E_{21}E_{12} > 0.$$

We thus conclude that under the assumption of gross substitutability, the system is stable. As demonstrated in Chap. 10, the equilibrium is globally stable.

Example The Walrasian-Marshallian adjustment process.[6]
Consider a one-input/one-output economy where a commodity Y is produced solely by means of labor L with a smooth production function $Y = f(L)$ (whose reverse is denoted by $L = L(Y)$). Suppose demand for the produced commodity can be represented by a smooth function $d = d(p, L^s, w,)$ with $\partial d / \partial p \leq 0$, p denoting the price and $w\ (=1)$ the nominal wage rate. Profits π are given by

$$\pi = pf(L) - wL.$$

Households' initial endowments consist of labor only, and labor supply L^s can be derived from the above demand function by means of Walras's Law $pd = L^s + \pi$. Owing to this relation, we can neglect the labor market in later analysis. The market equilibrium is expressed as

$$d(p^*, L(Y^*)) = Y^* > 0,$$
$$L'(Y^*) = p^* > 0. \tag{6.1.9}$$

---

6 This section is based on Chap. 3 in Flaschel, Franke, and Semmler (1997).

The conditions imply that demand equals supply and prices equal marginal wage costs. Assume that $L''(Y) \geq 0$ for the second derivative of the cost function $L(Y)$ at least in a neighborhood of Y^*. Out of equilibrium, the following type of the tâtonnement adjustment process is suggested by Mas-Colell[7]

$$\dot{p} = \beta_p[d(p, L(Y)) - Y],$$

$$\dot{Y} = \beta_Y[p - L'(Y)], \tag{6.1.10}$$

where $\beta_p > 0$ and $\beta_Y > 0$ are parameters. Prices are adjusted in the direction of the excess demand on the market for goods and goods supply is adjusted following the discrepancy between the current price for the goods and the marginal wage costs of producing the current supplies. The Jacobian J at the equilibrium point

$$J = \begin{bmatrix} \beta_p & 0 \\ 0 & \beta_Y \end{bmatrix} \begin{bmatrix} d_p & -1 \\ 1 & -L'' \end{bmatrix},$$

where we use $d_Y = d_\pi \pi'(L) L'(Y) = 0$ at the profit-maximization point. We are concerned only with the case of $|J| \neq 0$. Local asymptotic stability is guaranteed if either $d_p < 0$ or $L'' > 0$.

Example Tobin-Blanchard model.[8]
Let $q(t)$ denote the market value of equities as a ratio of the replacement cost.[9] Suppose aggregate expenditure, $e(t)$, is a function of income, $y(t)$, the market value of equities, $q(t)$, real (fixed) government expenditure, g, as follows

---

[7] Mas-Colell (1986: 53). Such a process of a proportional control of prices as well quantities was related to the works of Walras by, for instance, Morishima (1959, 1977), and Goodwin (1953, 1989), and Walker (1987).

[8] See Sec. 2 in Obsfeld and Rogoff (1999) and Sec. 10.8 in Shone (2002). The link between stock market behavior and income and interest rates was suggested by Tobin (1969).

[9] The variable q can be understood as follows. If all future returns (R) are equal and are discounted at the interest rate r, then the present value of equities (V) is equal to R/r. On the other hand, firms will invest until the replacement cost of any outstanding capital stock (RC) is equal to the return on investment (R/ρ, where ρ is the marginal efficiency of capital. Hence, $q = V/RC = \rho/r$. In the long term, $r = \rho$, i.e., $q = 1$. See Stevenson, Muscatelli and Gregory (1988: 156–59) for further explanation.

$$e = a_1 y(t) + a_2 q(t) + g, \quad 0 < a_1 < 1, \quad a_2 > 0.$$

The goods market adjusts with reaction coefficient $\sigma > 0$ as

$$\dot{y} = \sigma(e - y).$$

The money market is assumed to adjust instantaneously, which implies that the demand for real money balances is equal to the real money balances. That is

$$ky(t) - ur(t) = m_0.$$

Suppose that bonds and equities perfect substitutes. Hence, the rate of interest on bonds and the yield on equities should be equal

$$r(t) = \frac{b_1 y(t) + \dot{q}(t)}{q(t)},$$

where $b_1 y(t)$ is the firms' profits and $\dot{q}(t)$ is the capital gains. It is straightforward to demonstrate that the dynamics of the system which are given by the above four equations are reduced to two nonlinear nonhomogeneous differential equations

$$\dot{y} = \sigma(a_1 - 1)y + \sigma a_2 q + \sigma g,$$

$$\dot{q} = \left(\frac{kq}{u} - b_1\right)y - \frac{qm_0}{u}.$$

It can be shown that the system has a unique equilibrium point and it is a saddle point. For instance, specify the parameter values as

$$a_1 = 0.8, \quad a_2 = 0.2, \quad g = 7, \quad \rho = 2, \quad m_0 = 8, \quad k = 0.25,$$

$$u = 0.2, \quad b_1 = 0.1.$$

The equilibrium point is

$$(y^*, q^*) = (35.76, 0.76).$$

The Jacobian is

$$J = \begin{bmatrix} -0.4 & 0.4 \\ 0.85 & 4.7 \end{bmatrix}.$$

The two eigenvalues are respectively 4.7658 and -0.4658. Hence, the equilibrium point is a saddle point.

Example The Ramsey growth model.
We assume a one-sector economy in which 1 unit of output can be used to generate 1 unit of household consumption, or 1 unit of additional capital. Each household consists of one or more adults who are employed in the competitive labor market and receive wages for providing labor services. A household is imagined as an immortal extended family. The households receive interest income on assets, purchase goods for consumption and save by accumulating additional assets. Each household maximizes utility and incorporates a budget constraint over an infinite horizon. Denote $C(t)$ the total consumption at time t and $c(t) \equiv C(t)/N(t)$ is consumption per worker. It is assumed that the labor market clears at any point of time and each adult supplies 1 unit of labor services per unit of time. Households take the net rate of return $r(t)$ on assets and the wage rate $w(t)$ paid per unit of labor services as given in the competitive markets. Let $K(t)$ denote the capital existing at each time t and $N(t)$ the flow of labor services used at time t for production. The extended family is assumed to grow at an exogenously given rate n. Let the number of adults at time 0 be unity, the family size at time t is $N(t) = e^{nt}$. Each member supplies one unit of labor per unit time, without disutility.

We assume that production function $F(K(t), N(t))$ is neoclassical.[10] The marginal conditions are

$$r = F_K = f'(k), \quad w = F_N = f(k) - kf'(k),$$

where $k = K/N$. The household's preferences are expressed by an instantaneous utility function $u(c(t))$, where $c(t)$ is the flow of consumption per person, and a discount rate for utility, denoted by ρ

$$u(c) = \frac{c(t)^{1-\theta} - 1}{1-\theta}, \quad \theta > 0. \tag{6.1.11}$$

Assume that each household maximizes utility U as given by

$$U = \int_0^\infty u(c(t)) e^{nt} e^{-\rho t} dt, \quad c(t) \geq 0, \quad t \geq 0.$$

[10] See Sec. 4.1.

The household makes the decision subject to a lifetime budget constraint. We denote the net assets per household by $k(t)$ which is measured in units of consumables. The total income at each point of time is equal to $w + rk$. The flow budget constraint for the household is

$$\dot{k} = w + rk - c - nk = f - c - nk . \tag{6.1.12}$$

The equation means that the change rate of assets per person is equal to per capita income minus per capita consumption and the term, nk. It is assumed that the credit market imposes a constraint of borrowing, the present value of assets must be asymptotically nonnegative, that is

$$\lim_{t\to\infty}[k(t)\exp\{-\int_0^t(\rho-n)dv\}] \geq 0 .$$

The present-value Hamiltonian is given by[11]

$$J = u(c)e^{-(\rho-n)t} + \overline{\lambda}(w + rk - c - nk),$$

where $\overline{\lambda}$ is the present-value shadow price of income. The first-order conditions are

$$\frac{\partial J}{\partial c} = 0 \Rightarrow \overline{\lambda} = u'e^{-(\rho-n)t},$$

$$\frac{d\lambda}{dt} = -\frac{\partial J}{\partial k} \Rightarrow \frac{d\lambda}{dt} = -(\rho - n)\overline{\lambda} . \tag{6.1.13}$$

The transversality condition is given by

$$\lim_{t\to\infty}[\lambda(t)k(t)] = 0.$$

By Eqs. (6.1.13), we can derive

$$r = \rho - \frac{u''c}{u'}\left(\frac{1}{c}\frac{dc}{dt}\right). \tag{6.1.14}$$

This equation says that households choose consumption so as to equate the rate of return r to the rate of time preference ρ plus the rate of decrease of the marginal utility of consumption u' due to growing per capita consumption c. Inserting Eq. (6.1.11) into Eq. (6.1.14) yields

---

[11] Appendix A.6 introduces optimal control theory.

$$\dot{c}(t) = \frac{r(t) - \rho}{\theta} c(t) = \frac{f'(k) - \rho}{\theta} c(t). \qquad (6.1.15)$$

The trajectory of the economy is determined by Eqs. (6.1.12) and (6.1.15). The phase diagram in $c(t)$ and $k(t)$ is as shown in Fig. 6.1.3. Along the vertical line defined by

$$f'(k^*) = \rho,$$

change rate of the consumption per capita is equal to zero, i.e., $\dot{c}(t) = 0$. The consumption per capita increases to the left of the curve and falls to the right. Along the locus defined by,

$$c(t) = f(t) - nk(t),$$

change rate of the capital-labor ratio equals zero. The capital-labor ratio falls above the curve and increases below it. With the requirement $\rho > n$ (without which the utility becomes unbounded along feasible paths), the intersection of the two curves determines a unique steady state.

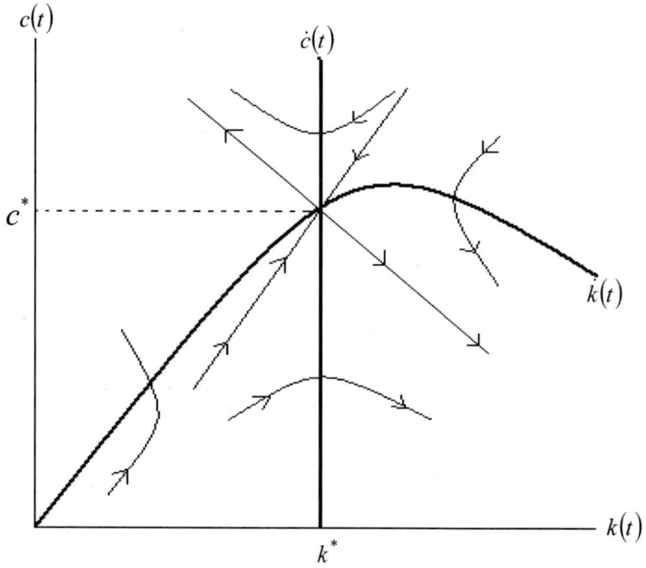

Fig. 6.1.3 The dynamics of the Ramsey model.

Local stability of the c - k system is determined by the characteristic roots of the following matrix of the coefficients of $k(t)$ and $c(t)$ equations linearized around the equilibrium point

$$J \equiv \begin{bmatrix} \dfrac{\partial \dot{k}}{\partial k} & \dfrac{\partial \dot{k}}{\partial c} \\ \dfrac{\partial \dot{c}}{\partial k} & \dfrac{\partial \dot{c}}{\partial c} \end{bmatrix} = \begin{bmatrix} f' - n & -1 \\ \dfrac{f''c}{\theta} & 0 \end{bmatrix}.$$

We have

$$\mathrm{tr}(J) = f' - n = \rho - n > 0, \quad |J| = \frac{f''c}{\theta} < 0. \qquad (6.1.16)$$

Therefore, the characteristic equation is

$$\phi^2 - \mathrm{tr}(J)\phi + |J| = 0.$$

Inserting Eq. (6.1.16) into the above equation yields

$$\phi^2 - (\rho - n)\phi + \frac{f''c}{\theta} = 0$$

$$\Rightarrow \phi_{1,2} = \frac{(\rho - n) \pm \sqrt{(\rho - n)^2 - 4f''c/\theta}}{2}.$$

The characteristic roots are real and opposite in sign. The equilibrium point is a saddle point.

Exercise 6.1

1 Consider the model

$$\ddot{x} + ax + bx^3 = 0,$$

in which a is positive and b can be positive or negative. Determine the equilibrium and analyze stability of the linearized system.

2 Do a phase plane analysis of the equation in the previous exercise. Compare to the results on stability obtained by linearization.

3 For the following questions, verify that $(0, 0)$ is a critical point, show that the system is almost linear, and discuss the type and stability of $(0, 0)$ by examining the corresponding linear system

(a) $\dot{x}_1 = x_1 - x_2^2, \ \dot{x}_2 = -x_1 - 2x_2 + x_1^2$;

(b) $\dot{x}_1 = -x_1 + x_2 + 2x_1x_2, \ \dot{x}_2 = -4x_1 - x_2 + x_1^2 - x_2^2$;

(c) $\dot{x}_1 = (1 + x_1)\sin x_2, \ \dot{x}_2 = 1 - x_1 - x_2 \cos x_2$.

4 For the following problems, (1) determine all critical points; (2) find the corresponding linear system near each critical point; (3) find eigenvalues of each linear system; (4) what conclusions can you draw about the nonlinear system from the eigenvalues.

(a) $\dot{x}_1 = (2 + x_1)(x_2 - x_1), \ \dot{x}_2 = (4 - x_1)(x_1 + x_2)$;

(b) $\dot{x}_1 = 1 - x_2, \ \dot{x}_2 = x_1^2 - x_2^2$.

5 For the following model of two competing species

$$\dot{x}_1 = x_1(a_{10} - a_{11}x_1 - a_{12}x_2),$$

$$\dot{x}_1 = x_1(a_{20} - a_{21}x_1 - a_{22}x_2),$$

where $a_{ij} > 0$ for all i and j. Discuss possible qualitative behavior of the system.

6 Discuss local stability of the following Keynesian adjustment process

$$\dot{p} = \beta_p[d(p, L(Y)) - Y],$$

$$\dot{Y} = \beta_Y[p - L'(Y)],$$

in which the parameters, variables, and functions are defined the same as in the Walrasian-Marshallian adjustment process, except that in the Keynesian procedure, quantities react to quantity discrepancies and prices to cost-price differentials.

7 We specify the parameter values of the Tobin-Blanchard model as

$$a_1 = 0.8, \ a_2 = 0.2, \ g = 7, \ \rho = 2, \ m_0 = 16, \ k = 0.5,$$

$$u = 0.25, \ b_1 = 0.15.$$

(a) Show that the system has a unique equilibrium points; and (b) Prove that the equilibrium point is a saddle point.

8 Discuss stability conditions of the housing market system[12]

$$\dot{P} = rP - R(h),$$

$$\dot{h} = g(P) - (\delta_h + n)h,$$

where

P = real price of a standardized housing unit
rP = operating cost of owning a home (r assumed constant)
R = real rental price
h $= H / N =$ housing per adult, where H is stock of housing and N adult population
δ_h = fixed rate of depreciation of stock of housing
n $= \dot{N} / N =$ fixed rate of population growth.

The function $R(h)$ (where $R' < 0$) is the inverse of the demand function for housing $(h =) H / N = f(R)$. The first condition, which may be rewritten as

$$rP = R(h) + \dot{P},$$

states that the operating cost of owning a home is equal to the real rental price plus the price change rate. The second equation results from the assumption that gross investment ($= \dot{H} + \delta_h H$) is an increasing function of the price of housing, $g(P)$ with $g' > 0$. That is

$$\dot{H} + \delta_h H = gN.$$

6.2 Liapunov Functions

We showed that the stability of a critical point of an almost linear system can usually be determined from a study of the corresponding linear system. Nevertheless, if the critical point is a center of the corresponding linear system, then no conclusion can be made about the nonlinear system. It is also important to investigate the *basin* of an asymptotically stable critical point, that is, the domain that all solutions starting within that domain approach the critical point. Since the theory of linearization is a local theory, it does not address this question. In this section, we

[12] The model is proposed by Mankiw and Weil (1989). See also Shone (2002: 358–363).

discuss another approach, known as *Liapunov's second method* or *direct method.*[13] The method determines the stability or instability of a critical point by constructing a suitable auxiliary function. The technique provides an estimation of the extent of basin of attraction of a critical point and it can also be applied to study systems of equations that are not almost linear.

The theory of Liapunov functions is a global approach toward determining asymptotic behavior of solutions. The previous theorem of local stability tells us that in the neighborhood of equilibrium, solution trajectories are attracted to the equilibrium if the eigenvalues of the linear part of the equation have negative real part. Basically, the method is a generalization of two physical principles for conservative systems, namely, (i) a rest position is stable if the potential energy is a local minimum, otherwise it is unstable, and (ii) the total energy is a constant during any motion. The Liapunov function shows that initial values from a large region converge to an equilibrium point. Let vector

$$x(t) = (x_1(t), x_2(t)),$$

be a solution of the 2-dimensional system

$$\dot{x}_j = f_j(x_1, x_2), \quad j = 1, 2,$$

or

$$\dot{x}(t) = f(x(t)). \tag{6.2.1}$$

For a function $V(t)$ of $x(t)$, we measure the time rate of change of "the energy of the system", $V(t)$, along a solution trajectory of Eq. (6.2.1) by taking the derivative of $V(t)$ with respect to t

$$\dot{V}(x) = \sum_{j=1}^{2} \frac{\partial V}{\partial x_j} \dot{x}_j = \sum_{j=1}^{2} \frac{\partial V}{\partial x_j} f_j(x). \tag{6.2.2}$$

The derivative of $V(t)$ can be calculated in terms of the differential equation itself – the solutions do not explicitly appear in this formula. This function can be used to determine the stability of equilibria of the differential equation.

[13] The method is referred to as a direct method because no knowledge of the solution of the system of differential equations is required.

Definition 6.2.1 Let U be an open subset of R^2 containing the origin. A real-valued C^1 function

$$V : U \to R;\ \ x \mapsto V(x)$$

is said to be *positive definite* on U if
(i) $V(0) = 0$;
(ii) $V(x) > 0$ for all $x \in U$ with $x \neq 0$.
A real-valued C^1 function V is said to be *negative definite* if $-V$ is positive definite.

For instance,

$$V(x_1, x_2) = x_1^2 + x_2^2$$

is positive definite on all of R^2, while

$$V(x_1, x_2) = x_1^2 + x_2^2 - x_2^3$$

is positive definite on only a sufficiently small strip about the x_1-strip. For

$$V(x_1, x_2, t) = x_1^2 + x_2^2 - 2\alpha x_1 x_2 \sin t,$$

if $|\alpha| < 1$, V is positive definite; if $|\alpha| = 1$, it is "positive semi-definite"; if $|\alpha| < 1$, it is "indefinite".

We note that if V is a positive definite, then V has a minimum at the origin. If this extreme point of V is isolated, then the surface given by the following equation

$$z = V(x_1, x_2),$$

representing the graph of V in R^3 near the origin has the general shape of a parabolic mirror pointing upward. Figure 6.2.1 depicts a graph of a positive definite function near the origin.

The following theorem, due to the Russian mathematician Alexander Mikhailovich Liapunov.

Theorem 6.2.1 (Liapunov)[14] Let $x^* = 0$ be an equilibrium point of

$$\dot{x} = f(x),$$

---

[14] A proof of the theorem is given in Hirsch and Smale (1974).

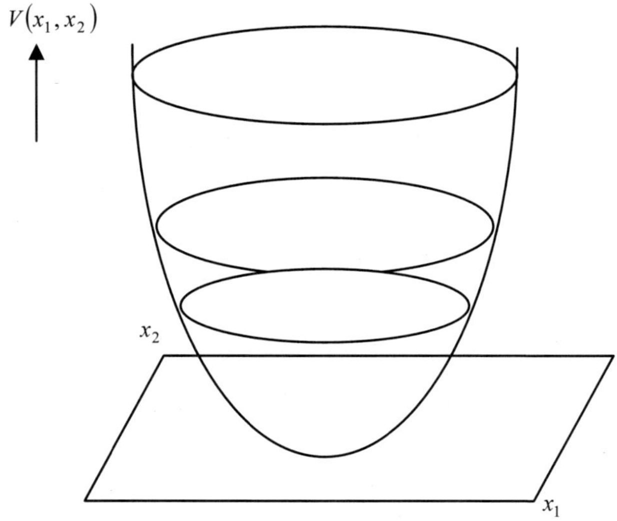

Fig. 6.2.1 A positive definite function V.

and V be a positive definite C^1 function on a neighborhood of 0.

(i) If $\dot{V} \le 0$ for all $x \in U - \{0\}$, then 0 is stable.

(ii) If $\dot{V} < 0$ for all $x \in U - \{0\}$, then 0 is asymptotically stable.

(iii) If $\dot{V} > 0$ for all $x \in U - \{0\}$, then 0 is unstable.

Definition 6.2.2 A positive definite function V on an open neighborhood U of the origin is said to be a *Liapunov function* for

$$\dot{x} = f(x),$$

if $\dot{V} \le 0$ for $x \in U - \{0\}$. When $\dot{V} < 0$ for all $x \in U - \{0\}$, the function V is called a *strict Liapunov function.*

Example By applying the above theorem with

$$V(x_1, x_2) = x_1^2 + x_2^2,$$

it is straightforward to demonstrate that $(0, 0)$ is an asymptotically stable equilibrium of the following differential equation

$$\dot{x}_1 = -x_1^3 + x_1 x_2, \quad \dot{x}_2 = -x_2^3 - x_1^2.$$

Example Consider a model of planar pendulum

$$\dot{x}_1 = x_2,$$

$$\dot{x}_2 = -\frac{g}{l}\sin x_1,$$

where g and l are positive parameters. The linearization of the differential equations at the origin has purely imaginary eigenvalues and the stability type of the equilibrium point at the origin cannot be deduced from the linear approximation. We introduce

$$V(x_1, x_2) = \frac{1}{2}x_2^2 + \frac{g}{l}(1 - \cos x_1).$$

V is positive definite in a sufficiently small neighborhood of the origin. Moreover, $\dot{V} \equiv 0$. Hence, the origin is stable.

In fact, $\dot{V} = 0$ holds for any solution of the pendulum model. This implies that for any solution

$$V(x_1(t), x_2(t)) = V(x_1(0), x_2(0)),$$

holds. Since V is 2π-periodic in x_1, we will confine our analysis to the vertical strip of the plane with

$$-\pi \le x_1 \le \pi .$$

Due to its periodicity, it is sufficient to take initial data on the x_2-axis. We now determine the shapes of the curves of

$$\frac{1}{2}x_2^2 + \frac{g}{l}(1 - \cos x_1) = \frac{1}{2}x_2^2(0).$$

For any $x_2(0)$, the curve defined by the above equation is symmetric with respect to the x_1-axis. Therefore, we need only examine

$$x_2 = \sqrt{x_2^2(0) - \frac{2g}{l}(1 - \cos x_1)}. \qquad (6.2.3)$$

With these observations, we can effectively construct the orbits of the pendulum by considering the values of $x_2(0)$. The directions of the orbits can be inferred from the vector field.

For $x_2(0) = 0$, Eq. (6.2.3) gives the equilibrium points $(-\pi, 0)$, $(0, 0)$, and $(\pi, 0)$. For $0 < x_2^2(0) < 4g/l$, the range of x_1 is an interval of length longer than 2π and symmetric about the origin. The curve defined by Eq. (6.2.3) yields a closed curve on the plane. Since there are no equilibrium points on it, this closed curve is a periodic orbit corresponding to the oscillation of the pendulum about the equilibrium position $(0, 0)$, see Fig. 6.2.2. For $x_2^2(0) = 4g/l$, the curve defined by Eq. (6.2.3) is a closed curve. However, on this closed curve, there are several orbits. In particular, the equilibrium points $(-\pi, 0)$ and $(\pi, 0)$ are on this curve. These equilibria correspond to the vertical position of the pendulum while the pendulum is "sitting on its head." There are two other special orbits: one whose α-limit set is $(-\pi, 0)$ and ω-limit set is $(\pi, 0)$, the other, which is the reflection of this orbit, whose α-limit set is $(\pi, 0)$ and ω-limit set is $(-\pi, 0)$. These special orbits are called heteroclinic orbits and they correspond to the motions of the pendulum from one equilibrium point to the other, in infinite time. We give the precise definition of heteroclinic orbit for future reference.

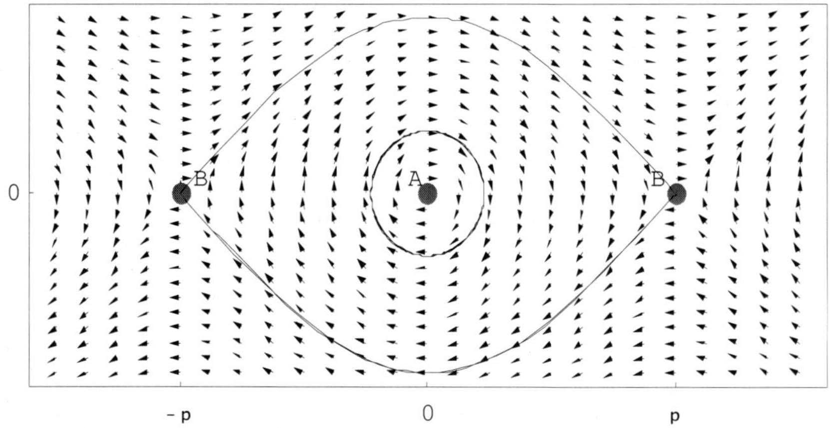

Fig. 6.2.2 Phase portraits of the pendulum on the plane.

Definition 6.2.3 An orbit whose α-limit set is an equilibrium point and ω-limit set is another equilibrium point is called a *heteroclinic orbit*.

If $x_2^2(0) > 4g/l$, then the range of x_1 is unrestricted and the curve defined by Eq. (6.2.3) is a 2π-graph over the x_1-axis. There is no equilibrium point on these curves and they correspond to the orbits of the motion of the pendulum with initial velocity so large that the pendulum revolves around without end.

Example Consider the second-order differential equation

$$\ddot{x} + q(x) = 0,$$

where the continuous function q satisfies $xq(x) > 0$ for $x \neq 0$, $q(0) = 0$. This differential equation can be written as

$$\dot{x}_1 = x_2,$$

$$\dot{x}_2 = -q(x_1).$$

The total energy of the system

$$V = \frac{x_2^2}{2} + \int_0^{x_1} q(s)\,ds,$$

serves as a Liapunov function for this system. It is straightforward to show that the origin is stable.

Example Consider

$$\dot{x}_1 = x_2,$$

$$\dot{x}_2 = -x_1 - 2ax_2 - x_1^3, \quad a > 0.$$

The origin is the only equilibrium point and the eigenvalues are $-a \pm ib$, where $b \equiv \sqrt{1 - a^2}$. Hence, the origin is asymptotically stable. We now apply Liapunov methods to estimate the basic of attraction of the origin. To put the origin system into the Real Jordan Normal form, introduce

$$y = Px, \quad P = \begin{bmatrix} 1 & 0 \\ -a & b \end{bmatrix} \Rightarrow P^{-1} = \frac{1}{b}\begin{bmatrix} b & 0 \\ a & 1 \end{bmatrix}.$$

Under the above transformation, the original system becomes

$$\dot{y}_1 = -ay_1 + by_2,$$

$$\dot{y}_2 = -by_1 - ay_2 - \frac{y_1^3}{b}.$$

Define

$$V(y_1, y_2) \equiv \frac{1}{2a}\left(y_1^2 + y_2^2\right).$$

Then

$$\dot{V} = -\left(y_1^2 + y_2^2\right) - \frac{1}{ab} y_1^3 y_2 .$$

We now determine the largest subset of R^2 containing the origin where $-\dot{V}$ is positive definite. To simplify the matter, we search for the largest such disk. From the symmetry of $-\dot{V}$, it is evident that the radius r_0 of the largest circle inside which $-\dot{V}$ is positive definite must satisfy

$$r_0^2 - \frac{1}{ab} r_0^4 = 0 .$$

That is,

$$r_0 = \sqrt{ab} .$$

Thus, every solution of $y(t)$ with $y(0)$ satisfying $\|y_0\| < r_0$ approaches the origin as $t \to \infty$. The circle of radius r_0 becomes an ellipse when transformed back to the original system. In fact, one can obtain a slightly larger basin of attraction of the origin.

Consider again the dynamic system

$$\dot{x} = A_{2\times 2} x + g(x), \quad x \in R^2 , \tag{6.2.4}$$

where A is real and $g(x)$ is of of magnitude smaller than Ax. We now construct explicit Liapunov functions for the linearized system $\dot{x} = Ax$ and show that they also work for the original system. Introduce

$$A = \begin{bmatrix} a & b \\ c & d \end{bmatrix}, \quad p \equiv Trace A \ (< 0) , \ q \equiv Det A > 0 .$$

As A is regular, we have

$$A^{-1} = \frac{1}{q} \begin{bmatrix} d & -b \\ -c & a \end{bmatrix} .$$

Theorem 6.2.2 Let $(0, 0)$ be an equilibrium point of Eqs. (6.2.4), where $g(x) = O\left(\|x\|^2\right)$ as $\|x\| \to 0$. Then the zero solution of Eqs. (6.2.4) is asymptotically stable when its linear approximation is asymptotically stable.

Proof: Define $V(x) = x^T Kx$ (where K is a 2×2 constant symmetry matrix to be determined) and differentiate V with regard to t for Eqs. (6.2.4)

$$\dot{V} = \dot{x}^T Kx + x^T K\dot{x} = x^T\left(A^T K + KA\right)x + 2g^T Kx\,. \qquad (6.2.5)$$

We now want to determine K such that $A^T K + KA = -I_{2\times 2}$ and V is positive definite for $p < 0$ and $q > 0$. We now show that there is a solution to $A^T K + KA = -I_{2\times 2}$ in the form of

$$K = m\left(A^T\right)^{-1} A^{-1} + nI\,,$$

where m and n are constants. Substitute this equation into $A^T K + KA = -I_{2\times 2}$

$$\frac{1}{q}\begin{bmatrix} 2md + 2naq & (nq - m)(b + c) \\ (nq - m)(b + c) & 2ma + 2ndq \end{bmatrix} = -\begin{bmatrix} 1 & 0 \\ 0 & 1 \end{bmatrix}.$$

This equation is satisfied if $m = -q/2p$ and $n = -1/2p$. Now, it is straightforward to show

$$K = -\frac{1}{2pq}\begin{bmatrix} c^2 + d^2 + q & -ac - bd \\ -ac - bd & a^2 + b^2 + q \end{bmatrix},$$

$$V = -\frac{\left(dx_1 - bx_2\right)^2 + \left(cx_1 - ax_2\right)^2 + q\left(x_1^2 + x_2^2\right)}{2pq}$$

$$< 0,\ \ \text{for } x \neq 0\,. \qquad (6.2.6)$$

Hence, K is symmetry and V is positive definite. Under K specified in Eqs. (6.2.6), Eq. (6.2.5) becomes

$$\dot{V} = -x_1^2 - x_2^2 + 2g^T Kx\,. \qquad (6.2.7)$$

For any p, q, there is clearly a neighborhood of the origin in which the terms $-x_1^2 - x_2^2$ predominates, that is to say, where $\dot{V}$ is negative definite.

Therefore, V is a strong Liapunov function for the original system for $p < 0$, $q > 0$.

Theorem 6.2.3 Let $(0, 0)$ be an equilibrium point of Eqs. (6.2.4), where $g(x) = O(\|x\|^2)$ as $\|x\| \to 0$. When the eigenvalues of A are different, nonzero, and at least one has positive real part, the zero solution of Eqs. (6.2.4) is unstable.

Proof: First consider the case that the two eigenvalues ρ_1 and ρ_2 are real and different. As the origin is unstable for the linearized system, at least one eigenvalue is positive. We know that there is a 2×2 matrix invertible C such that $C^{-1}AC = diag[\rho_j] \equiv D$. Introduce $x = CX$. Now Eqs. (6.2.4) are transformed into

$$\dot{X} = DX + C^{-1}g(CX). \tag{6.2.8}$$

Introduce

$$V(X) = X^T D^{-1} X = \frac{X_1^2}{\rho_1} + \frac{X_2^2}{\rho_2}.$$

Then $V(X) > 0$ at some point in every neighborhood of $X = 0$ (since instability requires ρ_1 or ρ_2 positive). As

$$\dot{V} = 2\left(X_1^2 + X_2^2 + \frac{X_1 g_1}{\rho_1} + \frac{X_2 g_2}{\rho_2} \right),$$

which is positive definite in a small enough neighborhood of the origin, we see the theorem holds when ρ_1 and ρ_2 are real and different.

Now examine the other case that the two eigenvalues ρ_1 and ρ_2 are conjugate complex with positive real part, $\rho_{1,2} = \alpha \pm i\beta$, $\alpha > 0$. In this case, there exists an invertible matrix G such that

$$G^{-1}AG = \begin{bmatrix} \alpha & -\beta \\ \beta & \alpha \end{bmatrix} \equiv \tilde{A}_{2\times 2}.$$

Introduce $x = GX$ and transform Eqs. (6.2.4) into

$$\dot{X} = \tilde{A}X + G^{-1}g(GX).$$

Define $V(X) = X^T X$. The function V is positive definite and $\dot{V} = 2\alpha X^T X$ is also positive definite. Hence, the origin is unstable.

We conclude this section with an embellishment of the theorem of Liapunov. The instability part of Theorem 6.2.1 has the deficiency of considering a full neighborhood of the origin and thus is not applicable to equilibria of saddle type. The following Četaev theorem remedies the shortcoming.[15]

Theorem 6.2.4 (Četaev) Let U be a sufficiently small open neighborhood of the origin. If there is an open region W and a C^1 function $V : \overline{W} \to R$ with the properties
(i) the origin is a boundary point of W;
(ii) $V(x) = 0$ for all x on the boundary points of W inside U;
(iii) $V(x) > 0$ and $\dot{V}(x) > 0$ for all $x \in W \cap U$,
then the origin is an unstable equilibrium point.

Example Instability with Četaev.
We consider the system of differential equations

$$\dot{x}_1 = x_1^3 + x_2 x_1^2,$$

$$\dot{x}_2 = -x_2 + x_1^2,$$

which has an equilibrium point at the origin. The eigenvalues of the linearized system at the origin are 0 and -1. To determine stability of the origin, introduce

$$V(x_1, x_2) = \frac{1}{2}(x_1^2 - x_2^2),$$

and the open region

$$W = \{(x_1, x_2) : x_1 > x_2 > -x_1\}.$$

Observe that $V(x) > 0$ for $x \in W$, and $V(x) = 0$ on the boundary. Next, we compute the derivative of V along the solutions of the differential equations above

$$\dot{V}(x_1, x_2) = x_1^4 - x_2(x_1^2 - x_1^3) + x_2^2.$$

In a sufficiently small neighborhood U of the origin, we estimate

$$\dot{V} \geq x_1^4 - (1 + \varepsilon)|x_2| x_1^2 + x_2^2 = (x_1^2 - |x_2|)^2 + (1 - \varepsilon)|x_2| x_1^2,$$

---

[15] See Sec. 9.3 in Hale and Koçak (1991).

where $\varepsilon \geq 0$ is small. It is easy to see that $\dot{V} > 0$ for x in a neighborhood of $x = 0$ and, in particular, for $x \in W$. Now, the conditions of the Četaev Theorem are satisfied and thus the origin is unstable.

Exercise 6.2

1 Consider

$$\dot{x} = y, \quad \dot{y} = -\sin y,$$

for each integer n, the system has a unique equilibrium, $(n\pi, 0)$. Show that

$$V(x, y) = y^2/2 + 1 - \cos x,$$

is a Liapunov function for the equilibria.

2 For the following equation

$$\frac{d^2u}{dt^2} + g(u) = 0,$$

where $g(0) = 0$, $g(u) > 0$ for $0 < u < k$, and $g(u) < 0$ for $-k < u < 0$. (1) Let

$$x_1 = u,$$

$$x_2 = \dot{u}.$$

Write the above equation as a system of two differential equations system and show that $x_1 = 0$ and $x_2 = 0$ is a critical point. (2) Show that

$$V(x_1, x_2) = \frac{x_2^2}{2} + \int_0^{x_1} g(s)ds, \quad -k < x_1 < k,$$

is a strict Liapunov function for $(0, 0)$.

3 Consider the system of equations

$$\dot{x}_1 = x_2 - x_1 f(x_1, x_2),$$

$$\dot{x}_2 = -x_1 - x_2 f(x_1, x_2),$$

where f is a real-valued C^1 function. Using

$$V(x_1, x_2) = x_1^2 + x_2^2,$$

show that if $f > 0$ in some open neighborhood of the origin, then the origin is asymptotically stable. What is the stability type of the origin if $f < 0$ in some open neighborhood of the origin?

4 Show that the origin is a stable equilibrium point of (i) and an unstable equilibrium point of (ii) in the following equations:

(i) $$\dot{x}_1 = -x_2 - 2x_1^3, \qquad \dot{x}_2 = 5x_1 - x_2^3;$$

(ii) $$\dot{x}_1 = x_1^3 + x_1 x_2, \qquad \dot{x}_2 = -x_2 + x_2^2 + x_1 x_2 - x_1^3.$$

6.3 Bifurcations in Planar Dynamical Systems

In Chap. 3, we discussed bifurcations in one-dimensional systems. We now examine bifurcations in planar differential equations. We first show a few bifurcations, which are essentially the same as those that we have studied in Sec. 3.3. In this section, it will always be assumed that functions have as many derivatives as necessary if this is not explicitly stated.[6]

Example Saddle-node (fold) bifurcation.
Consider the following product system

$$\dot{x}_1 = \lambda + x_1^2,$$

$$\dot{x}_2 = -x_2.$$

The second equation is linear with $x_2(t) \to 0$ as $t \to \infty$. Thus all the orbits of the system eventually approach x_1-axis where the dynamics of the system are governed by the first equation. However, this first equation is simply Eq. (3.3.3) in Sec. 3.3. The phase portraits of the flow of the system for various parameter values are now easy to construct. For $\lambda < 0$, there are two equilibrium points. One of these equilibrium points

---

[16] The examples below in this section are from Chap. 7 in Hale and Koçak (1991).

is a saddle point. The other equilibrium point is a node. At $\lambda = 0$ the two equilibria coalesce into one, and for $\lambda > 0$, the equilibrium point disappears. We depict vector fields of the system in Fig. 6.3.1.

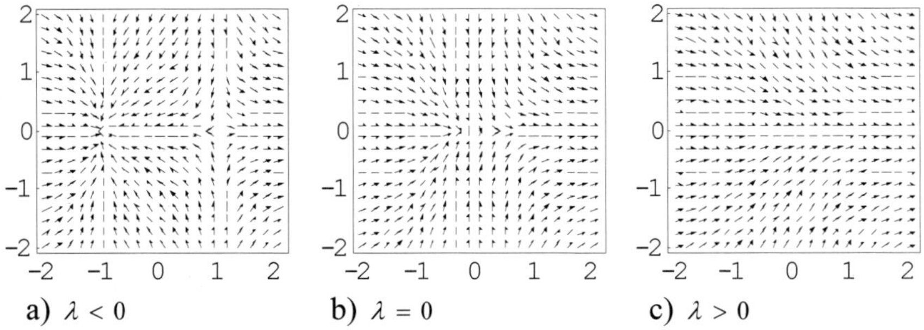

Fig. 6.3.1 Vector fields of the saddle-node bifurcation.

In fact, there are many planar dynamical systems which are equivalent to the above example. Consider a planar system

$$\dot{x} = f(x, \lambda), \quad x \in R^2, \quad \lambda \in R. \tag{6.3.1}$$

Assume that at $\lambda = 0$ it has the equilibrium at $x = 0$ with one eigenvalue $\rho_1 = 0$ and one eigenvalue $\rho_2 < 0$. It can be proved that Eq. (6.3.1) is locally topologically equivalent to the system[17]

$$\dot{x}_1 = \lambda + \sigma x_1^2,$$

$$\dot{x}_2 = -x_2,$$

where $\sigma = signa = \pm 1$.

Example Pitchfork bifurcation.
Consider

$$\dot{x}_1 = -\lambda x_1 - x_1^3,$$

$$\dot{x}_2 = -x_2. \tag{6.3.2}$$

As in the previous example, the dynamics of the system are contained in the first equation, which we have already analyzed in Sec. 3.3. The vector fields for three values of the parameter are depicted in Fig. 6.3.2.

---

[17] The proof is referred to Sec. 5.2 in Kuznetsov (1998).

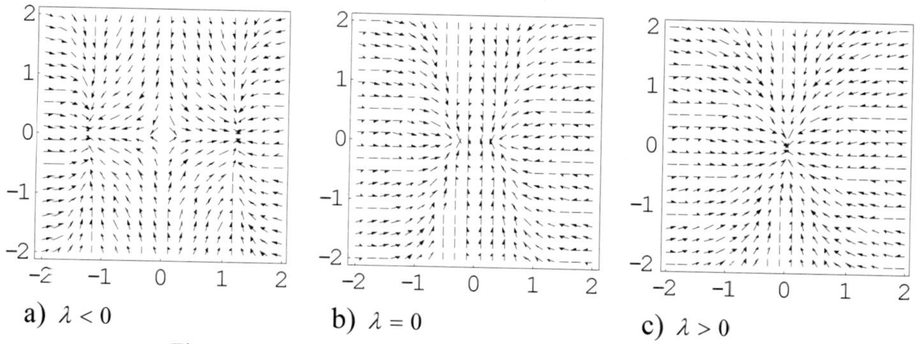

Fig. 6.3.2 Vector fields of the pitchfork bifurcation.

Example Vertical bifurcation.
Consider the following one-parameter perturbation of the harmonic oscillator

$$\dot{x}_1 = \lambda x_1 + x_2,$$

$$\dot{x}_2 = -x_1 + \lambda x_2,$$

when λ is a parameter. When the parameter satisfies $\lambda < 0$, we see that all the solutions spiral clockwise into the origin as t increases. For $\lambda = 0$, this is the harmonic oscillator and all the solutions are periodic so that the origin is a center. Since at this value of the parameter the number of periodic orbits changes from none to many, we consider $\lambda = 0$ a bifurcation point. For $\lambda > 0$ all solutions spiral out clockwise without bounds. We depict the vector fields for the vertical bifurcation as in Fig. 6.3.3.

Example The Poincaré-Andronov-Hopf bifurcation.
Consider

$$\dot{x}_1 = x_2 + x_1\left(\lambda - x_1^2 - x_2^2\right),$$

$$\dot{x}_2 = -x_1 + x_2\left(\lambda - x_1^2 - x_2^2\right),$$

where λ is a parameter.

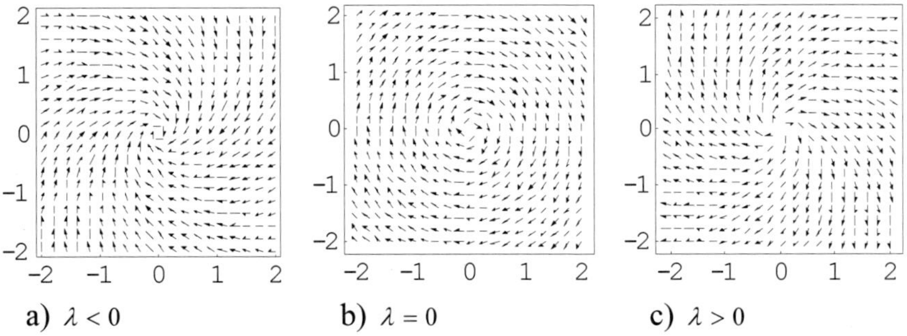

Fig. 6.3.3 Vector fields of the vertical bifurcation.

In polar coordinates, the system becomes

$$\dot{r} = r\left(\lambda - r^2\right),$$

$$\dot{\theta} = -1.$$

From this system, we see that for $\lambda \leq 0$, all solutions spiral clockwise to the origin with increasing time. When $\lambda > 0$, the origin becomes unstable and a periodic orbit of radius $\bar{r} = \sqrt{\lambda}$ appears. Furthermore, all the orbits, except the origin, spiral onto this periodic orbit. The birth or death of a periodic orbit through a change is the stability of an equilibrium point known as the Poincaré-Andronov-Hopf bifurcation. We depict the vector fields for the bifurcation in Fig. 6.3.4.

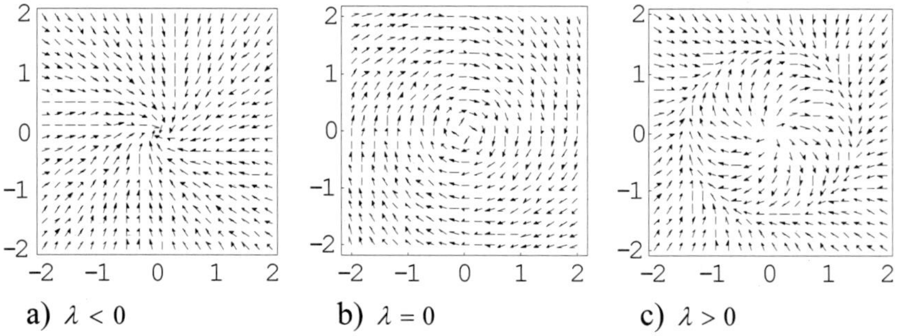

Fig. 6.3.4 Vector fields of the Hopf bifurcation.

Example Homoclinic or saddle-loop bifurcation.
Consider the planar system depending on a real parameter

$$\dot{x}_1 = x_2,$$

$$\dot{x}_2 = x_1 + \lambda x_2 - x_1^2. \tag{6.3.3}$$

For $\lambda = 0$, the system (6.3.3) is conservative with the first integral

$$H(x_1, x_2) = \frac{1}{2}(x_1^2 + x_2^2) + \frac{x_1^3}{3}.$$

The equilibrium point at $(1, 0)$ is a center locally surrounded by concentric periodic orbits. The equilibrium point at the origin, when viewed locally, is a saddle; when viewed globally, however, one of the orbits emanating from the origin terminates again at the origin after going around the other equilibrium. Indeed, the level set

$$H(x_1, x_2) = 0,$$

is rather special. It contains the equilibrium at the origin and the orbit whose α- and ω-limit sets are again the origin. Such orbits are called homoclinic orbits.

Definition 6.3.1 An orbit whose α- and ω-limit sets are both the same equilibrium point is called a *homoclinic orbit.*

When $\lambda \neq 0$, the center is destroyed but the saddle remains. The loop consisting of the homoclinic orbit and the equilibrium point at the origin is broken. We illustrate the manner in which the loop breaks depends on the sign of the parameter in Fig. 6.3.5.

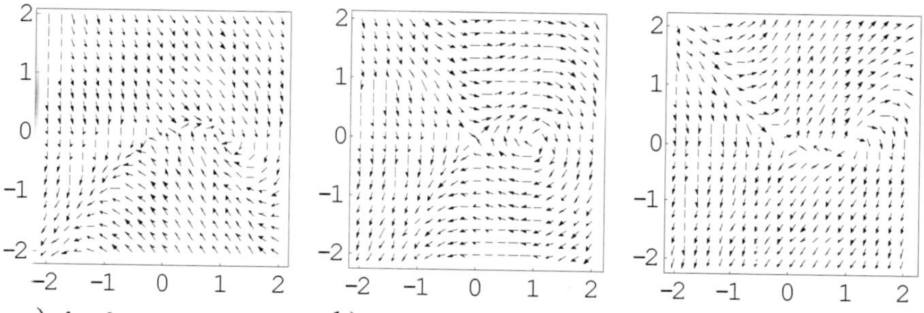

a) $\lambda < 0$ b) $\lambda = 0$ c) $\lambda > 0$

Fig. 6.3.5 Breaking a homoclinic loop bifurcation.

6.4 Periodic Solutions and Limit Cycles

We now consider possible existence of periodic solutions of second order autonomous systems

$$\dot{x}_j(t) = f_j(x_1, x_2), \quad j = 1, 2, \tag{6.4.1}$$

or in the vector form, $\dot{x} = f(x)$. Such solutions satisfy $x(t+T) = x(t)$ for all t and for some nonnegative constant T called the period. The corresponding trajectories are closed curves in the phase plane. A special case of a periodic solution is a critical point, which is periodic with any period. When we speak of periodic solutions, we exclude this case.

From Chap. 5, we know that a linear autonomous system

$$\dot{x} = A_{2\times 2} x,$$

has a periodic solution only if the eigenvalues of A are pure imaginary. The critical point is a center. If the eigenvalues are not pure imaginary, then the linear system has no periodic solution. Consider the following nonlinear system

$$\dot{x}_1 = x_1(1 - x_2),$$

$$\dot{x}_2 = \sigma x_2(x_1 - 1). \tag{6.4.2}$$

Dividing the two equations yields

$$\frac{dx_1}{dx_2} = \frac{x_1(1 - x_2)}{\sigma x_2(x_1 - 1)}.$$

The separation method solves the above equation as

$$G(x_1, x_2) = \sigma(x_1 - \ln x_1) + x_2 - \ln x_2 = A,$$

where A is a constant. Since $G(x_1, x_2)$ does not change as we move along a trajectory or solution curve of the equations, these trajectories are defined by the curves

$$G(x_1, x_2) = A$$

for different values of the constant A. We simulate the dynamics (6.4.2) with $\sigma = 1$. We see that $x_1(t)$ and $x_2(t)$ are periodic solutions as the trajectory is a closed curve.

We now study another dynamic system to demonstrate another way in which periodic solutions of nonlinear autonomous systems can occur.

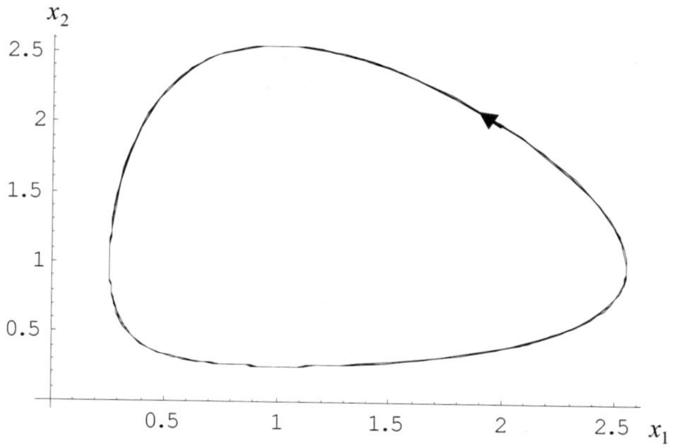

Fig. 6.4.2 A periodic solution of the system (6.4.2).

Example Consider a nonlinear system

$$\dot{x}_1 = x_1 + x_2 - x_1\left(x_1^2 + x_2^2\right),$$

$$\dot{x}_2 = -x_1 + x_2 - x_2\left(x_1^2 + x_2^2\right). \qquad (6.4.3)$$

The system has only one critical point $(0, 0)$ and it is almost linear in the neighborhood of the origin. The linearized system near the origin is

$$\begin{bmatrix} \dot{X}_1 \\ \dot{X}_2 \end{bmatrix} = \begin{bmatrix} 1 & 1 \\ -1 & 1 \end{bmatrix} \begin{bmatrix} X_1 \\ X_2 \end{bmatrix}.$$

The two eigenvalues are

$$\rho_{1,2} = 1 \pm i.$$

The origin is thus an unstable spiral point for both the linear system and the nonlinear system (6.4.3). Any solution that starts near the origin will spiral away from the origin. Since there are no other critical points, we might think that all solutions of the nonlinear system spiral out to infinity. But this is not correct as, demonstrated in Fig. 6.4.3, because far away from the origin the trajectories are directed inward.

We now analytically show the behavior of the system. We introduce polar coordinates r and θ

$$x_1 = r\cos\theta, \quad x_2 = r\sin\theta, \tag{6.4.4}$$

where $\lambda = 0$. First, we note that the system (6.4.3) contains the following relations between $x_1(t)$ and $x_2(t)$

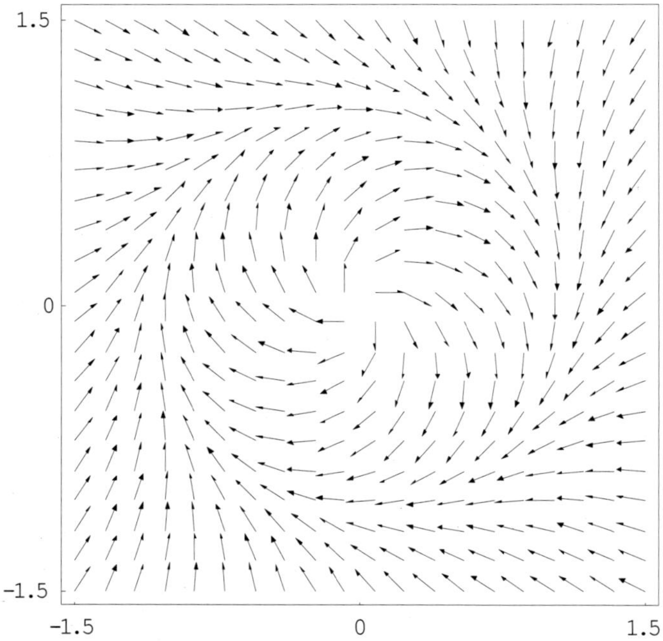

Fig. 6.4.3 Phase portraits of the system (6.4.3) near the origin.

$$x_1\dot{x}_1 + x_2\dot{x}_2 = \left(x_1^2 + x_2^2\right) - \left(x_1^2 + x_2^2\right)^2,$$

$$x_2\dot{x}_1 - x_1\dot{x}_2 = x_1^2 + x_2^2.$$

Substituting Eqs. (6.4.4) into the above equations yields

$$r\dot{r} = r^2\left(1 - r^2\right),$$

$$\dot{\theta} = -1. \tag{6.4.5}$$

The critical points of

$$r\dot{r} = r^2\left(1 - r^2\right),$$

are the origin and the point $r = 1$. The latter corresponds to the unit cycle in the phase plane. From the equation, we see that $\dot{r} > 0$ if $r < 1$, and $\dot{r} < 0$ if $r > 1$. Thus inside the unit circle the trajectories are directed outward, while outside the unit cycle they are directed inward. This is illustrated in Fig. 6.4.3. The solution of $\dot{\theta} = -1$ is

$$\theta(t) = -t + t_0,$$

where t_0 is an arbitrary constant. A solution to the system is

$$r(t) = 1, \ \theta(t) = -t + t_0. \tag{6.4.6}$$

As t increases, a point satisfying Eqs. (6.4.6) moves clockwise around the unit circle. Thus the nonlinear system has a periodic solution.

When $r \neq 1$, solutions of

$$r\dot{r} = r^2(1 - r^2)$$

are given by integrating

$$\frac{dr}{r(1 - r^2)} = dt \Rightarrow r(t) = (1 + c_0 e^{-2t})^{-1/2},$$

where c_0 is an arbitrary constant. Hence, the other solutions, except $r(t) = 1$, to Eqs. (6.4.5) are

$$r(t) = (1 + c_0 e^{-2t})^{-1/2}, \ \theta(t) = -t + t_0.$$

Given the initial conditions $r(0) = r_0$ and $\theta(0) = \theta_0$, we have the solution satisfying the initial conditions

$$r(t) = \{1 + (1/r_0^2 - 1)e^{-2t}\}^{-1/2}, \ \theta(t) = -t + \theta_0.$$

If $r_0 < 1$, then $r \to 1$ from the inside as $t \to +\infty$; If $r_0 > 1$, then $r \to 1$ from the outside as $t \to +\infty$. Thus in all cases the trajectories spiral toward the circle $r(t) = 1$ as $t \to +\infty$. The motion is illustrated in Fig. 6.4.4.

In this example, the circle $r = 1$ is a periodic solution. Also, all other nonclosed trajectories spiral toward it as $t \to +\infty$. In general, a closed trajectory in the phase plane is called a *limit cycle* if all other nonclosed trajectories spiral toward it, either from the inside or from the outside as $t \to +\infty$. If all trajectories that start near a closed trajectory (both inside and outside) spiral toward the closed trajectory as $t \to +\infty$, then the limit cycle is *asymptotically stable*. This kind of stability is called orbital stability as mentioned before. If the trajectories on one side spiral toward the closed trajectory, while those on the other side spiral away as $t \to +\infty$, then the closed trajectory is said to be *semistable*. If the trajectories on both sides of the closed trajectory spiral away as $t \to +\infty$, then the closed trajectory is *unstable*. If other trajectories neither approach nor depart from the closed trajectory, then it is called *stable*. The periodic solution to Eqs. (6.4.2) belongs to this type.

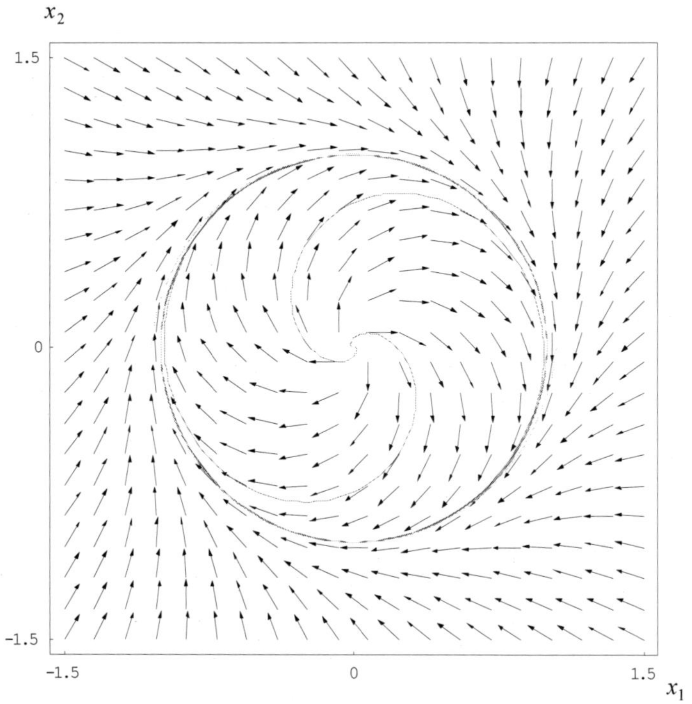

Fig. 6.4.4 A limit cycle generated by the system (6.4.4).

Example Semistable limit cycle.
Consider a nonlinear system

$$\dot{x}_1 = x_1\left(x_1^2 + x_2^2 - 1\right)^2 - x_2,$$

$$\dot{x}_2 = x_2\left(x_1^2 + x_2^2 - 1\right)^2 + x_1.$$

The system contains the following relations between $x_1(t)$ and $x_2(t)$

$$x_1\dot{x}_1 + x_2\dot{x}_2 = \left(x_1^2 + x_2^2\right)\left(x_1^2 + x_2^2 - 1\right)^2,$$

$$x_2\dot{x}_1 - x_1\dot{x}_2 = - x_1^2 - x_2^2 .$$

Under the polar coordinate transformation, the above system becomes

$$r\dot{r} = r^2\left(r^2 - 1\right)^2, \quad \dot{\theta} = 1 .$$

The critical points of

$$r\dot{r} = r^2\left(1 - r^2\right)$$

are the origin and the point $r = 1$. The latter corresponds to the unit cycle in the phase plane. From the equation, we see that $\dot{r} > 0$ for $r > 0$. Thus inside the unit circle the trajectories are directed outward, while outside the unit cycle they wind off on the outside. Figure 6.4.5 illustrates the cycle and two trajectories – one starting inside the cycle and the other outside the cycle.

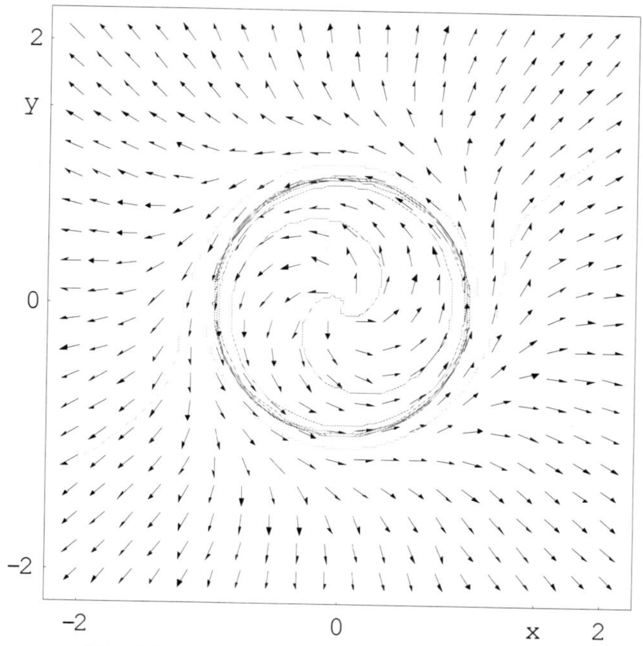

Fig. 6.4.5. An example of semistable limit cycles.

Example Multiple nested limit cycles.
Consider a nonlinear system

$$\dot{x}_1 = -x_2 + x_1\left(x_1^2 + x_2^2\right)\sin\left(\frac{1}{\sqrt{x_1^2 + x_2^2}}\right),$$

$$\dot{x}_2 = x_1 + x_2\left(x_1^2 + x_2^2\right)\sin\left(\frac{1}{\sqrt{x_1^2 + x_2^2}}\right).$$

The system contains the following relations between $x_1(t)$ and $x_2(t)$

$$x_1\dot{x}_1 + x_2\dot{x}_2 = \left(x_1^2 + x_2^2\right)^2 \sin\left(\frac{1}{\sqrt{x_1^2 + x_2^2}}\right),$$

$$x_2\dot{x}_1 - x_1\dot{x}_2 = -x_1^2 - x_2^2.$$

Under the polar coordinate transformation, the above system becomes

$$r\dot{r} = r^2 \sin\left(\frac{1}{r}\right), \quad \dot{\theta} = 1.$$

The critical points of $r\dot{r} = r^2 \sin(1/r)$ are the origin and the points

$$r = 1/n\pi, \quad i = 1, 2, 3 \cdots.$$

The latter corresponds to the cycles of radius $r = 1/n\pi$, which are limit cycles, unstable for odd n and stable for even n. To establish stability, introduce

$$\frac{1}{r(t)} = n\pi + h(t)$$

where

$$h(t) << n\pi.$$

Substituting

$$\frac{1}{r(t)} = n\pi + h(t)$$

into $\dot{r} = r\sin(1/r)$ yields

$$\dot{h} = -(n\pi + h)\sin(n\pi + h) = -(n\pi + h)\cos(n\pi)\sin(h).$$

When n is even, the above equation becomes

$$\dot{h} = -(n\pi + h)\sin(h).$$

When h is positive (negative), $\dot{h} < (>)\ 0$. Similarly, we can discuss temporary behavior of h when n is odd. We demonstrated the existence of multiple limit cycles as in Fig. 6.4.6.

Example The Brusselator model.
A hypothetical set of chemical reactions due to Prigogine and Lefever leads to the following Brusselator model[18]

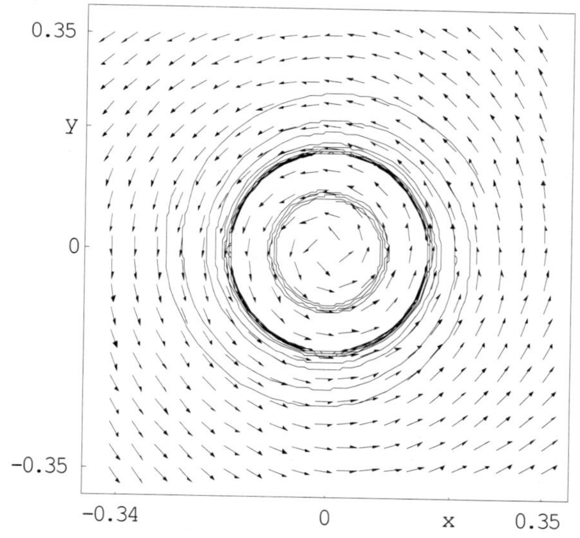

Fig. 6.4.6. Existence of multiple limit cycles.

$$\dot{x} = a - (1 + b)x + x^2 y,$$

$$\dot{y} = bx - x^2 y,$$

where x and y are nonnegative and a and b are positive. A stable limit cycle is demonstrated as in Fig. 6.4.7 with $a = 1$ and $b = 3$.

---

[18] Prigogine and Lefever (1968). Here, we only provide a simulation result to demonstrate the existence of limit cycles in the model.

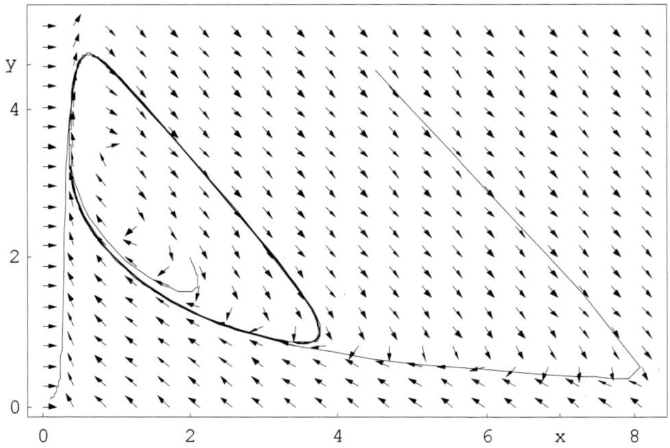

Fig. 6.4.7 A stable limit cycle in the Brusselator.

Exercise 6.4

1 Show that the following nonlinear system

$$\dot{x}_1 = -x_2 + \frac{x_1}{\sqrt{x_1^2 + x_2^2}}\left(1 - x_1^2 - x_2^2\right),$$

$$\dot{x}_2 = x_1 + \frac{x_2}{\sqrt{x_1^2 + x_2^2}}\left(1 - x_1^2 - x_2^2\right),$$

has a stable limit cycle.

2 Show that the following nonlinear system

$$\dot{x}_1 = x_2 + x_1\left(x_1^2 + x_2^2 - 1\right),$$

$$\dot{x}_2 = -x_1 + x_2\left(x_1^2 + x_2^2 - 1\right),$$

has an unstable limit cycle.

6.5 The Poincaré-Bendixson Theorem

The literature dealing with the existence and properties of limit cycles is vast. We now introduce a few important aspects of the literature. Consider

$$\dot{x}_j(t) = f_j(x_1, x_2), \quad j = 1, 2. \tag{6.5.1}$$

Theorem 6.5.1 Let functions $f_1(x)$ and $f_2(x)$ have continuous partial derivatives in a domain D of R^2. A closed trajectory of Eqs. (6.5.1) must necessarily enclose at least one critical point. If it encloses only one critical point, the critical point cannot be a saddle point.

Theorem 6.5.2 (Bendixson's negative criterion) Let functions $f_1(x)$ and $f_2(x)$ have continuous partial derivatives in a simply connected domain D of R^2.[19] If

$$\frac{\partial f_1}{\partial x_1} + \frac{\partial f_2}{\partial x_2},$$

has the same sign throughout D, then there is no closed trajectory of the system (6.5.1) lying entirely in D.

The theorem is also called *Bendixon's first theorem*. It is often used to establish the nonexistence of limit cycles of the two-dimensional first order differential equations.

Example Consider the system (6.4.3) again. We have

$$\frac{\partial f_1}{\partial x_1} + \frac{\partial f_2}{\partial x_2} = 2 - 4\left(x_1^2 + x_2^2\right) = 2\left(1 - 2r^2\right).$$

Hence

$$\frac{\partial f_1}{\partial x_1} + \frac{\partial f_2}{\partial x_2} > 0,$$

is positive for $0 \le r < \sqrt{2}$, so there is no closed trajectory in this circular disk. In fact, we have shown that there is no closed trajectory in the larger region $r < 1$. This implies that Theorem 6.5.2 may not give the best possible result. For $r > 1/\sqrt{2}$,

---

[19] A simply connected two-dimensional domain is one with no holes. Any closed curve or surface lying in the domain can be shrunk continuously to a point without passing outside of the domain.

$$\frac{\partial f_1}{\partial x_1} + \frac{\partial f_2}{\partial x_2} < 0 .$$

But the theorem is not applicable in this case because this annular region is not simply connected.

Example Consider

$$\dot{x}_1 = - g x_1 x_2 - \alpha x_1 ,$$

$$\dot{x}_2 = - g x_1 x_2 + \alpha x_2 ,$$

where $x_j > 0$, $j = 1, 2$. To apply the Bendixson's negative criterion, we calculate

$$\frac{\partial f_1}{\partial x_1} + \frac{\partial f_2}{\partial x_2} = - g\left(x_1 + x_2\right).$$

We conclude that $\partial f_1 / \partial x_1 + \partial f_2 / \partial x_2$ does not change its sign for $x_j > 0$, $j = 1, 2$. Hence, the system has no periodic solutions.

Theorem 6.5.3 (Poincaré-Bendixson theorem) Let functions $f_1(x)$ and $f_2(x)$ have continuous partial derivatives in a domain D of R^2. Let D_1 be a bounded subdomain in D, and let U be the region that consists of D_1 plus its boundary (all points of D_1 are in D). Suppose that U contains no critical point of the system (6.5.1). If there exists a constant t_0 such that $x_1 = \phi_1(t)$, $x_2 = \phi_2(t)$ is a solution of (6.5.1) that exists and stays in U for all $t \geq t_0$, then either $x_1 = \phi_1(t)$, $x_2 = \phi_2(t)$ is a periodic solution, or $x_1 = \phi_1(t)$, $x_2 = \phi_2(t)$ spirals toward a closed trajectory as $t \to +\infty$. In either case, the system (6.5.1) has a periodic solution in U.

This theorem is also referred to as the *second theorem of Bendixson.* If U does not contain a closed trajectory, then Theorem 6.5.1 shows that this trajectory must enclose a critical point. However, this critical point cannot be in U. Thus U cannot be simply connected; it must have a hole. Although this theorem gives the necessary and sufficient conditions for the existence of a limit cycle, it is often difficult to apply because it requires the knowledge of the nature of trajectories. It should be noted that the Poincaré-Bendixson theorem is restricted to two dimensions. Analogous theorems in higher dimensions do not exist.

When applying the Poincaré-Bendixson theorem, the following procedure is appropriate to a specific dynamic system in R^2: (i) Locate a fixed point of the dynamic system and examine its stability property; (ii) If the fixed point is unstable, search for an invariant set W enclosing the fixed point. When a closed orbit does not coincide with the boundary of W, the vector field described by the functions f_1 and f_2 must point into the interior of W. When the fixed point is unstable, trajectories starting in a neighboring of the fixed point will be repelled from it. The set U in the theorem can be considered as the subset of W which consists of W minus a neighbor of the fixed point. We now apply this procedure to the Kaldor model.

Example Periodic solutions in the Kaldor model.[20]
The Kaldor model is describe by

$$\dot{Y} = \alpha\{I(Y, K) - S(Y, K)\} \equiv \alpha F(Y, K),$$
$$\dot{K} = I(Y, K) - \delta K, \tag{6.5.2}$$

where variables and parameters are defined as

Y and K = output level and capital stock, respectively;
$I(Y, K)$ = investment function ($I_Y > 0$, $I_K < 0$);
$S(Y, K)$ = saving function ($0 < S_Y < 1$, $S_K > 0$)[21];
α and δ = a positive adjustment parameter and capital depreciation rate.

---

[20] The original model was proposed by Kaldor (1940, 1957, 1963). Kaldor's contribution was in conjunction with the work of Kalecki (1937, 1939), who investigated similar models but concentrated on different aspects of stability. The analysis below is based on Sec. 2.2 in Lorenz (1993). For the analysis and behavioral interpretations of the model, see also Chang and Smyth (1971) and Gabisch and Lorenz (1989).

[21] The assumption of $S_K > 0$ is not convincing. In Chang and Smyth (1971), it is assumed $S_K < 0$. As we require

$$I_K - S_K < 0,$$

the different signs do not affect our analytical conclusion.

Suppose that the system has at least one equilibrium point. The determinant and trace of the Jacobian at an equilibrium point (Y^*, K^*) are

$$\det J = \alpha F_Y (I_K - \delta) - \alpha F_K I_Y ,$$

$$trJ = \alpha F_Y + I_K - \delta. \tag{6.5.3}$$

The determinant must be positive in order to exclude the possibility of a saddle point. The fixed point is locally stable if the real parts of the eigenvalues are negative. This is guaranteed if the trace is negative. We are interested in the case that the fixed point is unstable, which is guaranteed by

$$\det J < 0, \quad trJ > 0 .$$

We now search a compact invariant set W such that the vector field (6.5.2) points inwards.

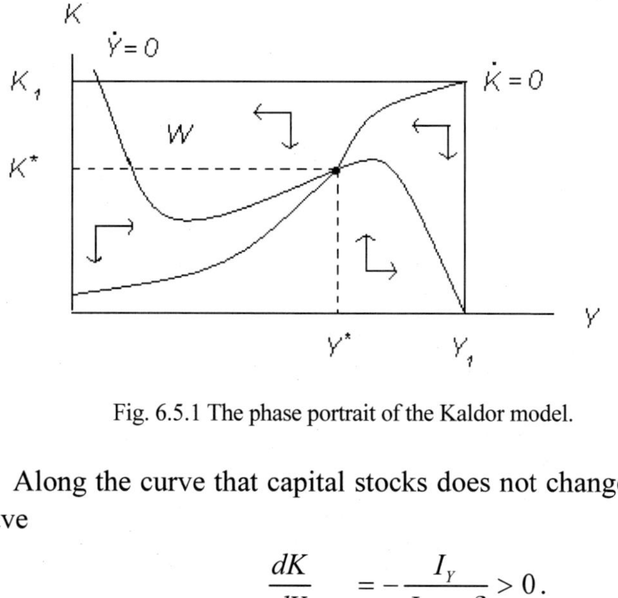

Fig. 6.5.1 The phase portrait of the Kaldor model.

Along the curve that capital stocks does not change (i.e., $\dot{K} = 0$), we have

$$\frac{dK}{dY}_{|\dot{K}=0} = -\frac{I_Y}{I_K - \delta} > 0 .$$

This implies that the locus of all points in the set

$$\{(Y, K) | \dot{K} = 0\},$$

is an upward sloping curve as shown in Fig. 6.5.1. For all K above (below) the curve for $\dot{K} = 0$, investment decreases (increases) because of $I_K - \delta < 0$.

Along the curve that output does not change (i.e., $\dot{Y} = 0$), we have

$$\frac{dK}{dY}_{|\dot{Y}=0} = \frac{S_Y - I_Y}{I_K - S_K} \begin{matrix} \geq \\ < \end{matrix} 0 .$$

Assume $I_K - S_K < 0$. The sign of dK / dY is dependent on the values of S_Y and I_Y. The difference $S_Y - I_Y$ is positive for low as well as for high levels of income and is negative in the neighborhood of the equilibrium point. It follows that the curve for $\dot{Y} = 0$ is negatively sloped for low and high values of Y and is positively sloped in a neighborhood of Y^*. Income increases (decreases) for all points below (above) the curve $\dot{Y} = 0$.

The subset

$$W = \{(Y, K) | \, 0 \leq Y \leq Y_1, \, 0 \leq K \leq K_1\},$$

is compact and the vector field points inwards the set on the boundary. Thus the requirements of the Poincaré-Bendixson theorem are satisfied. Therefore, the Kaldor model exhibits limit cycles.

Exercise 6.5

1 Apply the negative criterion to show no periodic solutions to

a) $\dot{x}_1 = -x_1 + x_2^2, \quad \dot{x}_2 = -x_2^3 + x_1^2$;

b) $\dot{x}_1 = -x_1 + 4x_2, \quad \dot{x}_2 = -x_1 - x_2^3$;

c) $\dot{x}_1 = -2x_1 e^{(x_1^2 + x_2^2)}, \quad \dot{x}_2 = -2x_2 e^{(x_1^2 + x_2^2)}$.

6.6 Lienard Systems

The Poincaré-Bendixson theorem can be used to establish the existence of limit cycles for certain planar systems. But it does not show us how to determine the exact number of limit cycles of a certain system or class of systems depending on parameters. This section represents some well-known results about the uniqueness of limit cycle for the so-called *Lienard system*

$$\dot{x}_1 = x_2 - F(x_1),$$
$$\dot{x}_2 = -g(x_1), \qquad (6.6.1)$$

under certain conditions on the functions F and g.[22]

Theorem 6.6.1 (Lienard's theorem) If $F,\, g \in C^1(R)$, and F and g are odd functions of x_1,[23]

$$x_1 g(x_1) > 0 \text{ for } x_1 \neq 0,\; F(0) = 0,\; F'(0) < 0,$$

F has a single positive zero at $x_1 = a$, and F increases monotonically to infinity for $x_1 \geq a$ as $x_1 \to \infty$, then the Lienard system (6.6.1) has exactly one limit cycle and it is stable.

Under the assumptions of the above theorem, we observe that the origin is the only critical point of the system (6.6.1); the flow on the positive x_2 - axis is horizontal and to the right, and the flow on the negative x_2 - axis is horizontal and to the left; the flow on the curve $x_2 = F(x_1)$ is vertical, downward for $x_1 > 0$ and upward for $x_1 < 0$; the system is invariant under $(x_1, x_2) \mapsto (-x_1, -x_2)$ and therefore if $(x_1(t), x_2(t))$ describes a trajectory of the system (6.6.1) so does $(-x_1(t), -x_2(t))$. We will apply this theorem to a modified Phillips model in the next chapter.

Example Consider the Van der Pol equation

$$\ddot{x} + \mu(x^2 - 1)\dot{x} + x = 0.$$

Show that for $\mu > 0$, the system has a unique limit cycle. The system can be rewritten as

$$\dot{x} + \mu\left(\frac{x^3}{3} - x\right) + \int_0^t x(\tau)d\tau = 0. \qquad (6.6.2)$$

[22] The system was first examined by the French physicist A. Lienard in 1928. Lienard studied this system in the equivalent form: $\ddot{x} + f(x)\dot{x} + g'(x) = 0$, where $f(x) = F'(x)$. The VdP equation, $\ddot{x} + \mu(x^2 - 1)\dot{x} + x = 0$, is a special case. The results about the Lienard system are referred to Sec. 3.8 in Perko (2000). See also Yeh (1986).

[23] A function is even if $f(x) = f(-x)$ and is odd if $-f(x) = f(-x)$.

Introduce

$$x_1 = x, \quad x_2 = -\int_0^t x(\tau)d\tau .$$

We rewrite Eq. (6.6.2) as

$$\dot{x}_1 = x_2 - \mu\left(\frac{x_1^3}{3} - x_1\right),$$

$$\dot{x}_2 = -x_1. \tag{6.6.3}$$

The system Eqs. (6.6.3) satisfies the conditions of Lienard's Theorem as

$$g(x_1) = x_1, \quad F(x_1) = \mu\left(\frac{x_1^3}{3} - x_1\right).$$

Hence, the system has a unique limit cycle.

The following theorem complements Lienard's Theorem. Lorenz applies this theorem proved by Levinson and Smith to guarantee the existence of a unique business cycle in a modified Phillips model.[24]

Theorem 6.6.3 (Levinson and Smith) The following Lienard system

$$\ddot{x} + f(x)\dot{x} + g(x) = 0$$

where $f = F'$ has a unique periodic solution if the following conditions are satisfied: (i) f and g are C^1; (ii) There exist $x_1 > 0$ and $x_2 > 0$ such that $f(x) < 0$ for $-x_1 < x < x_2$ and $f(x) > 0$ otherwise; (iii) $xg(x) > 0$, $x \neq 0$; (iv)

$$\int_0^\infty f(x)dx = \int_0^{\pm\infty} g(x)dx = \infty ;$$

and (v)

$$G(-x_1) = G(x_2)$$

[24] The following theorem is proved by Levinson and Smith (1942: 397). For economic applications to a modified Phillips model (1954), see Chap. 2 in Lorenz (1993).

where $G(x) \equiv \int_0^x g(s)ds$.

Theorem 6.6.3 (Zhang)[25] Under the assumptions that $a < 0 < b$, $F, g \in C^1(a, b)$, $x_1 g(x_1) > 0$ for $x_1 \neq 0$, $G(x_1) \to \infty$ as $x_1 \to a$ if $a = -\infty$ and $G(x_1) \to \infty$ as $x_1 \to b$ if $b = \infty$ (where $G(x_1) \equiv \int_0^{x_1} g(s)ds$), $f(x_1)/g(x_1)$ is monotone increasing on $(a, 0) \cup (0, b)$ and is not constant in any neighborhood of $x_1 = 0$, it follows that the system (6.6.1) has at most one limit cycle in the region $a < x < b$ and if it exists is stable.

Example As demonstrated by Perko,[26] the quadratic system

$$\dot{x}_1 = -x_2(1 + x_1) + \alpha x_1 + (1 + \alpha)x_1^2, \quad 0 < \alpha < 1,$$

$$\dot{x}_2 = x_1(1 + x_1),$$

has exactly one limit cycle and it is stable. Introduce a new independent variable τ by $d\tau = -(1 + x_1)dt$.
Then, the above system is transformed to a special case of Lienard's systems

$$\frac{dx_1}{d\tau} = x_2 - \frac{\alpha x_1 + (1 + \alpha)x_1^2}{(1 + x_1)}, \quad 0 < \alpha < 1,$$

$$\frac{dx_2}{d\tau} = -x_1.$$

Although it does not satisfy Lienard's Theorem, it satisfies Zhang's Theorem for $x > -1$. It is demonstrated that it has exactly one limit cycle and is stable.

Theorem 6.6.4 (Zhang[27]) If $g(x_1) = x_1$, $F \in C^1(R)$, $f(x_1)$ is an even function with exactly two positive $a_1 < a_2$ with $F(a_1) > 0$ and

---

[25] The theorem is proved in 1958 by the Chinese mathematician, Zhang Zhifen. Lorenz (1993: 57–60) applies Zhang's result to a simplified Kaldor model. We don't reproduce Lorenz's applications because they are done under very strong, if not unreasonable, assumptions from the economic point of view.

[26] See Perko (2000: 259).

$F(a_2) < 0$, and $f(x_1)$ (where $F(x_1) = \int_0^{x_1} f(s)ds$) is monotone increasing for $x_1 > a_2$, it follows that the system (6.6.1) has at most two limit cycles.

Example It can be demonstrated that the Lienard system with

$$g(x_1) = x_1, \quad F(x_1) = 0.32x_1^5 - \frac{4x_1^3}{3} + 0.8x_1,$$

has exactly two limit cycles.

Theorem 6.6.5 (Lins, de Melo and Pugh) If $g(x_1) = x_1$,

$$F(x_1) = a_1x_1 + a_2x_1^2 + a_3x_1^3,$$

and $a_1a_3 < 0$, then the system (6.6.1) has exactly one limit cycle. It is stable if $a_1 < 0$ and unstable if $a_1 > 0$.

Exercise 6.6

1 Show that the functions

$$F(x_1) = \frac{x_1^3 - x_1}{x_1^2 + 1},$$

and $g(x_1) = x_1$ satisfy the hypotheses of Lienard's Theorem.

6.7 The Andronov-Hopf Bifurcations in Planar Systems

Consider a two-dimensional dynamics including a parameter λ that is allowed to vary

$$\dot{x}_j(t) = f_j(x_1, x_2, \lambda), \quad j = 1, 2. \tag{6.7.1}$$

For each λ, suppose there is an isolated equilibrium point $x^* = (x_1^*, x_2^*)$ that depends on the choice of λ. Denote this dependence by $x^*(\lambda)$. The linearized system has a Jacobian matrix $A_{2\times2}$ that depends on λ. That is

[27] The theorem is proved in 1981 by Zhang (1981).

$$A(\lambda) = \begin{pmatrix} \dfrac{\partial f_1}{\partial x_1} & \dfrac{\partial f_1}{\partial x_1} \\ \dfrac{\partial f_1}{\partial x_1} & \dfrac{\partial f_1}{\partial x_1} \end{pmatrix},$$

evaluated at $x = x^*$. The two eigenvalues are given by $\rho_i(\lambda),\ i = 1, 2$. We assume that for some suitable range of λ values (we may choose near the origin without affecting the general conclusion), the eigenvalues are differentiable in λ and complex

$$\rho_{1,2} = \alpha(\lambda) \pm i\beta(\lambda).$$

The following *Andronov-Hopf bifurcation Theorem* in R^2 guarantees the existence of limit cycles.

Theorem 6.7.1 (The Andronov-Hopf birfurcation theorem)[28] Suppose that the equilibrium point $x^*(\lambda)$ is asymptotically stable for $\lambda < 0$ and unstable for $\lambda > 0$ and that $\alpha(0) = 0$. If $d\alpha(0)/d\lambda > 0$ and $\beta(0) \neq 0$ then for all sufficiently small $|\lambda|$, a closed orbit exists for λ either positive or negative. In particular, if $x^*(0)$ is locally asymptotically stable, then there is a stable limit cycle Γ about $x^*(\lambda)$ for all small $\lambda > 0$. Moreover, the amplitude of Γ grows as λ increases.

In Theorem 6.7.1, the real parts of the eigenvalues of $A(\lambda)$ cross the imaginary axis as λ moves past the origin $\lambda = 0$ from left to right (since $d\alpha(0)/d\lambda > 0$). In general, it is not readily determined whether or not the equilibrium $x^*(0)$ is locally asymptotically stable since $\alpha(0) = 0$ means that the equilibrium at $\lambda = 0$ is nonhyperbolic, which precludes any deduction from a knowledge of the linearized system. The linearized system is neutrally stable about $x^*(0)$ since the eigenvalues are imaginary. The theorem is somewhat ambiguous about the nature of the cycle.

---

[28] The Androvov-Hopf bifurcation was already studied by Poincaré. The systematic study of the conditions and the proof of the bifurcation theorem was done by Andronov and Leontovich in the two dimensional case in 1937. The proof of the bifurcation theorem in the n dimensional case is due to Hopf in 1942. The Andronov-Hopf bifurcation has a vast literature in mathematics as well as natural and social sciences.

The critical value of zero for the parameter λ is called a *Hopf bifurcation point*. It should be noted that $x^*(\lambda)$ can always be chosen as zero as for all λ simply by letting $z(t) = x(t) - x^*$ and rescaling Eqs. (6.7.1) to

$$\dot{z} = g(z, \lambda).$$

Example The predator-prey model.
We consider the following dynamics of two interacting populations whose levels at t are $x_1(t)$ and $x_2(t)$

$$\dot{x}_1 = rx_1\left(1 - \frac{x_1}{K}\right) - \frac{\beta x_1 x_2}{\alpha + x_1},$$

$$\dot{x}_2 = sx_2\left(1 - \frac{x_2}{vx_1}\right),$$

where α, r, s, v, and K are positive parameters. Ignoring the extinction equilibrium point in which $x_1 = K$ and $x_2 = 0$, there is one nontrivial rest state defined by

$$vx_1 = x_2, \quad r\left(1 - \frac{x_1}{K}\right) = \frac{\beta x_2}{\alpha + x_1}.$$

Denote this equilibrium point by x_1^* and x_2^* which are independent of the parameter s. The Jacobian matrix of the linearized system at the equilibrium is

$$A(\lambda) = \begin{pmatrix} k & -\beta x_1^* / (\alpha + x_1^*) \\ (k - \lambda)v & \lambda - k \end{pmatrix},$$

where $\lambda \equiv k - s$ and

$$k \equiv -\frac{rx_1^*}{K} + \frac{\beta v x_1^{*2}}{\alpha + x_1^{*2}}.$$

It follows that $\text{Det}A(\mu)$ is positive, independent of λ and $\text{Trace}A(\lambda) = \lambda$. The equilibrium point is stable for $\lambda < 0$ and unstable for $\lambda > 0$. The eigenvalues $\rho_i(\lambda)$ of $A(\lambda)$ are complex for all $|\lambda|$ small enough. At $\lambda = 0$, $\text{Re}\rho_i(0) = 0$ and $\text{Im}\rho_i(0) \neq 0$. We also have the following results

$$\frac{d\mathrm{Re}\rho_i(\lambda)}{d\lambda} = \frac{1}{2}\frac{d\mathrm{Trace}A(\lambda)}{d\lambda} = \frac{1}{2} > 0.$$

By the Hopf bifurcation theorem, a limit cycle exists for all $|\lambda|$ small.

Example The Van der Pol equation is given by

$$\dot{x}_1 = -x_2,$$

$$\dot{x}_2 = x_1 - \alpha\left(\frac{x_2^3}{3} - \lambda x_2\right), \quad \alpha > 0. \tag{6.7.2}$$

The equilibrium is $(0, 0)$. The Jacobian matrix of the linearized system is computed to be

$$A(\lambda) = \begin{pmatrix} 0 & -1 \\ 1 & \alpha\lambda \end{pmatrix}.$$

It can be seen that the determinant of the matrix is positive for all λ, i.e., $DetA(\lambda) > 0$; and the sign of $\mathrm{Trace}A(\lambda) = \alpha\lambda$ depends on the sign of λ. For $\lambda < 0$, the equilibrium is asymptotically stable and it is unstable for $\lambda > 0$.

At $\lambda = 0$, the eigenvalues are imaginary and no judgment about stability can be made. To further examine its stability, introduce

$$V(x_1, x_2) = \frac{x_1^2 + x_2^2}{2}.$$

Along orbits of Eqs. (6.7.2), we have

$$\dot{V} = x_1\dot{x}_1 + x_2\dot{x}_2 = -\alpha x_2^2\left(\frac{x_2^2}{3} - \lambda\right).$$

When $\lambda = 0$, $\dot{V}$ is negative except for $x_2 = 0$. However, the origin is the only invariant set in which $\dot{V} = 0$. Hence, we conclude that V is a Liapunov function. The origin is asymptotically stable.

The eigenvalues of $A(\lambda)$ are complex for $|\lambda|$ small enough with $\mathrm{Re}\,\rho_{1,2} = \alpha\lambda/2$ satisfying

$$\frac{d\,\mathrm{Re}\,\rho_{1,2}(\lambda)}{d\lambda} = \frac{\alpha}{2} > 0, \quad \mathrm{Re}\,\rho_{1,2}(0) = 0.$$

Also

$$\operatorname{Im} \rho_{1,2}(\lambda) \neq 0,$$

we conclude that a stable limit cycle exists about the origin for all $\lambda > 0$ small enough. The orbits repelled by the origin rend to the limit cycle. Just how small λ must be is not revealed by the Hopf bifurcation theorem. But it is known that a stable orbit exists for $0 < \lambda \leq 1$.

Example Apply the Hopf bifurcation theorem to the Kaldor model.
The Kaldor model is described by Eqs. (6.5.2)

$$\dot{Y} = \alpha\{I(Y, K) - S(Y)\} \equiv \alpha F(Y, K),$$

$$\dot{K} = I(Y, K) - \delta K, \tag{6.7.3}$$

where the variables and parameters are as in Sec. 6.4 and we neglect possible impact of wealth on saving.

Suppose that the system has at least one equilibrium point. The determinant and trace of the Jacobian at an equilibrium point (Y^*, K^*) are

$$\det J = \alpha F_Y(I_K - \delta) - \alpha F_K I_Y,$$

$$trJ = \alpha F_Y + I_K - \delta.$$

If

$$\det J > 0, \quad trJ = 0,$$

then the Jacobian has two complex conjugate eigenvalues. A Hopf bifurcation occurs if the complex conjugate roots cross the imaginary axis. If we choose α as the bifurcation parameter with the bifurcation value determined by

$$\alpha_0 = \frac{\delta - I_K}{F_Y}.$$

If $\alpha > \alpha_0$, the real parts of the eigenvalues are becoming positive. We will not provide expressions of the stability conditions as these complicated expressions provide few new insights.[29]

---

[29] They are provided in Sec. 3.2 in Lorenz (1993). They can also be calculated according to the method given by Iooss and Josephy (1980) which is applied to the Keynesian business model by the author.

We now state another form of the Andronov-Hopf bifurcation theorem.

Theorem 6.7.2 (Andronov-Hopf bifurcation theorem) Consider the system

$$\dot{x} = f(x, \lambda), \tag{6.7.4}$$

where

$$f \in C^{k+1}(R^2 \times R), k \geq 4, \ \ f(0, \lambda) \equiv 0.$$

Suppose that for small $|\lambda|$ the 2×2 matrix $f_x^{'}(0, \lambda)$ has a pair of complex conjugate eigenvalues

$$\alpha(\lambda) \pm i\omega(\lambda), \ \omega(\lambda) > 0, \ \alpha(0) = 0, \ \alpha'(0) > 0,$$

then

(i) there is a $\delta > 0$ and a function $\lambda \in C^{k-1}((-\delta, \delta), R)$ such that for $\varepsilon \in (-\delta, \delta)$ the system

$$\dot{x} = f(x, \lambda(\varepsilon)),$$

has a periodic solution $p(t, \varepsilon)$with period $T(\varepsilon) > 0$, also $T \in C^{k-1}$,

$$\lambda(0) = \lambda'(0) = 0, \ \ T(0) = \frac{2\pi}{\omega(0)}.$$

(ii) the origin $(x, \lambda) = (0, 0)$ of the space $R^n \times R$ has a neighborhood $U \subset R^n \times R$ that does not contain any periodic orbit of Eqs. (6.7.4) but those of the family $p(t, \varepsilon)$, $\varepsilon \in (-\delta, \delta)$.

(iii) if the origin $x = 0$ is a 3 - asymptotically stable (resp. 3 - unstable) 3 - unstable equilibrium of the system

$$\dot{x} = f(x, 0),$$

then $\lambda(\varepsilon) > 0$ (resp. $\lambda(\varepsilon) < 0$) for $\varepsilon \neq 0$, and the periodic solution $p(t, \varepsilon)$ is asymptotically orbitally stable (resp. unstable).[30]

Finally, similar to Theorem 3.6.1, we have a theorem for topological normal form for the Hopf bifurcation.

---

[30] For the concepts of h -asymptotical stability and instability see Chap. 7 in Farkas (1994).

Theorem 6.7.3 Any "generic" two-dimensional, one-parameter system

$$\dot{x} = f(x, \lambda)$$

at $\lambda = 0$ has the equilibrium $x = 0$ with eigenvalues

$$\rho_{1,2}(0) = \pm i\omega_0, \quad \omega_0 > 0,$$

is locally topologically equivalent near the origin to one of the following normal forms

$$\begin{bmatrix} \dot{y}_1 \\ \dot{y}_2 \end{bmatrix} = \begin{bmatrix} \beta & -1 \\ 1 & \beta \end{bmatrix} \begin{bmatrix} y_1 \\ y_2 \end{bmatrix} \pm \left(y_1^2 + y_2^2\right) \begin{bmatrix} y_1 \\ y_2 \end{bmatrix}.$$

It is straightforward to show that the following nonlinear dynamical system

$$\begin{bmatrix} \dot{y}_1 \\ \dot{y}_2 \end{bmatrix} = \begin{bmatrix} \beta & -1 \\ 1 & \beta \end{bmatrix} \begin{bmatrix} y_1 \\ y_2 \end{bmatrix} - \left(y_1^2 + y_2^2\right) \begin{bmatrix} y_1 \\ y_2 \end{bmatrix},$$

can be rewritten in the polar form as

$$\dot{r} = r\left(\beta - r^2\right),$$

$$\dot{\theta} = 1,$$

which is analyzed before. We can analyze the other case in the similar way.

Exercise 6.7

1 Applying Theorem 6.7.1 to the following fish harvesting model

$$\dot{N} = f(N) - vEN,$$

$$\dot{E} = \alpha(vpEN - cE),$$

where

$N(t)$ = population level at time t;
$E(t)$ = a measure of effort expended in fishing;
$f(N)$ = "natural growth" of the population (when $f(N) = rN(1 - N/K)$, the population growth is called logistical model);
v = a constant per-capita rate;

p = price of fish ($pvEN$ is the revenue from the harvest);
c = a constant cost of per unit effort (cE is the total cost);
α = a positive parameter.

Here. It is assumed that $f(N)$ is "well-behaved" and there are two positive numbers, defined by $\overline{N}$ $(= v/vp)$ and $\hat{N}$ where $\overline{N} > \hat{N}$, such that

$$\frac{f(N)}{N} \geq 0, \quad 0 < N < \overline{N},$$

$$\frac{d}{dN}\left(\frac{f(N)}{N}\right) > 0, \quad N < \hat{N}.$$

Chapter 7

Planar Dynamical Economic Systems

This chapter applies the concepts and theorems related to two-dimensional differential equations to various economic issues. Section 7.1 introduces the IS-LM model, one of the basic models in contemporary macroeconomics and examines its dynamic properties. Section 7.2 examines an optimal foreign debt model, maximizing the life-time utility with borrowing. In Sec. 7.3, we consider a dynamic economic system whose construction is influenced by Keynes' *General Theory*. Applying the Hopf bifurcation theorem, we demonstrate the existence of limit cycles in a simplified version of the Keynesian business model. Section 7.4 examines dynamics of unemployment within the framework of growth theory. In particular, we simulate the model to demonstrate how unemployment is affected by work amenity and unemployment policy. In Sec. 7.5, we establish a two-regional growth model with endogenous time distribution. We examine some dynamic properties of the dynamic systems. Section 7.6 models international trade with endogenous urban model formation. We show how spatial structures evolve in association of global growth and trade. In Sec. 7.7, we introduce a short-run dynamic macro model, which combines the conventional IS-LM model and Phillips curve. We also illustrate dynamics of the model under different financial policies. Section 7.8 introduces a growth model with public inputs. The public sector is treated as an endogenous part of the economic system. The system exhibits different dynamic properties examined in the previous two chapters.

7.1 The IS-LM Model

The IS-LM model is one of the main models in contemporary macroeconomics. In its static form, it composes two, IS and LM, curves. The IS curve denotes an equation for the relationship between real income and interest rate that leads to equilibrium in the goods market; while the LM curve for the relationship which leads to equilibrium in the money market.[1] Overall equilibrium is established where both the goods market and money market achieves equilibrium – a state determined at the intersection of the two curves.

First study the goods market. Real expenditure is the sum of consumer expenditure, investment expenditure and government expenditure. It is assumed that consumption expenditure is positively related to real income and the investment is positively related to real income but negatively related nominal interest rate. That is

$$e(t) = c(y(t)) + i(r(t),\ y(t)) + g,\ \ 0 < c_y < 1,\ \ i_r < 0,\ \ i_y > 0, \tag{7.1.1}$$

where e is real expenditure, c consumption expenditure, y real income, r nominal interest rate, and g government expenditure. Demand for real money balances, $m^d(t)$, is assumed to be positively related to real income and inversely related to the interest rate. That is

$$m^d(t) = l(y,\ r),\ \ l_y > 0,\ \ l_r < 0 .$$

The dynamics are in terms of excess demand in the goods market and excess demand for real money balances, i.e.

$$\dot{y} = \alpha(e - y) = \alpha[c(y) + i(r,\ y) + g - y],$$

$$\dot{r} = \beta(l(y,\ r) - m_0),\ \ \alpha,\ \beta > 0, \tag{7.1.2}$$

where we use Eq. (7.1.1) and m_0 is fixed supply of real money supply. The equilibrium in the goods market requires $\dot{y} = 0$; that is $c(y) + i(r,\ y) + g = y$. The equilibrium in the money market is obtained if $l(y,\ r) = m_0$.

Suppose that the two equations

---

[1] The model is explained in standard textbooks in macroeconomics (e.g., Abel and Bernanke, 1998, Blanchard, 1997).

$$c(y) + i(r, y) + g = y, \quad l(y, r) = m_0,$$

determines an overall equilibrium (y^*, r^*). Then the Jacobian at the equilibrium is

$$J = \begin{bmatrix} \alpha(c_y + i_y - 1) & \alpha i_r \\ \beta l_y & \beta l_r \end{bmatrix}. \tag{7.1.3}$$

We have

$$TrJ = \alpha(c_y + i_y - 1) + \beta l_r,$$

$$DetJ = \alpha\beta[(c_y + i_y - 1)l_r - l_y i_r].$$

We now show that if the IS curve is less steep than the LM curve, then $DetJ > 0$. For the IS curve we have

$$c(y) + i(r, y) + g = y.$$

Totally differentiating this equation with respect to y and r, we have the slope of the IS curve

$$\frac{dr}{dy} = \frac{1 - c_y - i_y}{i_r}.$$

Similarly, the slope of the IM curve is given by

$$\frac{dr}{dy} = -\frac{l_y}{l_r}.$$

If the IS curve is less steep than the LM curve, we have

$$\frac{1 - c_y - i_y}{i_r} < -\frac{l_y}{l_r} \Rightarrow (c_y + i_y - 1)l_r - l_y i_r > 0,$$

which guarantees $DetJ > 0$. This is certainly satisfied in the usual case of a negatively sloped IS curve and a positively sloped LM curve. The above analysis demonstrates that $DetJ > 0$ occurs when the two curves are positively sloped but the IS curve is less steep than the LM curve. Hence, if the trace is negative in sign, then the equilibrium is stable.

To simulate the model, we specify $c(t)$, $i(t)$ and $m^d(t)$ as

$$c(t) = a + b(1 - \tau)y(t),$$

$$i(t) = -hr(t) + jy(t),$$

$$m^d(t) = ky(t) - ur(t),$$

$$a,\ h,\ j,\ k,\ u > 0, \ \ 0 < b,\ \tau < 1, \tag{7.1.4}$$

where a, b, τ, h, j, k, and u are positive parameters. Here, a is autonomous expenditure, b marginal propensity to consume, τ marginal rate of tax, h and j are coefficients of investment in response respectively to r and y. We omit g as it can be technically included in a. Under Eqs. (7.1.4), Eqs. (7.1.2) become

$$\dot{y} = \alpha(a + a_1 y - hr),$$

$$\dot{r} = \beta(-m_0 + ky - ur), \ \ \alpha, \beta > 0, \tag{7.1.5}$$

where

$$a_1 \equiv b(1 - \tau) + j - 1,$$

which may be either positive or negative. The system has a unique equilibrium point

$$(y^*, r^*) = \left(\frac{au + m_0 h}{kh - a_1 u}, \frac{a_1 m_0 + ak}{kh - a_1 u} \right).$$

For $y^* > 0$, we should require $kh - a_1 u > 0$. The Jacobian is

$$J = \begin{bmatrix} \alpha a_1 & -\alpha h \\ \beta k & -\beta u \end{bmatrix}. \tag{7.1.6}$$

We have

$$TrJ = \alpha a_1 - \beta u,$$

$$DetJ = \alpha\beta[-ua_1 + kh] > 0.$$

If $a_1 < 0$, then $TrJ < 0$. The equilibrium is stable. In the case of $a_1 > 0$, if $TrJ = \alpha a_1 < (>)\ \beta u$, the equilibrium point is stable (unstable).

The IS curve is given by

$$a + a_1 y - hr = 0.$$

The slope of this curve is

$$\frac{dr}{dy} = \frac{a_1}{h}.$$

The IS curve has either a negative slope (when $a_1 < 0$) or a positive slope (when $a_1 > 0$). On the other hand, the IM curve is

$$-m_0 + ky - ur = 0.$$

The slope of the LM curve is positive because

$$\frac{dr}{dy} = \frac{k}{u} > 0.$$

Figure 7.1.1 shows that the system has a unique equilibrium when we specify the parameters and the initial conditions as follows

$$a = 38, \ b = 0.35, \ \tau = 0.3, \ h = 1.3, \ j = 0.3, \ k = 0.25,$$

$$u = 0.3, \ m_0 = 8, \ \alpha = 0.2, \ \beta = 0.4, \ y_0 = 90, \ r_0 = 50. \quad (7.1.7)$$

As the eigenvalues are equal to $-0.09 \pm 0.14i$, the stability is numerically confirmed.

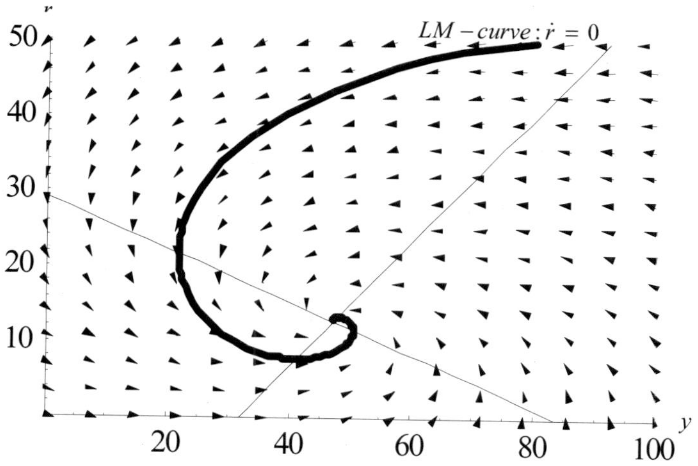

Fig. 7.1.1 The IS-LM model with stable equilibrium.

Figure 7.1.2 demonstrates instability when we specify the parameters and the initial conditions as follows

$$b = 0.6, \ j = 0.7, \ \alpha = 0.81, \ \beta = 0.3, \ y_0 = 50, \ r_0 = 20. \ (7.1.8)$$

The values of the parameters missed in (7.1.8) are specified as the same values of the corresponding parameters in (7.1.7). The eigenvalues are equal to $0.0036 \pm 0.265i$ with positive real parts.

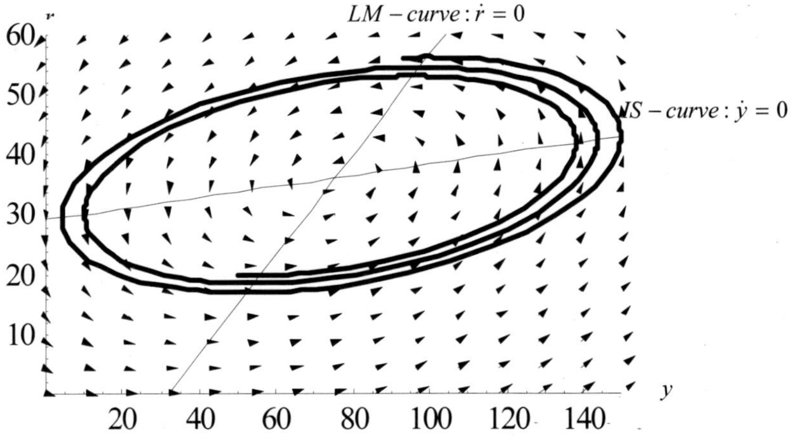

Fig. 7.1.2 The IS-LM model with unstable equilibrium.

Exercise 7.1

1 Consider the following IS-LM model

$$e = a + b(1 - \tau)y - hr + jy,$$

$$m^d = ky - ur,$$

$$\dot{y} = \alpha(e - y),$$

$$\dot{r} = \beta(m^d - m_0),$$

with the specified parameter values

$$a = 5, \quad b = 0.75, \quad \tau = 0.25, \quad h = 0.3, \quad j = 0.4, \quad k = 0.5,$$

$$u = 0.3, \quad m_0 = 10, \quad \alpha = 0.25, \quad \beta = 0.4.$$

(a) Find the equilibrium values of y and r; (b) What are the equations for the IS and the LM curve? and (c) Discuss the stability.

7.2 An Optimal Foreign Debt Model

Pitchford proposes an optimal model with foreign debt in a representative-agent type macroeconomy.[2] Treating other income y as given and assuming that the rate of interest $r(t)$ depends on borrowing to depend on the level of debt $A(t)$ of the borrower, we can describe the motion of debt by

$$\dot{A}(t) = R(A) + y - c(t), \tag{7.2.1}$$

where c is consumption and $R(A) \equiv r(A)A$. It is required

$$R'(A) = r'A + r > 0, \quad R'' < 0, \quad R \begin{cases} > 0 & \text{if } A > 0, \\ < 0 & \text{if } A < 0, \\ = 0 & \text{if } A = 0. \end{cases}$$

It is assumed that there exists some (bankruptcy) level $\overline{A}$ (< 0) of A at which consumption is forced to zero. This assumption prevents the agent from borrowing without limit. The optimal problem is to maximize

$$\underset{c}{Max} \int_0^\infty U(c) e^{-\rho t} \, dt,$$

subject to Eq. (7.2.1). We require:

$$U_c(c) > 0, U_{cc}(c) < 0, U_c(0) = \infty.$$

The Hamiltonian for the problem is

$$H = U(c) + \lambda(R(A) - c + y).$$

The necessary conditions are

$$H_c = U_c - \lambda = 0,$$

$$\dot{\lambda} = \rho\lambda - H_A = \rho\lambda - R'\lambda.$$

Taking derivatives of $U_c - \lambda = 0$ with respect to t yields

$$U_{cc}\dot{c} = \dot{\lambda}.$$

Substituting $U_{cc}\dot{c} = \dot{\lambda}$ and $U_c = \lambda$ into the above second equation yields

[2] Refer Pitchford (1991) for explanation in detail and extensions of the model.

$$\dot{c} = [\rho - R'(A)]\frac{U_c}{U_{cc}}. \tag{7.2.2}$$

Hence, the dynamics is given by Eq. (7.2.1) and Eq. (7.2.2). The stationary state of the system is determined by

$$c = R(A) + y,$$

$$\rho = R'(A) = r' A + r. \tag{7.2.3}$$

In the case of $c = 0$, $c = R(A) + y$ becomes

$$r(A_1)A_1 = - y < 0.$$

We see that A_1 is negative. The equilibrium value A^* of A is given by

$$\rho = r'(A^*)A^* + r(A^*).$$

As $R'' < 0$, as long as the subjective discount rate ρ is greater than $r(0)$, A^* we will negative. If ρ were less than $r(0)$, then A^* would be positive. The phase arrows are given by $\partial \dot{A} / \partial c < 0$ and $\partial \dot{c} / \partial A < 0$. The phase diagram is illustrated in Fig. 7.2. Check that the equilibrium is a saddle point.

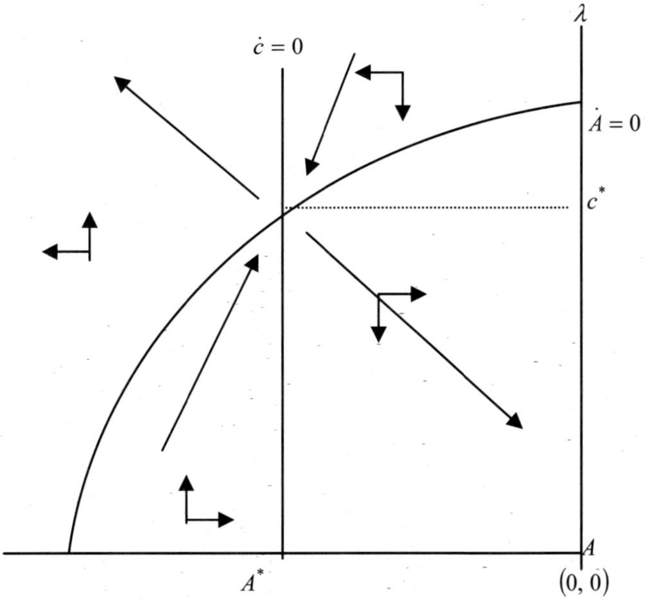

Fig. 7.2.1 Phase diagram for the model of foreign debt.

The figure illustrates the case that the optimal strategy for the economy is to remain permanently in debt. For instance, if

$$A(0) = 0,$$

then the optimal consumption involves borrowing heavily initially, and gradually reducing consumption over time to achieve the equilibrium. At the equilibrium, income is just sufficient to cover interest payments and maintain a steady consumption level.

Exercise 7.2

1 Consider Forster's model of pollution control [3]

$$\underset{c}{Max} \int_0^\infty [U(C) - V(P)] e^{-\rho t}\, dt, \text{ s.t.: } \dot{P} = Z(C) - \delta P,$$

where $C(t)$ and $P(t)$ are respectively consumption and pollution, ρ is the subjective discount rate, and δ represents the environment's natural capacity to restore itself. The utility function U and disutility function V satisfy

$$U_C > 0, \;\; U_{CC} < 0, \;\; U_C(0) = \infty,$$

$$V(0) = 0, \;\; V_P > 0, \;\; V_{PP} > 0, \;\; U_P(0) = \infty.$$

The function Z is defined by

$$Z(C) = g(C) - h(Y - C)$$

where $g(C)$ represents the pollution generated by the act of consumption and $h(Y - C)$ represents the effectiveness of anti-pollution activity with the activity intensity $Y - C$. The two functions satisfy

$$g(0) = 0, \;\; g_C > 0, \;\; g_{CC} > 0 \;\text{ for } C > 0, \;\; g_C(0) = 0,$$

$$h(0) = 0, \;\; h' > 0, \;\; h'' < 0 \;\text{ for }\; Y - C > 0, \;\; h'(0) = +\infty.$$

Find optimal conditions (in terms of two-dimensional differential equations with the pollution and costate as variables), illustrate phase diagram, and discuss stability.

---

[3] See Forster (1977) or Ferguson and Lim (1998).

7.3 The Simplified Keynesian Business Cycle Model

We now consider a dynamic economic system whose construction is essentially influenced by Keynes' *General Theory*. The simplified Keynesian business model is described by

$$\dot{Y} = \alpha\{I(Y, R) - S(Y, R)\} \equiv \alpha F(Y, R),$$
$$\dot{R} = \beta\{L(Y, R) - L_s\}, \tag{7.3.1}$$

on which parameters and variables are defined as

Y = output level;
R = rate of interest;
$I(Y,R)$ = investment function ($I_Y > 0,\ I_R < 0$);
$S(Y,R)$ = savings function ($S_Y > 0,\ S_R > 0$);
$L(Y,R)$ = total demand function for money ($L_Y > 0,\ L_R < 0$);
L_s = the fixed supply of money;
α, β = positive adjustment parameters.

The dynamic system states that if investment is larger than savings, then output level tends to increase, and vice versa; if the money demanded is larger than that supplied, the rate of interest tends to increase. Here, the requirements

$$I_Y > 0,\ I_R < 0,\ S_Y > 0,\ S_R > 0,\ L_Y > 0,\ L_R < 0,$$

imply that investment is positively related to output level, and negatively related to rate of interest; an increase in output or rate of interest will make people save more; more money is demanded if output increases or rate of interest falls.

The existence of a positive equilibrium (Y_0, R_0) determined by the intersection of $I(Y, R) = S(Y, R)$ and $L(Y,R) = L_s$, is assumed. It is sufficient to limit the discussion to a local domain. Torre first noticed the existence of business cycles in this system.[4] Our examination is based on the application of the Hopf bifurcation theorem to the model by Zhang.[5]

---

[4] See Torre (1977).
[5] See Zhang (1991: 78-81).

We apply the Hopf bifurcation theorem to identify the existence of limit cycles. We first find the conditions for the existence of a pair of purely imaginary eigenvalues and identify the loss of stability of the equilibrium. Referring to Torre, we know that these conditions are established if

$$\alpha_0 = \frac{\beta L_R}{F_Y},$$

$$F_Y L_R - F_R L_Y > 0, \quad F_Y > 0, \tag{7.3.2}$$

hold at (Y_0, R_0). As α is meaningful at any point in R^+, there is a value of the parameter α such that the first equality in (7.3.2) is valid. As $F_Y = I_Y - S_Y$, we interpret $F_Y > 0$ as that the marginal investment in the product is larger than the marginal savings with regard to output. Further interpretation of (7.3.2) is referred to Torre. The following theorem identifies the Hopf bifurcation in the system.

Theorem 7.3.1 (The existence of business cycles in the simplified Keynesian model) Let (7.3.2) hold at (Y_0, R_0). Then there exist limit cycles – Hopf bifucations – around (Y_0, R_0). The critical value of bifurcation parameter α is α_0. The bifurcated cycle of period $2\pi / \omega(\varepsilon)$ is approximately given by

$$Y(\varepsilon, t) = Y_0 + 2\varepsilon \alpha_0 F_R \cos[\omega(\varepsilon)t] + O(\varepsilon^2),$$

$$R(\varepsilon, t) = R_0 - 2\varepsilon \{z_0 \sin[\omega(\varepsilon)t] + \alpha_0 F_R \cos[\omega(\varepsilon)t]\} + O(\varepsilon^2), \tag{7.3.3}$$

where

$$z_0 \equiv \{\alpha_0 \beta (F_Y L_R - F_R L_Y)\}^{1/2},$$

ε is the expansion amplitude parameter, and

$$\alpha = \alpha_0 + \varepsilon^2 x_2 + O(\varepsilon^4),$$

$$\omega(\varepsilon) = z_0 + \varepsilon^2 \omega_2 + O(\varepsilon^4), \tag{7.3.4}$$

where x_2 and ω_2 are parameters. Moreover, if x_2 is positive, the periodic solutions are stable, while if x_2 is negative, they are unstable.

Proof: This theorem is proved by Zhang, applying the bifurcation method of Iooss and Joseph.[6] We now apply this method to approximately calculate the periodic solutions.

We denote by x small perturbations of α from α_0 as follows

$$x \equiv \alpha - \alpha_0 .$$

To write the system in a local form, introduce

$$U_1(t) = Y(t) - Y_0, \ \ U_2(t) = R(t) - R_0 .$$

Substituting this transformation into Eqs. (7.3.1) yields

$$\dot{U} = J(x)U + N(x, U, U) + O(U^3), \tag{7.3.5}$$

where $U = (U_1, U_2)$ and J is the Jacobian at the equilibrium

$$J(x) = \begin{bmatrix} \alpha F_Y & \alpha F_R \\ \alpha L_Y & \alpha L_R \end{bmatrix},$$

and N is the quadratic terms of U. The two eigenvalues are

$$z_{1,2}(x) = \frac{\beta L_R + \alpha F_Y}{2} \pm \left\{ \frac{(\beta L_R + \alpha F_Y)^2}{2} - \alpha\beta(F_Y L_R - L_Y F_R) \right\}^{1/2} . \tag{7.3.6}$$

As the conditions (7.3.2) hold at $x = 0$, we see that there are a pair of purely imaginary eigenvalues, $\pm iz_0$. If we denote $z(x)$ the eigenvalue which equals iz_0 at $x = 0$, then $\mathrm{Re}[z_x(0)]$ is not equal to zero. Thus the loss of stability of the equilibrium is guaranteed. We have identified the conditions for Hopf bifurcations.

To obtain an approximate expression of the periodic solutions, we calculate the eigenvector X and adjoint eigenvector X^* with regard to $z(x)$ from

$$JX = z(x)X,$$

$$J^T X^* = \bar{z}(x)X^*, \tag{7.3.7}$$

which satisfy

$$\langle X, X^* \rangle = 1, \langle \bar{X}, X^* \rangle = 0,$$

[6] See Iooss and Joseph (1980).

where $\langle\, , \rangle$ is the product operator in C^2. The solution to the above equations is given by

$$X = \begin{bmatrix} \alpha_0 F_R \\ -iz_0 - \alpha_0 F_Y \end{bmatrix}, \quad X^* = \begin{bmatrix} -\dfrac{i\beta L_Y}{2\alpha_0 z_0 F_Y} \\ \dfrac{-z_0 + i\alpha_0 F_Y}{2\alpha_0 z_0 F_Y} \end{bmatrix}. \tag{7.3.8}$$

As X and $\overline{X}$ are independent, U can be expressed as a combination of them in the form of

$$U(t) = \sigma(t)X + \overline{\sigma}(t)\overline{X}, \tag{7.3.9}$$

in which function $\sigma(t)$ is to be determined. Substituting Eqs. (7.3.9) into Eqs. (7.3.5), multiplying the resulted equations by $\overline{X}^*$, and then adding the equations, we have

$$\dot{\sigma} = z(x)\sigma + r_0\sigma^2 + 2r_1|\sigma|^2 + r_2\sigma^2, \tag{7.3.10}$$

where r_0 and r_1 are imaginary numbers. Here, we should not explicitly give r_0 and r_1. As shown by Iooss and Josephy, the solution to Eq. (7.3.10) can be constructed by the following series

$$\begin{bmatrix} g(s, \varepsilon) \\ \omega(\varepsilon) - z_0 \\ x(\varepsilon) \end{bmatrix} = \sum_{i=1}^{\infty} \varepsilon^i \begin{bmatrix} g_i(\varepsilon) \\ \omega_i \\ x_i \end{bmatrix}, \tag{7.3.11}$$

where

$$\sigma(t) = \sigma(s, \varepsilon), \quad s = \omega(\varepsilon)t, \quad \omega(0) = z_0, \quad x = x(\varepsilon),$$

$$\varepsilon = \int_0^{2\pi} \exp(-is)\sigma(s, \varepsilon)ds.$$

The coefficients of low orders with respect to ε are determined by

$$x_i = \omega_i = 0, \quad i = 1, 3, \cdots, 2n-1, \cdots,$$

$$\sigma_1(s) = \exp(is),$$

$$\sigma_2(s) = \frac{r_0 \exp(i2s) - 2r_1 - (r_2/3)\exp(-i2s)}{iz_0},$$

$$i\omega_2 - z_x(0)x_2 = 2i\frac{r_0 r_1 - 2|r_1|^2 - |r_2|^2/3}{z_0}.$$

From these equations, we can explicitly solve x_i, ω_i $(i = 1, 2, 3)$, σ_1 and σ_2.

Define a real number D as

$$D \equiv -\left[\varepsilon \operatorname{Re}\{z_x(0)\} + O(\varepsilon^3)\right]\frac{dx}{d\varepsilon},$$

where

$$x = \frac{x_2 \varepsilon^2}{2} + O(\varepsilon^3).$$

According to the factorization theorem,[7] if D is positive, the cycle is unstable; if D is negative, it is stable. Thus we have proved the stability conditions in Theorem 7.3.1.

We will not explicitly give the values of r_i, ω_i, x_i and other parameters because their expressions are too complicated. The behavior of the system near the equilibrium is illustrated in Fig. 7.3.1. The cycle size is dependent on the bifurcation parameter. As the parameter is further away from its critical value, the radius of the cycle becomes larger.

The rate of interest is sometimes higher than its point equilibrium value and sometimes lower. Although it may arrive at R_0, the rate of interest cannot stay there permanently. As soon as it arrives at this equilibrium value, it tends to move away from it. It is driven by the nonlinear interdependent forces in the system. From Eqs. (7.3.3), we have

$$\delta R(t) = -\alpha \sin[\omega(\varepsilon)t] - \frac{F_Y}{F_R}\delta Y(t) + O(\varepsilon),$$

where

$$\delta R(t) \equiv \frac{R(t) - R_0}{2\varepsilon}, \quad \delta Y(t) \equiv \frac{Y(t) - Y_0}{2\varepsilon}.$$

[7] See Chap. 7 in Iooss and Joseph (1980).

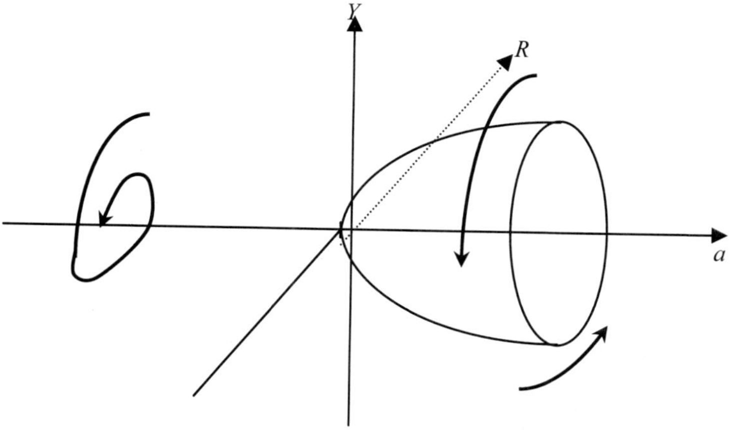

Fig. 7.3.1 Bifurcated cycles in the Keynesian model.

As $\delta Y(t)$ is a periodic function which is independent of $\delta R(t)$, we see that the interactions between the two variables may appear very complicated over time.

7.4 The Welfare Economy with Unemployment

We are now concerned with dynamics of unemployment in a welfare economy. The idea is that if society offers a generous welfare for unemployment, people have incentives to 'remain unemployed' – earning welfare payments and enjoying leisure. Marx and Keynes provided different reasons for the existence of unemployment. Neither Marx nor Keynes was concerned with unemployment issues in welfare economies in which an unemployed person may be paid 'well'.

We are now concerned with an economic system similar to the OSG model. Let the population be fixed and denoted by N. Let subscript indexes, i and u, denote respectively employed and unemployed. We denote $N_i(t)$ and $N_u(t)$ respectively the number of workers employed by the production sector and the number of persons unemployed. We have

$$N_i(t) + N_u(t) = N .$$

The production function of the economy is specified with the Cobb-Douglas form

$$F(t) = K^{\alpha}(zN_i)^{\beta}, \;\; \alpha + \beta = 1,$$

where z is the level of human capital, $K(t)$ is the capital stocks of the economy and $F(t)$ is the total output at time t. The marginal conditions are given by

$$r(t) = \frac{(1 - \tau(t))\alpha F(t)}{K(t)}, \;\; w_i(t) = \frac{(1 - \tau(t))\beta F(t)}{N_i(t)},$$

where $r(t)$ is the rate of interest, $w_i(t)$ is the wage rate and $\tau(t)$ is the tax rate at time t. Let each unemployed people be paid $w_u(t)$ amount of money by the government at time t. Suppose w_u is at least not higher than the wage rate, i.e., $0 < w_u(t) < w_i(t)$. It is assumed that the unemployment payment rate is related to the wage rate as follows

$$w_u(t) = \varpi w_i(t), \;\; 0 < \varpi < 1.$$

Let income per person j be denoted by $y_j(t)$. The current incomes are

$$y_j(t) = r(t)k_j(t) + w_j(t), \;\; j = i, u,$$

in which k_j is the level of capital stocks owned by per person j. Let $s_j(t)$ denote the savings made by per person j at time t. The utility level that a person j obtains is dependent on $c_j(t)$ and $s_j(t)$

$$U_j(t) = A_j c_j(t)^{\xi_j} s_j(t)^{\lambda_j}, \;\; \xi_j, \lambda_j > 0, \;\; j = i, u,$$

in which the parameters, ξ_j and λ_j, are person j's propensities to consume goods and to own wealth, respectively. In $U_j = A_j c_j^{\xi_j} s_j^{\lambda_j}$, A_j is the amenity level of person j. We assume A_i to be constant. Maximizing $U_j(t)$ subject to the budget constraint

$$c_j(t) + s_j(t) = y_j(t) + \delta k_j(t) \equiv \hat{y}_j(t)$$

(where $\delta \equiv 1 - \delta_k$) yields

$$c_j(t) = \xi_j \hat{y}_j(t), \;\; s_j(t) = \lambda_j \hat{y}_j(t).$$

Person j's capital accumulation is given by $\dot{k}_j(t) = s_j(t) - k_j(t)$. Substituting $s_j(t) = \lambda_j(t)\hat{y}_j(t)$ into these equations yields

$$\dot{k}_j(t) = \lambda_j \hat{y}_j(t) - k_j(t), \quad j = i, u . \tag{7.4.1}$$

The government's budget constraint is

$$w_u(t)N_u(t) = \tau(t)F(t) .$$

People stop changing their current situation when their utility levels are the same for either of the two types of life, i.e.

$$U_u(t) = U_i(t), \quad \text{if } N_u(t) > 0 .$$

The output is either consumed or invested, i.e.

$$(c_i + s_i - k_i + \delta_k k_i)N_i + (c_u + s_u - k_u + \delta_k k_u)N_u = F . \tag{7.4.2}$$

By the definitions, we have

$$k_i(t)N_i(t) + k_u(t)N_u(t) = K(t) .$$

We try to find differential equations governing the dynamic system. Substituting $c_j = \xi_j \hat{y}_j$ and $s_j = \lambda_j \hat{y}_j$ into $U_u = U_i$ yields

$$\frac{\hat{y}_i(t)}{\hat{y}_u(t)} = A \equiv \frac{A_u \xi_u^{\xi_u} \lambda_u^{\lambda_u}}{A_i \xi_i^{\xi_i} \lambda_i^{\lambda_i}} .$$

The disposable personal income of the employed is proportional to that of the unemployed with a constant ratio. From $w_u N_u = \tau F$, $w_u = \varpi w_i$, and $w_i = (1 - \tau)\beta F / N_i$, we have

$$\tau(t) = \frac{1}{\omega_0 / u(t) + 1} , \tag{7.4.3}$$

where

$$\omega_0 \equiv \frac{1}{\beta \varpi}, \quad u(t) \equiv \frac{N_u(t)}{N_i(t)} .$$

Substituting $w_u = \varpi w_i$ and $F = K^\alpha (zN_i)^\beta$ into $w_u N_u = \tau F$ yields

$$\varpi w_i N_u = \tau K^\alpha (zN_i)^\beta .$$

From the above equation and $k_i N_i + k_u N_u = K$, we know

$$w_i(t) = (k_i(t) + k_u(t)u(t))^\alpha \frac{z^\beta \tau(t)}{\varpi u(t)} .$$

Inserting Eq. (7.4.3) into the above equation yields

$$w_i(t) = \left(k_i(t) + k_u(t)u(t)\right)^{\alpha} \frac{z^{\beta}}{\left(\omega_0 + u(t)\right)\overline{\omega}}. \tag{7.4.4}$$

The wage rate $w_i(t)$ is a function of $u(t)$, $k_i(t)$ and $k_i(t)$. Inserting

$$\hat{y}_j = y_j + \delta k_j, \; y_j = rk_j + w_j$$

into $\hat{y}_i = A\hat{y}_u$, we get

$$(k_i - Ak_u)r + (1 - \overline{\omega}A)w_i = \delta(Ak_u - k_i), \tag{7.4.5}$$

where we use $w_u = \overline{\omega}w_i$. The marginal conditions lead to

$$\frac{r}{w_i} = \frac{\alpha}{(k_i + k_u u)\beta},$$

where

$$k_i N_i + k_u N_u = K,$$

is applied. Application of the above equation to Eq. (7.4.5) results in

$$\left[\frac{\alpha}{\beta}\frac{k_i - Ak_u}{k_i + k_u u} + (1 - \overline{\omega}A)\right]w_i = \delta(Ak_u - k_i).$$

With the above equation and Eq. (7.4.4), we have

$$\Phi(u) \equiv \Phi_1(u) + \delta(k_i - Ak_u) = 0, \tag{7.4.6}$$

where

$$\Phi_1(u) \equiv \frac{\alpha(k_i - Ak_u) + \beta(1 - \overline{\omega}A)(k_i + k_u u)}{(\omega_0 + u)(k_i + k_u u)^{\beta}}\omega_0 z^{\beta}.$$

We now show that for $k_i(t)$ and $k_u(t)$, equation $\Phi(u) = 0$ has solutions. We note that $u(t)$ is meaningful for $0 < u(t) < +\infty$. If $k_i - Ak_u > 0$ and $1 - \overline{\omega}A > 0$, $\Phi(u) = 0$ has no meaningful solution. These conditions say that the unemployment payment rate is low, the amenity level of unemployment is relatively low (in comparison with the amenity level of work), and the wealth associated with work life style is relatively high. Life with unemployment is so disagreeable that no one wants to be at leisure. The opposite case is $k_i - Ak_u < 0$ and $1 - \overline{\omega}A < 0$. This situation implies that to work is so loathsome that no one wants to have a job. Hence, for the problem to have a meaningful solution, we have to require either of the following two combinations: (i)

$k_i - Ak_u < 0$ and $1 - \overline{\varpi}A > 0$ or (ii) $k_i - Ak_u > 0$ and $1 - \overline{\varpi}A < 0$. Condition (i) says that to work is not to earn a lot of money but to work is relatively pleasant; condition (ii) says that to work is to enjoy a large amount of wealth, but to work is also to lose leisure enjoyed in unemployment. These conditions tell that for co-existence of the two life styles among the workers, no life style provides everything desirable. For convenience of discussion, we are concerned with the case of

$$k_i - Ak_u > 0, \ 1 - \overline{\varpi}A < 0.$$

The other case can be similarly discussed. We have

$$\Phi(0) = [\alpha(k_i - Ak_u) + \beta k_i(1 - \overline{\varpi}A)]k_i^{-\beta} z^{\beta} + \delta(k_i - Ak_u),$$

$$\Phi(+\infty) = \delta(k_i - Ak_u) > 0.$$

The equation has at least one solution in the case of $\Phi(0) < 0$. Since the sign of $d\Phi / du$ is ambiguous, it is possible that the problem has multiple equilibria.

Once we find a solution, $u(t) = u(k_i(t), k_u(t))$, of $\Phi(u) = 0$ (as a function of $k_i(t)$ and $k_i(t)$), we can express all the other variables as functions of $k_i(t)$ and $k_i(t)$. By the definition of u and $N_i + N_u = N$, we solve

$$N_i = \frac{N}{1 + u}, \quad N_u = \frac{uN}{1 + u}.$$

It is straightforward to check that K, F, τ, r, w_j, y_j, $\hat{y}_j$, c_j, s_j, U_j as unique functions of $k_i(t)$ and $k_i(t)$. We thus conclude that the system is governed by the two-dimensional differential equations

$$\dot{k}_j(t) = \lambda_j \hat{y}_j(k_i(t), k_u(t)) - k_j(t), \quad j = i, u. \tag{7.4.7}$$

Since it is difficult to provide a comprehensive analysis of the system, we omit examining it. In particular, it is complicated to analyze the dynamics when $\Phi(u) = 0$ has multiple equilibria. By Eqs. (7.4.1), in equilibrium, we have

$$\lambda_j \hat{y}_j = k_j, \quad j = i, u.$$

Substituting $\lambda_j \hat{y}_j = k_j$ into $c_j = \xi_j \hat{y}_j$ and $s_j = \lambda_j \hat{y}_j$ yields

$$c_j = \frac{\xi_j k_j}{\lambda_j}, \quad s_j = k_j, \quad j = i, u. \tag{7.4.8}$$

Substituting Eqs. (7.4.8) into $U_j = A_j c_j^{\xi_j} s_j^{\lambda_j}$ and then using $U_u = U_i$, we get $k_u = ak_i$ where

$$a \equiv \left(\frac{A_i}{A_u}\right)\left(\frac{\xi_i}{\lambda_i}\right)^{\xi_i}\left(\frac{\lambda_u}{\xi_u}\right)^{\xi_u}.$$

In the case of $\xi_u = \xi_i$, we have $k_u = k_i A_i / A_u$. By $c_j = \xi_j k_j / \lambda_j$, the definitions of y_j and $\hat{y}_j$, we get

$$w_j = (\delta_j - r)k_j, \quad j = i, u,$$

where

$$\delta_j \equiv \frac{\xi_j}{\lambda_j} + \delta_k.$$

By the above equation, $k_u = ak_i$ and $w_u = \varpi w_i$, we have

$$\varpi = \frac{\delta_u - r}{\delta_i - r} a.$$

In the case of $\lambda_i = \lambda_u$, we have $\varpi = A_i / A_u$. It says that if consumers have the same propensities in the two life styles, the unemployment payment rate is equal to the ratio of the levels of amenity-at-work and amenity-in-leisure. As we assume that amenity levels are invariant, this requirement is hardly satisfied. We neglect this case by requiring $\lambda_i \neq \lambda_u$. In general, it is reasonable to assume that A_i and A_u are dependent on wage rates and unemployment policy. For instance, when wage rate becomes higher, it is reasonable to expect that A_i becomes higher (with A_u being kept constant).

By $\varpi = a(\delta_u - r)/(\delta_i - r)$, we solve

$$r = \frac{a\delta_u - \varpi\delta_i}{a - \varpi}. \tag{7.4.9}$$

The following two cases guarantee $r > 0$, $\delta_u > r$, and $\delta_i > r$

Case 1: $a/\varpi > \delta_i / \delta_u > 1$ or
Case 2: $a/\varpi < \delta_i/\delta_u < 1$.

In case 1, we have $\lambda_i < \lambda_u$. That is, the consumer in leisure has a higher level of the propensity to hold wealth than the consumer at work. For convenience of discussion, let $\delta_k = 0$. The condition of $a/\varpi > \delta_i/\delta_u$ in case 1 becomes

$$\frac{A_i}{\varpi A_u} > \frac{(1/\lambda_i - 1)^{\lambda_i}}{(1/\lambda_u - 1)^{\lambda_u}}.$$

We see that the ratio $A_i/\varpi A_u$ has to be large in the case of $\lambda_i < \lambda_u$. Otherwise, the problem has no meaningful solution. In the remainder of this section, we are only concerned with case 1.

Assumption 7.4.1 In the remainder of this section, we assume $a/\varpi > \delta_i/\delta_u > 1$.

Under Assumption 7.4.1, $r > 0$, $\delta_u > r$ and $\delta_i > r$ are guaranteed. We solve r by

$$r = \frac{a\delta_u - \varpi\delta_i}{a - \varpi}.$$

We now solve the other variables. By $w_i = (1-\tau)\beta F/N_i$, $w_u = \varpi w_i$, and $w_u N_u = \tau F$, we get

$$\frac{N_u}{N_i} = \frac{\tau}{(1-\tau)\beta\varpi}.$$

By this equation and $N_i + N_u = N$, we have

$$N_i = \frac{\beta\varpi N}{\beta\varpi + \tau/(1-\tau)}, \quad N_u = \frac{\tau N}{\beta\varpi(1-\tau) + \tau}. \tag{7.4.10}$$

Substituting Eqs. (7.4.8) into Eq. (7.4.2) yields

$$\delta_i K_i + \delta_u K_u = F,$$

where

$$K_j = k_j N_j, \quad j = i, u.$$

By

$$\delta_i K_i + \delta_u K_u = F, \quad F = \frac{rK}{\alpha(1-\tau)}, \quad K_i + K_u = K,$$

we solve

$$K_u = \frac{\delta_i - r/\alpha(1-\tau)}{\bar{\delta}} K,$$

$$K_i = \frac{-\delta_u + r/\alpha(1-\tau)}{\bar{\delta}} K, \tag{7.4.11}$$

where

$$\bar{\delta} \equiv \delta_i - \delta_u = \frac{1}{\lambda_i} - \frac{1}{\lambda_u} > 0.$$

By $F = rK/\alpha(1-\tau)$ and $F = K^{\alpha}(zN_i)^{\beta}$, we have

$$K = zN_i\left\{(1-\tau)\frac{\alpha}{r}\right\}^{1/\beta}.$$

By $w_i = (\delta_i - r)k_i$, $w_i = (1-\tau)\beta F/N_i$ and $F = rK/\alpha(1-\tau)$, we have

$$\beta rK = (\delta_i - r)\alpha K_i\,.$$

Substituting K_i in Eqs. (7.4.11) and Eq. (7.4.9) into this equation yields

$$\frac{1}{1-\tau} = \frac{\beta\bar{\delta}}{(\delta_i - r)} + \frac{\alpha\delta_u}{r} > 0. \tag{7.4.12}$$

By Eqs. (7.4.9) and (7.4.12), we can show that $1/(1-\tau) > 1$ holds if

$$\left(\alpha + \frac{\beta\varpi}{a}\right)\left(\frac{\delta_i}{\delta_u}\right) > 1.$$

Since $\alpha + \beta\varpi/a < 1$ and $\delta_i/\delta_u > 1$ by Assumption 7.4.1, we see that $(\alpha + \beta\varpi/a)(\delta_i/\delta_u) > 1$ may not be guaranteed by Assumption 7.4.1. For $1/(1-\tau) > 1$, it is necessary to require

$$(\alpha + \beta\varpi/a)\delta_i > \delta_u\,.$$

From $(\alpha + \beta\varpi/a)(\delta_i/\delta_u) > 1$ and Assumption 7.4.1, we get

$$\delta_i - \frac{r}{(1-\tau)\alpha} = \left(\alpha + \frac{\beta\varpi}{a} - \frac{\delta_u}{\delta_i}\right)\frac{\delta_i}{a} > 0,$$

$$-\delta_u + \frac{r}{(1-\tau)\alpha} = \frac{-\varpi\delta_i + a\delta_u}{a\alpha}\beta > 0.$$

By the above equations and Eqs. (7.4.11), we have $K_u > 0$ and $K_i > 0$. Summarizing the discussion in this section, we have the following proposition.

Proposition 7.4.1 If Assumption 7.4.1 and

$$\left(\alpha + \frac{\beta\varpi}{a}\right)\delta_i > \delta_u$$

are satisfied, then the dynamic system has a unique equilibrium. We solve the equilibrium values of the variables by the following procedure: r by Eq. (7.4.9) $\rightarrow \tau$ by Eq. (7.4.12) $\rightarrow N_j$, $j = i, u$, by Eqs. (7.4.10) $\rightarrow K = zN_i\{\alpha(1-\tau)/r\}^{1/\beta} \rightarrow F = K^{\alpha}(zN_i)^{\beta} \rightarrow w_i = (1-\tau)\beta F / N_i \rightarrow w_u = \varpi w_i \rightarrow K_j$ by Eqs. (7.4.11) $\rightarrow k_j = K_j / N_j \rightarrow c_j$ and s_j by Eqs. (7.4.8) $\rightarrow y_j = c_j + s_j \rightarrow U_j = A_j c_j^{\xi_j} s_j^{\lambda_j}$.

We specify the parameter values as

$$\alpha = 0.3, \quad A \equiv \frac{A_i}{A_u} = 1.1, \quad \lambda_i = 0.6, \quad \lambda_u = 0.75, \quad N = 1, \quad z = 12. \tag{7.4.13}$$

The population is unity (equaling 0.1 billion) and the level of human capital is fixed at 12. The consumers, who are unemployed, have a higher propensity to own wealth higher than these consumers, who are employed. The ratio of the amenities is larger than unity. Figure 7.4.1 demonstrates how the economy is affected by the unemployment policy under (7.4.13). The unemployment payment rate is between 40 percent to 65 percent. It can be shown that when the parameter lies beyond

$$0.4 \le \varpi \le 0.65,$$

the system does not have a meaningful solution under (7.4.13). As the figure demonstrates, the unemployment rate rises as the unemployment policy is 'improved'. In association with the rise in unemployment rate, the tax rate is also increased. As ϖ rises, the national income, the total wealth, and the total consumption fall. The wage rate falls, and the rate of interest decreases slightly. The wealth, income, and consumption per capital all fall for the workers and the unemployed.

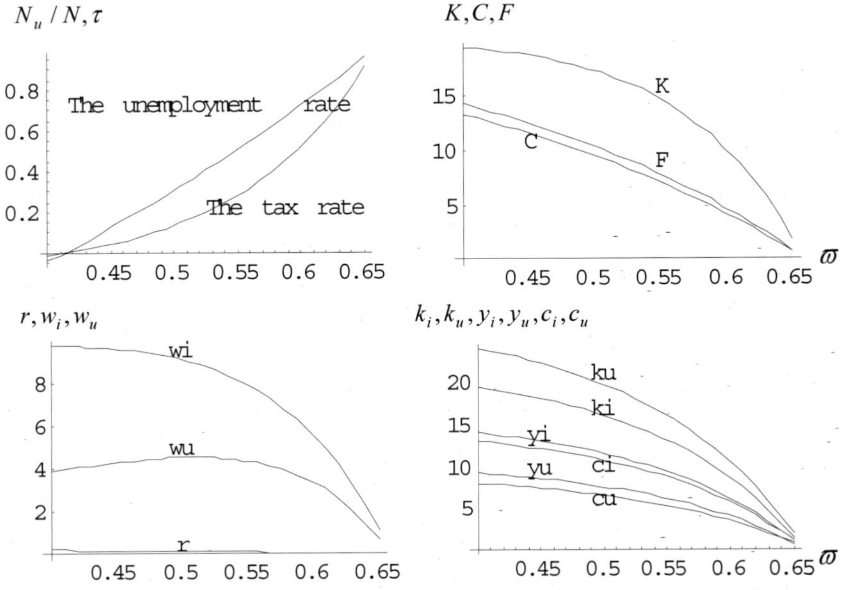

Fig.7.4.1 The equilibrium values and the unemployment policy, $0.4 \le \varpi \le 0.65$.

We now fix $\varpi = 0.45$ and still accept (7.4.13), except for the value of A. We consider that A change from 0.76 to 1.2. The simulation results are shown in Fig. 7.4.2. As the job amenity is improved - such as working conditions and social respect for work are changed, the unemployment rate and the tax rate decline. The national product, the total wealth, and the total consumption are increased. It should be noted that the change rate is reasonably low if we limit the discussion to the domain $1.0 \le A \le 1.2$. The national output, the total capital, and the total capital are increased. The rate of interest slightly rises, and the wage rates increase.

We now examine effects of change in the workers' propensity to save. We fix $\varpi = 0.45$ and still accept (7.4.13), except for the value of λ_i. We assume λ_i to change from 0.39 to 6.1. The simulation results are shown in Fig. 7.4.3. As the propensity to own wealth rises, the unemployment rate and tax rate fall. The national output, the total capital, and the total capital are increased. The rate of interest slightly rises, and the wage rates increase.

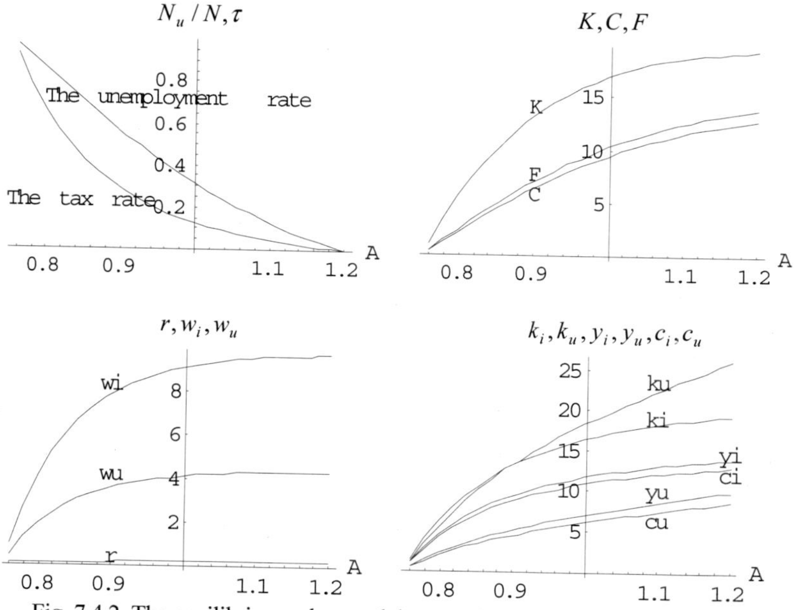

Fig. 7.4.2 The equilibrium values and the amenities, $0.76 \leq A \leq 1.2$.

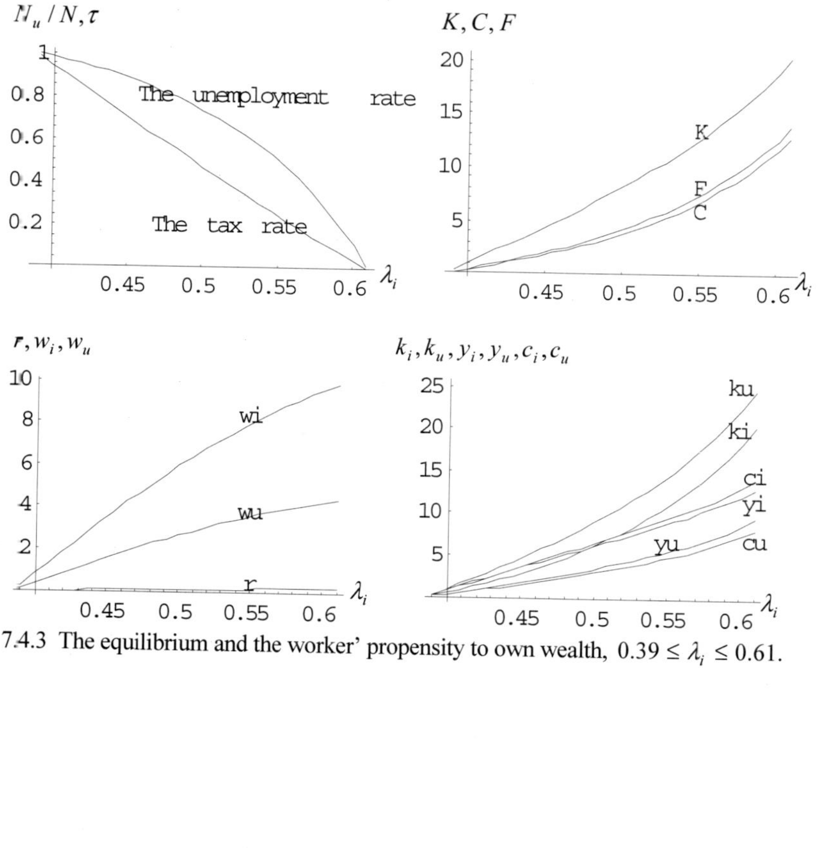

Fig. 7.4.3 The equilibrium and the worker' propensity to own wealth, $0.39 \leq \lambda_i \leq 0.61$.

7.5 Regional Growth with Endogenous Time Distribution

We consider an economic system which consists of two regions, indexed by 1 and 2, respectively.[8] The two regions' product is qualitatively homogeneous and is either consumed or invested. We assume a homogeneous population. A person is free to choose where he works and where he lives. We assume that any person chooses the same region where he works and lives. Each region has fixed land. It is assumed that land quality, climates and environment are homogenous within each region, but climates and environment may be different between the two regions. The land is used only for housing. We select commodity to serve as numeraire, with all the other prices being measured relative to its price.

N = the fixed population of the economy;
L_j = the fixed territory size of region j, $j = 1, 2$;
$K(t)$ = total capital stocks of the economy at time t;
$F_j(t)$ = output levels of region j's industrial sector at time t;
$K_j(t)$ = capital stocks employed by region j's production sector;
$N_j(t)$ = labor force employed by region j's production sector;
$c_j(t)$ = per capita's consumption level of commodity in region j;
$s_j(t)$ = net savings of per capita in region j;
$l_j(t)$ = lot size per capita in region j;
$T_j(t)$ = working time in region j;
$T_j^h(t)$ = leisure time in region j;
$y_j(t)$ = net income per capita in region j;
$r(t)$ = rate of interest;
$w_j(t)$ = region j's wage rate per unity of working time;
$R_j(t)$ = region j's land rent.

We specify the production functions of the two regions as follows

$$F_j(t) = K_j^{\alpha}\left(z_j T_j N_j\right)^{\beta}, \quad \alpha, \beta > 0, \quad \alpha + \beta = 1, \quad j = 1, 2, \qquad (7.5.1)$$

[8] This section is based on Chap. 3 in Zhang (2003).

in which z_j are working efficiency index of region j's labor force and $z_j T_j N_j$ is region j's total efficient labor force. We require z_j to be constant and $z_1 > z_2$. The marginal conditions are given by

$$r = \frac{\alpha F_j}{K_j}, \quad w_j = \frac{\beta F_j}{T_j N_j}, \quad j = 1, 2. \qquad (7.5.2)$$

Suppose that each worker owns L_1 / N amount of land in region 1 and L_2 / N in region 2 and it is impossible to sell land but it is free to rent one's own land to others. The land revenue, R_0, per worker is given by

$$R_0 = \frac{L_1 R_1 + L_2 R_2}{N}. \qquad (7.5.3)$$

If we denote $k_j(t)$ capital stocks owned by per capita in region j, the interest payment per capita is given by $r(t)k_j(t)$. Under the specified land ownership, the current income per capita, $y_j(t)$, in region j consists of the wage income, $T_j(t)w_j(t)$, land revenue, R_0, and interest payment $r(t)k_j(t)$, i.e.

$$y_j = T_j w_j + r k_j + R_0. \qquad (7.5.4)$$

A typical person's utility level, $U_j(t)$, in region j is dependent on the leisure time, $T_j^h(t)$, lot size, $l_j(t)$, consumption level, $c_j(t)$, of community, and the net savings, $s_j(t)$ in the following way

$$U_j = A_j T_{hj}^{\sigma} l_j^{\eta} c_j^{\xi} s_j^{\lambda}, \quad \sigma, \eta, \xi, \lambda > 0, \quad \sigma + \eta + \xi + \lambda = 1, \qquad (7.5.5)$$

in which σ, η, ξ and λ are respectively the household's propensity to use leisure time, to utilize lot size, to consume the commodity, and to hold wealth, and A_j is region j's amenity level. Let T_0 denote the total available time. The time constraint requires that the amounts of time allocated to each specific use add up to the time available

$$T_j + T_j^h = T_0, \quad j = 1, 2. \qquad (7.5.6)$$

The budget constraints are given by

$$l_j R_j + c_j + s_j = \hat{y}_j, \quad j = 1, 2, \qquad (7.5.7)$$

where the disposable income

$$\hat{y}_j = y_j + k_j - \delta_k k_j,$$

where δ_k is the fixed depreciation rate of physical capital. By Eqs. (7.5.4) and (7.5.6), we rewrite Eqs. (7.5.7) as follows

$$l_j R_j + c_j + s_j + T_j w_j = R_0 + T_j^h w_j + rk_j + (1 - \delta_k)k_j, \quad j = 1, 2. \tag{7.5.8}$$

A typical person maximizes the person's utility subject to the budget constrain. The optimal solution is

$$T_j^h w_j = \sigma\Omega_j, \quad l_j R_j = \eta\Omega_j, \quad c_j = \xi\Omega_j, \quad s_j = \lambda\Omega_j,$$
$$j = 1, 2, \tag{7.5.9}$$

where

$$\Omega_j \equiv R_0 + T_0 w_j + rk_j + \delta k_j, \quad \delta \equiv 1 - \delta_k.$$

According to the definitions of k_j and s_j, the capital accumulation of a typical person in region j is given by

$$\dot{k}_j = s_j - k_j.$$

Substituting s_j in Eqs. (7.5.9) into these two equations yields

$$\dot{k}_j = \lambda\Omega_j - k_j, \quad j = 1, 2. \tag{7.5.10}$$

As households are freely mobile between the two regions, they should have the same level of the utility, irrespective of where they live. That is

$$U_1(t) = U_2(t). \tag{7.5.11}$$

By the definitions of K, k_j and N_j

$$K = k_1 N_1 + k_2 N_2. \tag{7.5.12}$$

The assumption that labor force, capital stocks and land are fully employed is represented by

$$N_1 + N_2 = N, \quad K_1 + K_2 = K, \quad l_j N_j = L_j, \quad j = 1, 2. \tag{7.5.13}$$

We have thus built the model. The system has 29 variables, N_j, K_j, F_j, c_j, s_j, k_j, l_j, T_j, T_j^h, y_j, U_j, w_j, R_j ($j = 1, 2$), K, r and R_0. We now examine conditions for the existence of equilibria of the dynamic system.

We are now concerned with the conditions for existence of economic equilibria. By Eqs. (7.5.10), at equilibrium we have

$$\lambda \Omega_j = k_j, \quad j = 1, 2. \tag{7.5.14}$$

Substituting Eqs. (7.5.14) into Eqs. (7.5.9) yields

$$T_j^h w_j = \frac{\sigma k_j}{\lambda}, \quad l_j R_j = \frac{\eta k_j}{\lambda}, \quad c_j = \frac{\xi k_j}{\lambda}, \quad s_j = \delta_k k_j,$$

$$j = 1, 2. \tag{7.5.15}$$

Substituting Eqs. (7.5.15) into Eqs. (3.1.14) yields

$$\delta_1 K = F_1 + F_2, \tag{7.5.16}$$

where we use Eq. (7.5.12) and

$$\delta_1 \equiv \frac{\xi}{\lambda} + \delta_k.$$

By $r = \alpha F_1 / K_1 = \alpha F_2 / K_2$ in Eqs. (7.5.2) and Eq. (7.5.16), we have

$$r = \alpha \delta_1, \tag{7.5.17}$$

where we use

$$K_1 + K_2 = K.$$

By Eqs. (7.5.1), $r = \alpha F_1 / K_1 = \alpha F_2 / K_2$ in Eqs. (7.5.2) and Eq. (7.5.17), we get

$$z_j T_j N_j = \delta_1^{1/\beta} K_j, \quad j = 1, 2. \tag{7.5.18}$$

By Eqs. (7.5.18) and Eqs. (7.5.2)

$$w_j = \frac{\beta z_j}{\delta_1^{\alpha/\beta}}, \quad j = 1, 2. \tag{7.5.19}$$

Substituting $l_j = L_j / N_j$ and Eqs. (7.5.15) into (7.5.5) and then using Eq. (7.5.11), we obtain

$$\frac{k_1}{k_2} = A \Lambda^{\eta \upsilon}, \tag{7.5.20}$$

where we use $w_1 / w_2 = z$ and

$$\Lambda \equiv \frac{N_1}{N_2}, \quad z \equiv \frac{z_1}{z_2}, \quad \upsilon \equiv \frac{1}{1 - \eta}, \quad A \equiv z^{\sigma \upsilon} \left(\frac{A_2}{A_1} \right)^{\upsilon} \left(\frac{L_2}{L_1} \right)^{\upsilon \eta}.$$

By the definition of Λ and $N_1 + N_2 = N$, we have

$$N_1 = \frac{\Lambda N}{1 + \Lambda}, \quad N_2 = \frac{N}{1 + \Lambda}. \tag{7.5.21}$$

By $l_j R_j = \eta k_j / \lambda$ in Eqs. (7.5.15), $l_j = L_j / N_j$ and Eq. (7.5.3), we get

$$R_0 = \frac{\eta K}{\lambda N}, \tag{7.5.22}$$

where we use (7.5.12). By the definitions of Ω_j and Eqs. (7.5.14), we get

$$R_0 + T_0 w_j = \delta_2 k_j, \quad j = 1, 2, \tag{7.5.23}$$

where

$$\delta_2 \equiv \frac{1}{\lambda} - \delta - r = \frac{\sigma + \eta + \beta\xi}{\lambda} + \beta\delta_k > 0.$$

By Eqs. (7.5.23), we have

$$k_1 - k_2 = w_0 \equiv (z - 1)\frac{T_0 w_2}{\delta_2} > 0. \tag{7.5.24}$$

Region 1's wage rate is higher than region 2's wage rate, i.e., $w_1 > w_2$ and the level of capital stocks owned by per capita in region 1 is higher than that in region 2, i.e., $k_1 > k_2$. By Eqs. (7.5.20) and (7.5.24), we solve

$$k_1 = \frac{w_0 A\Lambda^{\eta\upsilon}}{A\Lambda^{\eta\upsilon} - 1}, \quad k_2 = \frac{w_0}{A\Lambda^{\eta\upsilon} - 1}. \tag{7.5.25}$$

It is necessary to require $A\Lambda^{\eta\upsilon} > 1$. Substituting Eq. (7.5.22) and k_2 in Eqs. (7.5.25) into

$$R_0 + T_0 w_2 = \delta_2 k_2,$$

in Eqs. (7.5.23), we have

$$K = \frac{\lambda N T_0 w_2}{\eta(A\Lambda^{\eta\upsilon} - 1)}(z - A\Lambda^{\eta\upsilon}). \tag{7.5.26}$$

Since $A\Lambda^{\eta\upsilon} > 1$, for $K > 0$ it is necessary to require:

$$z > A\Lambda^{\eta\upsilon}.$$

Since $k_j > 0$ are guaranteed by $A\Lambda^{\eta\upsilon} > 1$, by $T_j^h w_j = \sigma k_j / \lambda$ in Eqs. (7.5.15), we see that $T_0 > T_j^h > 0$ are satisfied if

$$T_0 > \frac{\sigma k_j}{\lambda w_j}, \quad j = 1, 2.$$

By Eqs. (7.5.25), we see that these two inequalities are satisfied if

$$A\Lambda^{\eta\upsilon} > \frac{\lambda\delta_2 z}{\lambda\delta_2 z - \sigma(z-1)}, \quad A\Lambda^{\eta\upsilon} > \frac{\sigma}{\lambda\delta_2}(z-1)+1.$$

Since

$$z > \frac{\sigma}{\lambda\delta_2}(z-1)+1 > \frac{\lambda\delta_2 z}{\lambda\delta_2 z - \sigma(z-1)} > 1,$$

we should require Λ to satisfy

$$z > A\Lambda^{\eta\upsilon} > \frac{\sigma}{\lambda\delta_2}(z-1)+1. \tag{7.5.27}$$

Under (7.5.27), we have $K > 0$, $k_j > 0$ and $T_0 > T_j^h > 0$.

Substituting Eqs. (7.5.21) and (7.5.25) into Eq. (7.5.12) and then using Eq. (7.5.26), we get

$$\Phi(\Lambda) \equiv \left(A\Lambda^{\eta} + 1\right)(z-1) - \left(z - A\Lambda^{\eta\upsilon}\right)(1+\Lambda)\frac{\lambda\delta_2}{\eta}. \tag{7.5.28}$$

It is shown that the function, $\Phi(\Lambda)$, has the following properties

$$\Phi = \frac{\beta\xi + \beta\lambda\delta_k - (z-1)\sigma\eta/\lambda\delta_2}{\eta}(1+\Lambda)(z-1) - \frac{\sigma}{\lambda\delta_2}(z-1)^2,$$

$$at \ A\Lambda^{\eta\upsilon} = \frac{\sigma}{\lambda\delta_2}(z-1)+1,$$

$$\Phi = (1+z\Lambda)(z-1) \ \ at \ A\Lambda^{\eta\upsilon} = z,$$

$$\Phi^* \equiv \frac{d\Phi}{d\Lambda} = \upsilon A\Lambda^{\eta\upsilon}(z-1) + \upsilon\lambda\delta_2 A\Lambda^{\eta\upsilon-1}(1+\Lambda) - \frac{z\lambda\delta_2}{\eta}$$

$$+ \frac{\lambda\delta_2 A\Lambda^{\eta\upsilon}}{\eta} > \left\{\frac{\sigma\upsilon}{\lambda\delta_2}(z-1) + \upsilon + \frac{\sigma\upsilon}{\eta} - \frac{\lambda\delta_2}{\eta}\right\}(z-1) + \upsilon\lambda\delta_2,$$

$$\text{for } z > A\Lambda^{\eta\upsilon} > \frac{\sigma}{\lambda\delta_2}(z-1)+1. \tag{7.5.29}$$

Since $\Phi > 0$ at $A\Lambda^{\eta\upsilon} = z$, we see that the equation, $\Phi(\Lambda) = 0$, has at least one solution satisfying (7.5.27) if $\Phi < 0$ at

$$A\Lambda^{\eta\upsilon} = (z-1)\sigma / \lambda\delta_2 + 1 .$$

By the first equation in (7.5.29), we see that $\Phi < 0$ at

$$A\Lambda^{\eta\upsilon} = (z-1)\sigma / \lambda\delta_2 + 1 ,$$

is held if

$$\xi + \lambda\delta_k > (z-1)\frac{\sigma\eta}{\lambda\beta\delta_2} .$$

As

$$\sigma\eta / \lambda\delta_2 < \min\{\sigma, \eta\},$$

we see that if $z-1$ is not large, the requirement is satisfied. If $\Phi^* > 0$ under (7.5.27), the equation has a unique solution in the interval. By (7.5.29), we see that it is acceptable to require $\Phi^* > 0$. For instance, if

$$\eta + (z-1)\left(\frac{\sigma+\eta}{\lambda\delta_2}\right) > \frac{z-1}{\upsilon} ,$$

then $\Phi^* > 0$ under (7.5.27) is held.

Proposition 7.5.1 We assume

$$\xi + \lambda\delta_k > (z-1)\frac{\sigma\eta}{\lambda\beta\delta_2}, \quad \eta + \frac{\sigma+\eta}{\lambda\delta_2}(z-1) > \frac{z-1}{\upsilon} .$$

The dynamic system has a unique equilibrium. The unique equilibrium values of the variables are given by the following procedure: r by Eq. (7.5.17) $\rightarrow$ w_j, $j = 1, 2$, by Eqs. (7.5.19) $\rightarrow$ R_0 by Eq. (7.5.28) $\rightarrow$ N_j by Eqs. (7.5.21) $\rightarrow$ k_j by Eqs. (7.5.25) $\rightarrow$ K by Eq. (7.5.26) $\rightarrow$ T_j^h by Eq. (7.5.22) $\rightarrow$ T_j, $0 < T_j^h < T_0$, by Eqs. (7.5.15) $\rightarrow$ $T_j = T_0 - T_j^h$ $\rightarrow$ K_j by Eqs. (7.5.18) $\rightarrow$ F_j by Eqs. (7.5.1) $\rightarrow$ $l_j = L_j / N_j \rightarrow$ R_j, c_j and s_j by Eqs. (7.5.15) $\rightarrow$ y_j by Eqs. (7.5.4) $\rightarrow$ U_j by (7.5.5).

The assumptions in Proposition 7.5.1 are satisfied if the difference between the two regions' levels of working efficiency is appropriately small. It should be remarked that the dynamic system might have a unique equilibrium even if the requirements in Proposition 7.5.1 are not satisfied. In the remainder of this chapter, we examine effects of changes in some parameters. We accept the assumptions in Proposition 7.5.1 in the remainder of the chapter.

We now examine the effects of changes in region 1 's working efficiency, z_1, on the economic structure. Taking derivatives of Eq. (7.5.28) with respect to z_1 yields

$$z_1\Phi^* \frac{d\Lambda}{dz_1} = (1 - z)\sigma\upsilon A\Lambda^{\upsilon} - zA\Lambda^{\upsilon} + \frac{\sigma + \beta\xi + \beta\lambda\delta_k}{\eta} z + \frac{z\lambda\delta_2\Lambda}{\eta}$$

$$- \frac{\sigma\lambda\delta_2\upsilon A\Lambda^{\eta\upsilon}}{\eta} - \frac{\sigma\lambda\delta_2\upsilon A\Lambda^{\upsilon}}{\eta} > \frac{\sigma + \beta\xi + \beta\lambda\delta_k - \sigma\lambda\delta_2\upsilon}{\eta} z$$

$$+ \left\{(1 - \sigma\upsilon)\frac{z\lambda\delta_2}{\eta} + \sigma\upsilon - \sigma\upsilon z^2 - z^2\right\}\Lambda, \qquad (7.5.30)$$

in which we use (7.5.27) to get the right-hand side of (7.5.30) and $\Phi^* > 0$ under Proposition 7.5.1. We have

$$\frac{\sigma + \beta\xi + \beta\lambda\delta_k - \sigma\lambda\delta_2\upsilon}{\upsilon} =$$

$$(1 - \sigma - 2\eta)\sigma + (1 - \sigma - \eta)\beta\lambda\delta_k + (1 - \sigma - \eta)\beta\xi .$$

As $\sigma + \eta + \xi + \lambda = 1$, the above term is positive if $\xi + \lambda > \eta$. By (7.5.30), we see that if the term

$$(1 - \sigma\upsilon)\frac{z\lambda\delta_2}{\eta} + \sigma\upsilon - \sigma\upsilon z^2 - z^2 > 0,$$

then $d\Lambda / dz_1 > 0$. As

$$(1 - \sigma\upsilon)\frac{z\lambda\delta_2}{\eta} + \sigma\upsilon - \sigma\upsilon z^2 - z^2$$

$$= \left\{(\xi + \lambda)\frac{\upsilon\lambda\delta_2}{\eta} - z\right\}z - \sigma\upsilon(1 + z)(z - 1),$$

we see that the term is positive if $z - 1$ and η are sufficiently small. Otherwise, it is difficult to judge the sign of the term. Summarizing the above discussion, we conclude that if $z - 1$ and η are sufficiently small, it is reasonable to have $d\Lambda / dz_1 > 0$. In the remainder of this section, we assume $d\Lambda / dz_1 > 0$. We have

$$\frac{1}{N_1}\frac{dN_1}{dz_1} = \frac{1}{(1 + \Lambda)\Lambda}\frac{d\Lambda}{dz_1} > 0,$$

$$\frac{1}{N_2}\frac{dN_2}{dz_1} = -\frac{1}{1 + \Lambda}\frac{d\Lambda}{dz_1} < 0.$$

If the regional working efficiency difference, $z - 1$ and the propensity η to consume housing are appropriately small, some of region 2 's population will migrate to region 1 as region 1's working efficiency is improved.

By Eqs. (7.5.7) and (7.5.15),

$$y_j = \left(\frac{\eta}{\lambda} + \frac{\eta}{\lambda} + \delta_k\right)k_j .$$

By Eqs. (7.5.25) and these two equations, we have

$$\frac{1}{y_1}\frac{dy_1}{dz_1} = \frac{1}{k_1}\frac{dk_1}{dz_1} = \frac{1}{(z - 1)z_2} - \frac{\sigma\upsilon}{(A\Lambda^{\eta\upsilon} - 1)z_1} - \frac{\eta\upsilon}{(A\Lambda^{\eta\upsilon} - 1)\Lambda}\frac{d\Lambda}{dz_1},$$

$$\frac{1}{y_2}\frac{dy_2}{dz_1} = \frac{1}{k_2}\frac{dk_2}{dz_1} = \frac{1}{(z - 1)z_2} - \frac{\sigma\upsilon A\Lambda^{\eta\upsilon}}{(A\Lambda^{\eta\upsilon} - 1)z_1} - \frac{\eta\upsilon A\Lambda^{\eta\upsilon-1}}{(A\Lambda^{\eta\upsilon} - 1)\Lambda}\frac{d\Lambda}{dz_1}.$$

It is difficult to explicitly judge the signs of dk_j / dz_1. By Eq. (7.5.17) and Eqs. (7.5.19), we have

$$\frac{dr}{dz_1} = \frac{dw_2}{dz_1} = 0, \quad \frac{dw_1}{dz_1} = \frac{w_1}{z_1} > 0.$$

By Eq. (7.5.22) and Eqs. (7.5.23), we get

$$\frac{1}{R_0}\frac{dR_0}{dz_1} = \frac{1}{K}\frac{dK}{dz_1} = \frac{\lambda\delta_2 N}{\eta K}\frac{dk_2}{dz_1}.$$

By $l_j = L_j / N_j$

$$\frac{1}{l_1}\frac{dl_1}{dz_1} = -\frac{1}{N_1}\frac{dN_1}{dz_1} < 0,$$

$$\frac{1}{l_2}\frac{dl_2}{dz_1} = -\frac{1}{N_2}\frac{dN_2}{dz_1} > 0.$$

By Eqs. (7.5.15) and $T_j + T_j^h = T_0$, we get

$$\frac{1}{c_j}\frac{dc_j}{dz_1} = \frac{1}{k_j}\frac{dk_j}{dz_1}, \quad \frac{dT_j}{dz_1} = -\frac{1}{k_j}\frac{dT_j^h}{dz_1}, \quad j = 1, 2,$$

$$\frac{1}{T_1^h}\frac{dT_1^h}{dz_1} = \frac{1}{k_1}\frac{dk_1}{dz_1} - \frac{1}{z_1}, \quad \frac{1}{T_2^h}\frac{dT_2^h}{dz_1} = \frac{1}{k_2}\frac{dk_2}{dz_1}.$$

By Eqs. (7.5.1), (7.5.18), $R_j = \eta K_j / \lambda L_j$ and $K_j = k_j N_j$, we have

$$\frac{1}{R_j}\frac{dR_j}{dz_1} = \frac{1}{F_j}\frac{dF_j}{dz_1} = \frac{1}{K_j}\frac{dK_j}{dz_1} = \frac{1}{k_j}\frac{dk_j}{dz_1} + \frac{1}{N_j}\frac{dN_j}{dz_1}, \quad j = 1, 2.$$

7.6 Growth with International Trade and Urban Pattern Formation

This section proposes a simple two-country and one-commodity trade growth model with free capital mobility and urban pattern formation to provide some insights into relationships between commodity prices, factor prices, land values, production, preferences and trade volumes.[9] The growth aspects of our model is based on the international macroeconomic one-sector growth model with perfect capital mobility.[10]

The system consists of two countries, indexed by j, $j = 1, 2$ and only one good is produced in the system. The good is assumed to be composed of homogeneous qualities, and to be produced by employing two factors of production - labor and capital. Perfect competition is assumed to prevail in goods markets both within each country and between the countries, and commodities are traded without any barriers such as transport costs or tariffs. The households achieve the same utility

[9] The model is based Chap. 10 in Zhang (2002).

[10] See Oniki and Uzawa (1965), Rodriguez (1975), Frenkel and Razin (1987), Ruffin (1979), Buiter (1981), Wang (1990), Ikeda and Ono (1992), Devereux and Shi (1991), Turnovsky (1997), and Rauch (1991).

level regardless of where they locate.[11] The industrial production is similar to that in the OGM. To describe the industrial sector, we introduce

N_j = the fixed population of country j, $j = 1, 2$;

$K_{ij}(t)$ = the capital stocks employed by country j's production sector at time t;

$w_j(t)$ = country j's wage rate;

$r(t)$ = rate of interest;

$F_j(t)$ = output of country j's production sector;

$C_j(t)$ = country j's net consumption level;

$K_j(t)$ = capital stocks owned by country j;

$E(t) > (<)\ 0$ = country 2's (1's) capital stocks utilized by country 1 (2); and

$K(t)$ = the world capital stock.

We assume that production is carried out by combination of capital and labor force in the form of

$$F_1 = zK_{i1}^{\alpha} N_1^{\beta}, \; F_2 = K_{i2}^{\alpha} N_2^{\beta}, \; \alpha + \beta = 1, \; \alpha, \beta > 0, \tag{7.6.1}$$

where α and β are parameters. Here, we call z the efficiency parameter of economic production. The parameter may be simply interpreted as a measurement of working efficiency difference between the two countries. If $z > 1$, we say that country 1's workers work more effectively than country 2's workers. The marginal conditions are given by

$$w_j = \frac{\beta F_j}{N_j}, \; r = \frac{\alpha F_j}{K_{ij}}, \; j = 1, 2. \tag{7.6.2}$$

The world capital is equal to the sum of the capital stocks owned by the two countries, i.e. $K_1 + K_2 = K$. According to the definitions of K_{ij}, K_j and E, we have

---

[11] The spatial aspects of this model are referred to the urban model in Chap. 4 of this book.

$$K_{i1} = K_1 + E, \;\; K_{i2} = K_2 - E. \tag{7.6.3}$$

The above equations state that the capital stocks utilized by each country is equal to the capital stocks owned by the country plus the foreign capital stocks.

We now describe housing production and behavior of households. First, we introduce

L_j = the fixed (territory) length of country j, $j = 1, 2$;
ω_j = distance from the CBD to a point in the residential area in country j;
$R_j(\omega_j, t)$ = land rent per household at location ω_j;
$k_j(\omega_j, t)$ = capital stocks owned by the household at location ω_j;
$c_j(\omega_j, t)$ = consumption level of the household at location ω_j;
$y_j(\omega_j, t)$ = net income of the household at location ω_j;
$n_j(\omega_j, t)$ = residential density of the household at location ω_j;
$L_{hj}(\omega_j, t)$ = lot size of the household at location ω_j.

According to the definitions of L_{hj} and n_j, we have

$$n_j(\omega_j, t) = \frac{1}{L_{hj}(\omega_j, t)}, \;\; 0 \le \omega_j \le L_j, \;\; j = 1, 2. \tag{7.6.4}$$

We assume the public land ownership. The total land revenue is given by

$$\overline{R}_j(t) = \int_0^L R_j(\omega_j, t) d\omega_j, \;\; j = 1, 2.$$

The income from land per household is given by

$$\bar{r}_j(t) = \frac{\overline{R}_j(t)}{N_j}, \; j = 1, 2.$$

The current income is given by

$$y_j(\omega_j, t) = r(t) k_j(\omega_j, t) + w_j(t) + \bar{r}_j(t), \;\; j = 1, 2. \tag{7.6.5}$$

As in Sec. 4.4, we specify $A_j(\omega_j, t)$ and $T_{hj}(\omega_j, t)$ as

$$A_j(\omega_j, t) = \frac{\mu_{1j}}{n_j^{\mu_j}(\omega_j, t)}, \quad T_{hj}(\omega_j) = T_{0j} - \upsilon_j \omega_j,$$

$$\mu_{1j}, \mu_j, T_{0j}, \upsilon_j > 0. \tag{7.6.6}$$

The utility level $U_j(\omega_j, t)$ of the household at location ω_j is given by

$$U_j(\omega_j, t) = A_j T_{hj}^{\sigma_j} c_j^{\xi_j} L_{hj}^{\eta_j} S_j^{\lambda_j}, \quad \sigma_j, \xi_j, \eta_j, \lambda_j, \quad j = 1, 2, \tag{7.6.7}$$

where $S_j(\omega_j, t)$ is the savings of the household at ω_j, σ_j, ξ_j, η_j and λ_j are respectively interpreted as country j's propensities to use leisure time, to consume the commodity, to use lot size and to own wealth. As the population is homogeneous within each country, we have

$$U_j(\omega_{j1}, t) = U_j(\omega_{j2}, t), \quad 0 \le \omega_{j1}, \omega_{j2} \le L_j.$$

The budget constraint is given by

$$c_j(\omega_j, t) + R_j(\omega_j, t) L_{hj}(\omega_j, t) + S_j(\omega_j, t) = \hat{y}_j(\omega_j, t),$$

where

$$\hat{y}_j(\omega_j, t) = y_j(\omega_j, t) + (1 - \delta_k) k_j(\omega_j, t).$$

Maximizing U_j subject to the budget constraint yields

$$c_j(\omega_j) = \xi_j \rho_j \hat{y}_j(\omega_j), \quad R_j(\omega_j) L_{hj}(\omega_j) = \eta_j \rho_j \hat{y}_j(\omega_j),$$

$$S_j(\omega_j) = \lambda_j \rho_j \hat{y}_j(\omega_j), \tag{7.6.8}$$

where

$$\rho_j = \frac{1}{\xi_j + \eta_j + \lambda_j}.$$

According to the definition of $S_j(\omega_j, t)$, we have the following capital accumulation for the household at location ω_j

$$\dot{k}_j(\omega_j) = S_j(\omega_j) - k_j(\omega_j), \quad 0 \le \omega_j \le L_j.$$

Substituting $S_j(\omega_j)$ in Eqs. (7.6.8) into the above equation yields

$$\dot{k}_j(\omega_j) = s_j y_j(\omega_j) - \delta_j k_j(\omega_j), \quad 0 \le \omega_j \le L_j, \tag{7.6.9}$$

where

$$s_j \equiv \lambda_j \rho_j, \ \ \delta_j = \delta_0 + (\xi_j + \eta_j)\rho_{0j}, \ \rho_{0j} \equiv (1 - \delta_k)\rho_j.$$

As there is no migration, the following population constraints are held

$$\int_0^{L_j} n_j(\omega_j) d\omega_j = N_j, \ \ j = 1, 2. \tag{7.6.10}$$

By the definitions of n_j, c_j and C_j, we have

$$\int_0^{L_j} c_j(\omega_j) n_j(\omega_j) d\omega_j = C_j, \ \ j = 1, 2.$$

The product is either invested or consumed. That is

$$F_1 + F_2 = C_1 + \overline{S}_1 + C_2 + \overline{S}_2,$$

where $\overline{S}_j(t)$ is given by

$$\overline{S}_j \equiv \int_0^{L_j} [s_j(\omega_j) - (1 - \delta_k) k_j(\omega_j)] n_j(\omega_j) d\omega_j, \ \ j = 1, 2.$$

By the definitions of n_j, k_j and K_j, we have

$$\int_0^{L_j} k_j(\omega_j) n_j(\omega_j) d\omega_j = K_j, \ \ j = 1, 2.$$

The total income $Y_j(t)$ of country j is equal to the sum of incomes of its population, i.e.

$$Y_j \equiv \int_0^{L_j} y_j(\omega_j) n_j(\omega_j) d\omega_j.$$

We have thus built the two-country trade model with economic growth and economic geography under perfectly competitive institution. The system has 18 space-time-dependent variables, k_j, c_j, L_{hj}, S_j, n_j, A_j, U_j, R_j, y_j ($j = 1, 2$), and 21 time-dependent variables, F_j, K_{ij}, K_j, Y_j, C_j, w_j, $\overline{S}_j$, $\overline{R}_j$, $\overline{r}_j$ ($j = 1, 2$), r, K, and E_j. It contains 39 independent equations. We now examine the properties of the dynamic system.

We now examine dynamic properties of the system. First, we show that the dynamics can be described by the motion of two variables, K_1 and K_2. By Eqs. (7.6.5), (7.6.8) and (7.6.9) we see that the capital stocks owned by per household and the net income are identical over space within each country at any point of time. Hence, we have

$$K_j = k_j N_j, \ Y_j = y_j N_j, \ Y_j = rK_j + w_j N_j + \bar{R}_j. \qquad (7.6.11)$$

We rewrite the dynamics, Eqs. (7.6.9), in aggregated terms as follows

$$\dot{K}_j = s_j Y_j - \delta_j K_j, \ j = 1, 2. \qquad (7.6.12)$$

Our problem is to show that we can express $Y_j(t)$ as functions of K_1 and K_2.

Multiplying all the equations in Eqs. (7.6.8) by $n_j(\omega_j, t)$ and then integrating the resulted equations from 0 to L_j with respect to ω_j, we obtain

$$C_j = \xi_j \rho_j Y_j + \xi_j \rho_{0j} K_j, \ \bar{R}_j = \eta_j \rho_j Y_j + \eta_j \rho_{0j} K_j,$$

$$\bar{S}_j = \lambda_j \rho_j Y_j + \lambda_j \rho_{0j} K_j. \qquad (7.6.13)$$

Substituting $\bar{R}_j$ in Eqs. (7.6.11), and r and w_j in Eqs. (7.6.2) into Y_j in Eqs. (7.6.11), we have

$$Y_j = \frac{\alpha K_j F_j / K_{ij} + \beta F_j + \eta_j \rho_{0j} K_j}{1 - \eta_j \rho_j}. \qquad (7.6.14)$$

From

$$r = \frac{\alpha F_1}{F_{i1}} = \frac{\alpha F_2}{F_{i2}}$$

and Eqs. (7.6.1), we have

$$\frac{K_{i1}}{F_{i2}} = \frac{z^{1/\beta} N_1}{N_2}.$$

From this equation and Eqs. (7.6.3), we have

$$E = z_1 K_2 - z_2 K_1, \ K_{i1} = z_1 K, \ K_{i2} = z_2 K, \qquad (7.6.15)$$

where

$$z_1 \equiv \frac{z^{1/\beta} N_1}{z^{1/\beta} N_1 + N_2}, \ z_2 \equiv \frac{N_2}{z^{1/\beta} N_1 + N_2}.$$

We can thus express K_{ij} and F_j as functions of K_1 and K_2. By Eqs. (7.6.12), (7.6.15) and (7.6.14), we have

$$\dot{K}_1 = \frac{(\alpha K_1 / z_1 K + \beta)\lambda_1 z z_1^{\alpha} N_1^{\beta} K^{\alpha} - (\xi_1 + \delta_k \lambda_1) K_1}{\xi_1 + \lambda_1},$$

$$\dot{K}_2 = \frac{(\alpha K_2 / z_2 K + \beta)\lambda_2 z_2^{\alpha} N_2^{\beta} K^{\alpha} - (\xi_2 + \delta_k \lambda_2) K_2}{\xi_2 + \lambda_2}. \tag{7.6.16}$$

As $K = K_1 + K_2$, we see that the dynamic system (7.6.16) is only dependent on $K_1(t)$ and $K_2(t)$. Accordingly, the above two differential equations determine the capital stocks owned by the two countries, independent of the other variables in the system. We can show that all the other variables in the system are uniquely determined as functions of K_j and ω_j ($0 \le \omega_j \le L_j, j = 1, 2$) at any point of time.

Proposition 7.6.1 For any given (positive) level of the capital stocks, $K_1(t)$ and $K_2(t)$, at any point of time, all the other variables in the system are uniquely determined as functions of $K_1(t)$ and $K_2(t)$. The dynamics of $K_1(t)$ and $K_2(t)$ are determined by Eqs. (7.6.16).

Proof: We already uniquely determined k_j, y_j, E, K, F_j, K_{ij} and Y_j as functions of K_1 and K_2. From Eqs. (7.6.13) and (7.6.8), we directly determine $\overline{R}_j$, C_j, $\overline{S}_j$, c_j and S_j. The income from land ownership per household is given by: $\overline{r} = \overline{R} / N$. The rate of the interest r, and the wage rate w_j are uniquely determined by Eqs. (7.6.2). Substituting Eqs. (7.6.4) and (7.6.6) into $U_j(\omega_j)$ in (7.6.7) yields

$$U_j(\omega_j) = \mu_{1j} \left(T_{0j} - \upsilon_j \omega_j \right)^{\sigma_j} c_j^{\xi_j} n_{hj}^{-\mu_j - \eta_j} S_j^{\lambda_j}. \tag{7.6.17}$$

Substituting (7.6.17) into $U_j(0) = U_j(\omega_j)$, we have

$$n_j(\omega_j) = n_j(0) \left(1 - \frac{\upsilon_j \omega_j}{T_{0j}} \right)^{\sigma_j / (\mu_j + \eta_j)}. \tag{7.6.18}$$

Substituting Eqs. (7.6.18) into Eqs. (7.6.10) and then integrating the resulted equation from 0 to L_j, we have

$$n_j(0) = \frac{\upsilon_j B_j N_j T_{0j}^{-1}}{1 - \left(1 - \upsilon_j L_j / T_{0j}\right)^{B_j}}, \quad j = 1, 2. \tag{7.6.19}$$

We assume

$$1 - \frac{\upsilon_j L_j}{T_{0j}} > 0.$$

This simply means that the available time is sufficient to travel from the boundary of the country to the CBD. By Eqs. (7.6.18) and (7.6.19), we determine the residential density at any location in the two countries. The lot size per household is given by:

$$L_{hj}(\omega_j) = \frac{1}{n_j(\omega_j)}.$$

From Eqs. (7.6.8), we have the land rent $R_j(\omega_j)$. The local amenity $A_j(\omega_j)$ is given by Eqs. (7.6.6). The utility level $U_j(\omega_j)$ is given by (7.6.7). We thus showed how to determine all the variables in the system as unique functions of K_1 and K_2.

Before further analyzing the dynamic properties of the system, we examine how the differences in values of some variables between the two countries are determined at any point of time. Substituting K_{i1} and K_{i2} in Eqs. (7.6.15) into Eqs. (7.6.1) and (7.6.2), we have

$$\frac{F_1}{F_2} = \frac{w_1}{w_2} = z^{1/\beta}, \quad r = \alpha z \left(\frac{N_1}{z_1 K} \right)^{\beta},$$

where we assume $N_1 = N_2$. If country 1 works more effectively than country 2, both country 1's output and wage rate are higher than country 2. In the free trade system, if the world capital stocks K is increased, the interest rate r is reduced. From Eqs. (7.6.19) and (7.6.18), we get the ratio of the residential densities as follows

$$\frac{n_1(0)}{n_2(0)} = \frac{\upsilon \left[1 - (1 - \upsilon_2 L_2 / T_{02})^{B_2}\right]}{\left[1 - (1 - \upsilon_1 L_1 / T_{01})^{B_1}\right]},$$

$$\frac{n_1(\omega_1)}{n_2(\omega_2)} = \frac{n_1(0)(1 - \upsilon_1 \omega_1 / T_{01})^{B_1 - 1}}{n_2(0)(1 - \upsilon_2 \omega_2 / T_{02})^{B_2 - 1}}, \quad 0 \le \omega_j \le L_j, \quad j = 1, 2,$$

where

$$\upsilon \equiv \frac{\upsilon_1 B_1 N_1 T_{01}}{\upsilon_2 B_2 N_2 T_{02}}, \quad B_j \equiv \frac{\sigma_j}{\mu_j + \eta_j} + 1.$$

We see that the ratios are determined by the differences in the population, the transportation conditions, the available times, the crowding effects, the preferences and the territory sizes between the two countries. For instance, when the two countries are identical in the transportation condition, the available time, the population, the residential distribution is different if the two countries have different preferences, i.e., $B_1 \neq B_2$. It should be remarked that as we have already explicitly solved all the variables as functions of $K_1(t)$ and $K_2(t)$ at any point of time, it is direct to compare all the variables in the system between the two countries.

We now examine whether or not the dynamic system has equilibrium. By Eqs. (7.6.16) equilibrium of the dynamic systems is given by

$$\left(\frac{\alpha K_1}{z_1 K} + \beta\right)\lambda_1 z z_1^{\alpha} N_1^{\beta} K^{\alpha} = (\xi_1 + \delta_k \lambda_1) K_1,$$

$$\left(\frac{\alpha K_2}{z_2 K} + \beta\right)\lambda_2 z_2^{\alpha} N_2^{\beta} K^{\alpha} = (\xi_2 + \delta_k \lambda_2) K_2. \tag{7.6.20}$$

It can be shown that the above equations have a unique solution.

Proposition 7.6.2 The dynamic trade system has a unique stable equilibrium.

Proof: We first show that the system has a unique equilibrium. Dividing the first equation by the second one in Eqs. (7.6.20), we have

$$\frac{K_1}{K_2} = \Lambda, \tag{7.6.21}$$

where Λ is defined in Eqs. (7.6.24). Substituting Eq. (7.6.21) into the first equation in Eqs. (7.6.20), we have

$$K = \overline{\Lambda} \equiv \left[\frac{\alpha / z_1 + \beta / \Lambda + \beta}{\xi_1 + \delta_k \lambda_1} \lambda_1 z z_1^{\alpha} N_1^{\beta}\right]^{1/\beta}, \tag{7.6.22}$$

where $\overline{\Lambda}$ is a constant. By $K_1 + K_2 = K$, Eqs. (7.6.21) and (7.6.22), we have

$$K_1 = \frac{\Lambda\bar{\Lambda}}{1+\Lambda}, \quad K_2 = \frac{\bar{\Lambda}}{1+\Lambda}.$$

We obtained a unique equilibrium. We now provide stability conditions for the equilibrium.

It is easy to calculate that the two eigenvalues, ϕ_1 and ϕ_2, are given by

$$\phi^2 - (a_1 + b_1)\phi + a_1 b_2 - a_2 b_1 = 0, \tag{7.6.23}$$

where

$$a_1 \equiv -\frac{\lambda_1 z z_1^{\alpha} K_2 N_1^{\beta} K^{-\beta} / K_1^2 + (\xi_1 + \delta_k \lambda_1)/K}{\xi_1 + \lambda_1} \beta K_1 < 0,$$

$$a_2 \equiv \frac{\lambda_1 z z_1^{\alpha} N_1^{\beta} K^{-\beta} / K_1 - (\xi_1 + \delta_k \lambda_1)/K}{\xi_1 + \lambda_1} \beta K_1,$$

$$b_1 \equiv \frac{\lambda_2 z_2^{\alpha} N_2^{\beta} K^{-\beta} / K_2 - (\xi_2 + \delta_k \lambda_2)/K}{\xi_2 + \lambda_2} \beta K_2,$$

$$b_2 \equiv -\frac{\lambda_2 z_2^{\alpha} K_1 N_2^{\beta} K^{-\beta} / K_2^2 + (\xi_2 + \delta_k \lambda_2)/K}{\xi_2 + \lambda_2} \beta K_2 < 0.$$

As a_1 and b_2 are negative, we see that the system is stable if

$$a_2 b_1 - a_1 b_2 < 0.$$

From Eq. (7.6.23), we directly obtain

$$a_2 b_1 - a_1 b_2 =$$

$$\frac{\lambda_1 z z_1^{\alpha} N_1^{\beta} K^{1+\alpha} (\xi_2 + \delta_k \lambda_2)/K_1^2 + \lambda_2 z z_2^{\alpha} N_2^{\beta} K^{1+\alpha} (\xi_1 + \delta_k \lambda_1)/K_2^2}{(\xi_1 + \lambda_1)(\xi_2 + \lambda_2) K^2}.$$

Accordingly, the unique equilibrium is stable.

The ratio of the capital stocks owned by the two countries and the foreign capital stocks E at equilibrium are given by

$$\frac{K_1}{K_2} = \Lambda \equiv \Lambda_1 + \left(\Lambda_1^2 + \Lambda_0\right)^{1/2},$$

$$E = \left(1 - \frac{N_2 \Lambda}{z^{1/\beta} N_1}\right) z_1 K_2, \tag{7.6.24}$$

where

$$\Lambda_0 \equiv \frac{(\xi_2 / \lambda_2 + \delta_k) z^{1/\beta} N_1}{(\xi_1 / \lambda_1 + \delta_k) N_2}, \quad \Lambda_1 \equiv \frac{\alpha / \beta z_1 + 1}{2} \Lambda_0 - \frac{\alpha / \beta z_2 + 1}{2},$$

where we use Eq. (7.6.21) and the definitions of z_1 and z_2. Here, if $\xi_1 / \lambda_1 > (<)\ \xi_2 / \lambda_2$, we say that country 1's net propensity to own wealth is lower (higher) than country 2. From (7.6.7), we see that when $\xi_1 / \lambda_1 > \xi_2 / \lambda_2$, country 1's propensity to own wealth is lower than country 2. From Eqs. (7.6.24), we see that it is not easy to explicitly determine the sign of E. The trade direction is affected by the differences in the population, preferences and working efficiency between the two countries. To examine the sign of E, we examine a few special cases. In the case that the two countries have an identical working efficiency and net propensity to own wealth, i.e., $z = 1$ and $\xi_1 / \lambda_1 = \xi_2 / \lambda_2$, we have:

$$\frac{K_1}{K_2} = \frac{N_1}{N_2}, \quad E = 0.$$

The country with larger population has more capital stocks but trade is in balance.

We now examine the case that the two countries have identical population and working efficiency and country 1's net propensity to own wealth is higher than country 2, i.e., $N_1 = N_2$, $z = 1$, and $\xi_1 / \lambda_1 < \xi_2 / \lambda_2$. In this case, we have

$$\Lambda_0 = \frac{\xi_2 / \lambda_2 + \delta_k}{\xi_1 / \lambda_1 + \delta_k} > 1, \quad \Lambda_1 = \frac{(1 + \alpha)(\Lambda_0 - 1)}{2\beta} > 0.$$

By Eqs. (7.6.24), we have

$$\frac{K_1}{K_2} = \Lambda_1 + \left(\Lambda_1^2 + \Lambda_0\right)^{1/2} > 1, \quad E = (1 - \Lambda) z_1 K_2 < 0.$$

When country 1's net propensity to own wealth is higher than country 1, country 1 has more capital stocks than country 2 and some of country 1's capital stocks are utilized by country 2.

Another case is that the two countries have identical net propensity to own wealth and identical population, but different working efficiency, i.e., $N_1 = N_2$, $\xi_1 / \lambda_1 = \xi_2 / \lambda_2$ and $z \neq 1$. As

$$\Lambda_0 = z^{1/\beta}, \ \Lambda_1 = \frac{z^{1/\beta} - 1}{2},$$

the country which has a higher working efficiency owns more capital stocks and some of its capital stocks is utilized by the other country, i.e., $K_1 / K_2 > (<) \, 1$ and $E < (>) \, 0$ in the case of $z > (<) \, 1$. In the case that the two countries have identical population and country 1 works more effectively than country 2 and country 1's net propensity to own wealth is higher than country 2, i.e.

$$N_1 = N_2, \ \frac{\xi_1}{\lambda_1} < \frac{\xi_2}{\lambda_2}, \ z > 1,$$

it is obvious to check:

$$\Lambda_1 > 0, \ \Lambda_0 > 1, \ \frac{K_1}{K_2} > 1, \ E < 0.$$

7.7 A Dynamic Macro Model with Monetary Policy[12]

We now introduce a (short-run) dynamic macro model. Capital accumulation is neglected. The model is a combination of the conventional IS-LM model and Phillips curve. The formal model is described by the following set of equations

$$Y(t) = D(Y^D(t), r(t) - \pi(t), A(t)) + G,$$

$$0 < D_1 < 1, \ D_2 < 0, \ D_3 > 0, \tag{7.7.1}$$

$$Y^D = Y - T + rb(t) - \pi A, \tag{7.7.2}$$

$$A = m(t) + b, \tag{7.7.3}$$

$$m = L(Y, r, A), \ L_1 > 0, \ L_2 < 0, \ 0 \le L_3 \le 1, \tag{7.7.4}$$

[12] This section is based on Chap. 2 in Turnovsky (2000). We will not examine the model. Interpretations are referred to Turnovsky.

$$p(t) = \alpha(Y - \bar{Y}) + \pi, \quad \alpha > 0, \tag{7.7.5}$$

$$\dot{\pi} = \gamma(p - \pi), \quad \gamma > 0, \tag{7.7.6}$$

$$\dot{A} = \dot{m} + \dot{b} = G - T + rb - p(m + b), \tag{7.7.7}$$

where

Y = real output, or national income;
$\bar{Y}$ = the fixed capacity of production;
D = real private expenditure;
Y^D = real disposable income;
C = consumption expenditure by the private sector;
A = real wealth of the private sector;
I = gross private domestic investment;
G = total government purchases of goods and services
P = price level of output;
p = rate of inflation ($p \equiv \dot{P}/P$);
r = nominal rate of interest;
π = the expected rate of inflation;
T = (exogenous) net real tax payments;
M = nominal stock of outside money, assumed the liability of central bank;
B = nominal stock of government bonds;
m = real stocks of money ($m \equiv M/P$, where);
b = real stocks of bonds ($b \equiv B/P$).

Here, financial wealth defined by Eq. (7.7.3) consists of the stock of money plus government bonds outstanding. The first five equations, Eqs. (7.7.1) – (7.7.5), comprise an instantaneous set of relationships, in which the five variables, Y, Y^D, r, p, and m or b (depending on the government financing policy) are determined instantaneously in terms of the predetermined values of π, A, and b or m. The dynamics of the system are described by Eqs. (7.7.6) and (7.7.7).

Equation (7.7.7) is the budget constraint, expressed in real terms with rb being the interest payments on the outstanding government debt and

$-p(m+b)$ described the "inflation tax" on government debt.[13] To complete the model, government policy needs to be specified. There is only one constraint on the government budget but the government has four policy instruments, debt, money, taxes, and government expenditure. Only three of the four can be chosen independently. As G and T are fixed in this model, the government concerns with the mix between bond financing and money financing of the deficit. We are now concerned with three government policies.

(i) Fixed real stock of money policy

The monetary policy is to maintain the real stock of money constant, i.e., $m = \overline{m}$. As $m = M/P$, this policy can also be expressed as $M(t) = \overline{m}P(t)$. Under $m = \overline{m}$, the system becomes

$$Y = D(Y - T + r(A - \overline{m}) - \pi A,\ r - \pi,\ A) + G,$$

$$\overline{m} = L(Y, r, A),$$

$$p = \alpha(Y - \overline{Y}) + \pi,$$

$$\dot{\pi} = \gamma(p - \pi),$$

$$\dot{A} = G - T + r(A - \overline{m}) - pA.$$

Since the monetary policy fixes the real stock of money and the accumulation of real wealth takes the form of real bonds, the government deficit is referred as *bond-financed*. Denote an equilibrium point by (π^*, A^*). The Jacobian at the point is

$$J = \begin{bmatrix} \gamma(p_\pi - 1) & p_A \\ (A - \overline{m})r_\pi - Ap_\pi & (r - p) + (A - \overline{m})(r_A - Ap_A) \end{bmatrix},$$

---

[13] If we assume price of government bonds $P_b = 1$, then the government budget constraint in nominal terms is given by: $\dot{M} + \dot{B} = P(G - T) + rB$. The right-hand side equals the nominal value of government expenditures on goods and services, plus the nominal value of interest payments on outstanding governments bonds, less revenue raised by taxes. This deficit is financed either by issuing additional money or selling more bonds, or by some combinations of the two. Substituting $\dot{M} = m\dot{P} + \dot{m}P$ and $\dot{B} = b\dot{P} + \dot{b}P$ into the constraint yields (7.6.7).

where p_π and r_π are obtained by solving the linear equations

$$Y_\pi = D_1(Y_\pi + r_\pi(A - \overline{m}) - A) + D_2(r_\pi - 1),$$

$$L_1 Y_\pi + L_2 r_\pi = 0,$$

$$p_\pi = \alpha Y_\pi + 1,$$

and p_A and r_A are obtained by solving the linear equations

$$Y_A = D_1(Y_A + r_A(A - \overline{m}) + r - \pi) + D_2 r_A + D_A,$$

$$L_1 Y_A + L_2 r_A + L_3 = 0,$$

$$p_A = \alpha Y_A .$$

The analysis is straightforward, even though behavioral interpretations are tedious.

(ii) Constant rate of nominal monetary growth policy

The monetary policy maintains a constant rate of nominal monetary growth

$$\dot{M} = \mu M ,$$

where μ is a constant. As $M = mP$, this policy can also be written in the form of

$$\dot{m} = (\mu - p)m . \tag{7.7.8}$$

The system now becomes

$$Y = D(Y - T + r(A - m) - \pi A, r - \pi, A) + G,$$

$$m = L(Y, r, A),$$

$$p = \alpha(Y - \overline{Y}) + \pi,$$

$$\dot{\pi} = \gamma(p - \pi),$$

$$\dot{A} = G - T + r(A - m) - pA,$$

together with Eq. (7.7.8). We see that the dynamics are now a third-order system.

Since the monetary policy fixes the real stock of money and the accumulation of real wealth takes the form of real bonds, the government deficit is referred as *bond-financed.*

(iii) Fixed real stock of government bonds policy
The monetary policy fixes a real stock of government bonds

$$b(t) = \bar{b},$$

where $\bar{b}$ is constant. This policy can also be expressed in terms of the nominal stock of bonds B by $B(t) = \bar{b}P(t)$.

Under the monetary policy, the system is described by

$$Y = D(Y - T + r\bar{b} - \pi A,\ r - \pi,\ A) + G,$$

$$A - \bar{b} = L(Y, r, A),$$

$$p = \alpha(Y - \bar{Y}) + \pi,$$

$$\dot{\pi} = \gamma(p - \pi),$$

$$\dot{A} = G - T + r\bar{b} - pA.$$

The instantaneous variables, Y, p, and r can now be solved at each point in time in terms of G, π, A, and $\bar{b}$.

7.8 Economic Growth with Public Services

We now introduce a growth model with public inputs.[14] The economy is populated with a continuum of infinitely lived agents, whose measure is normalized to one. The representative agent supplies one unit of labor services inelastically. The agent chooses consumption so as to maximize the discounted sum of utility

$$\int_0^\infty \frac{c^\sigma(t) g^\eta(t)}{\sigma} e^{-\rho t} dt, \quad \sigma < 1, \quad \eta \le 1, \tag{7.8.1}$$

[14] The model is proposed by Zhang (2000jx), influenced by Barro (1990), Abe (1995), and Futagami, Morita, and Shibata (1993). Zhang's work is of the global nature and generates much richer dynamics than the traditional works.

subject to the following constraint

$$c + \dot{k} = (1 - \tau) f(k, g), \tag{7.8.2}$$

where ρ is the subjective discount rate, c is private consumption, τ is the government's tax rate on income, g is government consumption, and f is the income. The agent takes government consumption as given.

The production function of the private sector is specified as

$$f(k, g) = k^{\alpha}(t) g^{\beta}(t), \;\; \alpha + \beta > 1, \;\; \beta < 1,$$

where k is capital stock. Here, $\alpha + \beta > 1$ implies increasing returns to scale in private production. The case of $\beta < 1$ is referred to as mild increasing returns, while the case of $\beta > 1$ as strong increasing returns. Suppose that the economy's total capital stock per person is $\bar{k}$. Consequently, the domain for the capital stock is restricted to $[0, \bar{k}]$. The government's budget is balanced every instant, so that

$$g = \tau f(k, g) = \tau k^{\alpha} g^{\beta}. \tag{7.8.3}$$

The current-value Hamiltonian for this problem is

$$H = \frac{c^{\sigma} g^{\eta}}{\sigma} + \lambda[(1 - \tau) f - c],$$

where λ is costate variable. The first-order necessary conditions include

$$\lambda = c^{\sigma - 1} g^{\eta}, \tag{7.8.4}$$

$$\dot{\lambda} = \lambda[\rho - \alpha \tau_0 k^{\alpha_0 \beta_0}], \tag{7.8.5}$$

where

$$\beta_0 \equiv \frac{1}{1 - \beta}, \;\; \tau_0 \equiv (1 - \tau) \tau^{\beta \beta_0}, \;\; \alpha_0 \equiv \alpha + \beta - 1,$$

and the associated transversality condition

$$\lim_{t \to \infty} \lambda k e^{-\rho t} = 0. \tag{7.8.6}$$

Substituting $g = \tau k^{\alpha} g^{\beta}$, $f = k^{\alpha} g^{\beta}$, and $\lambda = c^{\sigma - 1} g^{\eta}$ into Eq. (7.8.2) yields

$$\dot{k} = \tau_0 k^{\alpha \beta_0} - \frac{\tau^{\eta / \gamma} k^{\alpha \eta / \gamma}}{\lambda^{\sigma_0}}, \tag{7.8.7}$$

where $\sigma_0 = 1/(1-\sigma)$ and $\gamma \equiv 1/\beta_0\sigma_0$. Equations (7.8.5) and (7.8.7) jointly define a planar dynamic system in (λ, k) on $[0, \infty)\times[0, \bar{k}]$. The system has a unique equilibrium point

$$k^* = \left(\frac{\rho}{\alpha\tau_0}\right)^{1/\alpha_0\beta_0}, \quad \lambda^* = \frac{\alpha\tau^{\eta\beta_0}}{\rho}k^{*\alpha\eta\beta_0 - 1+\sigma}.$$

The Jacobian at equilibrium is

$$J = \begin{bmatrix} 0 & -\rho\alpha_0\beta_0\dfrac{\lambda^*}{k^*} \\ \dfrac{\rho\sigma_0}{\alpha}\dfrac{\lambda^*}{k^*} & (1-\sigma-\eta)\dfrac{\rho}{\gamma} \end{bmatrix}.$$

We have

$$DetJ = \frac{\rho^2\alpha_0}{\alpha\gamma} > 0, \quad TrJ = (1-\sigma-\eta)\frac{\rho}{\gamma}.$$

The sign of the trace is the same as that of $(1-\sigma-\eta)$.

Case A: $\sigma + \eta > 1$

In this case, the trace is negative, the equilibrium is a sink. Calculate the discriminant (as a function of η)

$$\Delta(\eta) \equiv (TrJ)^2 - 4DetJ = \left[\frac{(1-\sigma-\eta)^2}{\gamma} - \frac{4\alpha_0}{\alpha}\right]\frac{\rho^2}{\gamma}.$$

There exists a value of η, denoted by η_s, such that $\Delta(\eta_s) = 0$. As $\sigma + \eta > 1$, [15] we solve

$$\eta_s = 1 - \sigma + 2\sqrt{\frac{\alpha_0\gamma}{\alpha}}.$$

If $\eta_s < \eta < 1$, the equilibrium point is a stable node; if $\eta_s > \eta$, the equilibrium point is a stable focus.

Case B: $\sigma + \eta < 1$

---

[15] It is not difficult to show that $\eta < 1$ can be generally warranted under a wide range of parameter values.

In this case, the trace is positive, the equilibrium is a source. There exists a value of η, denoted by η_u, such that $\Delta(\eta_u) = 0$.

As $\sigma + \eta < 1$, we solve $\Delta(\eta_u) = 0$ as

$$\eta_u = 1 - \sigma - 2\sqrt{\frac{\alpha_0 \gamma}{\alpha}}.$$

If $\eta_c < \eta < 1$, the equilibrium point is an unstable focus; if $\eta_c > \eta$, the equilibrium point is an unstable node.

Case C: $\sigma + \eta = 1$

In this case, the Jacobian has a pair of pure imaginary eigenvalues. Denote $\eta_0 = 1 - \sigma$ and the two eigenvalues are $\pm i\delta_0$, where

$$\delta_0 \equiv 2\rho\sqrt{\frac{\alpha_0}{\alpha\gamma}}$$

at η_C. Moreover, it is straightforward to show that at $\eta = \eta_0$, the real part of the derivative of eigenvalue with respect to η does not vanish. Therefore, the conditions of the Hopf bifurcation theorem are satisfied. The dynamic system exhibits limit cycles.[16]

7.9 Endogenous Population Growth in the Ramsey Framework

In studying interdependence between population and economic growth, it is necessary to take account of effects of economic factors on fertility and mortality. Many empirical studies have found effects of economic variables, such as per capita income, age rate, levels of female and male education, social welfare on fertility and mortality. We introduce endogenous birth rate in the Ramsey model.[17]

The production sides are described by the Cobb-Douglas production function

$$\hat{y} = A\hat{k}^{\alpha},$$

where

[16] See Zhang (2000) for further investigation.

[17] Refer to Sec. 9.2.2 in Barro and Sala-i-Martin (1995).

$$0 < \alpha < 1, \ \hat{k} = \frac{K}{\hat{N}}, \ \hat{y} = \frac{Y}{\hat{N}},$$

and $\hat{N} \equiv Ne^{xt}$ is the effective labor input with $x \geq 0$ being the rate of exogenous, labor-augmenting technological progress in the economy. If capital depreciates at the constant rate δ_k, then the marginal conditions are given by

$$r = \alpha A\hat{k}^{\alpha - 1} - \delta_k, \ w = (1 - \alpha)A\hat{k}^{\alpha} e^{xt}.$$

Let $n \geq 0$ be a family's birth rate; a choice variable of households at every point in time t. Let $d > 0$ stand for the mortality rate. For simplicity, assume d to be constant. According to the definitions, we have

$$\dot{N}(t) = (n(t) - d)N(t). \tag{7.9.1}$$

The formation of household utility is

$$U = \int_0^\infty \frac{e^{-\rho t}}{1 - \theta}\left\{\left[N(t)^{\psi} c(t)(n(t) - d)^{\phi}\right]^{1-\theta} - 1\right\}dt. \tag{7.9.2}$$

The term $e^{-\rho t}$ is the 'altruism factor'. Here, we can think of the pure rate of time preference as 0 in the present context. To explain

$$\frac{\left[N(t)^{\psi} c(t)(n(t) - d)^{\phi}\right]^{1-\theta} - 1}{1 - \theta},$$

we consider that the temporary utility that generation t obtains is given by

$$U^*(t) = N(t)^{1-\varepsilon} u(c(t), n(t) - d), \tag{7.9.3}$$

where ε is a positive parameter and $N(t)$ is the number of adult descendants in generation t. The term $u(c, n - d)$ represents the utility generated during adulthood from consumption and the presence of (net) children. The condition $\varepsilon > 0$ measures the degree of altruism between parents and children. We assume that utility function, $u(c, n - d)$, is taken on the following form that the elasticity of marginal utility with respect to c and $n - d$ is constant, i.e.

$$u(t) = \frac{\left[c(t)(n(t) - d)^{\phi}\right]^{1-\theta} - 1}{1 - \theta},$$

where $\phi > 0$ and $\theta > 0$. We also require $\phi(1-\theta) < 1$. This condition guarantees diminishing marginal utility with respect to n. Substituting the above equation into Eq. (7.9.3) yields

$$U^*(t) = \frac{\left[N(t)^{\psi} c(t)(n(t)-d)^{\phi}\right]^{1-\theta}}{1-\theta},$$

where

$$\psi \equiv \frac{1-\varepsilon}{1-\theta}.$$

We see that

$$U = \int_0^{\infty} e^{-\rho t} U^*(t)\, dt,$$

gives Eq. (7.9.2).

Assume that each child costs an amount η for the birth and rearing. For tractability of analysis, we consider η to be spent entirely at the time of birth, even though in reality the cost should be spent over many years. Per unit of time the number of births is $n(t)N(t)$. The total expenditures on child rearing per unit of time is equal to $\eta n(t)N(t)$; the expenditure per capita is $\eta n(t)$. Here, the child-bearing cost is assumed to be proportional to the number of children. In reality, there may be scale-effects in cost determination. Moreover, the setup cost for a family to have its first child suggests possible existence of a range in which the cost per child diminishes with the number of children. Eventually, the costs would increase more than linearly with the number, because the parents are very old when they have children. It should be noted that η is related to many other variables such as the value of parents' time and children's quality. It is argued that the cost η tends to rise with parents' wage rates or with other measures of the opportunity costs of parental time. Greater educational attainment of adults, for instance, tends to increase η. More generally, η tends to rise in per capital consumption $c(t)$ and per capita asset $k(t)$. For simplicity of analysis, it is assumed a linear relation between η and k as follows

$$\eta(t) = b_0 + bk(t), \tag{7.9.4}$$

where b_0 and b are non-negative parameters. We interpret b_0 as the goods cost of child bearing and $bk(t)$ as the part of the cost that increases with the capital-labor ratio. In the following discussion, we assume $b_0 = 0$.

We now construct the family budget constraint. Each family member is assumed, for convenience of analysis, to receive the same wage rate, irrespective of his/her age and human capital. The family's assets earn the rate of return $r(t)$. The budget constraint can be expressed as

$$\dot{k} = w + (r - n + d)k - bnk - c. \tag{7.9.5}$$

It is assumed that each household takes as given the path of the wage rate and the rate of return.

The household's optimization is to choose the path of the control variables $c(t)$ and $n(t)$ to maximize U in (7.9.2). The problem is subject to the initial assets $k(0)$; the transition equations for the two state variables, $N(t)$ and $n(t)$, given by Eqs. (7.9.1) and (7.9.5), $c(t) \geq 0$ and $n(t) \geq 0$. The Hamiltonian for the problem is

$$J = \int_0^\infty \frac{e^{-\rho t}}{1-\theta}\left\{\left[N^\psi c(n-d)^\phi\right]^{1-\theta} - 1\right\} + \mu(n-d)N,$$

$$+ \upsilon[w + (r - n + d)k - bnk - c],$$

where υ and μ are the shadow prices associated with the two state variables, $k(t)$ and $N(t)$. The first-order conditions for maximization are

$$\frac{\partial J}{\partial c} = \frac{\partial J}{\partial n} = 0,$$

$$\dot{\upsilon} = -\frac{\partial J}{\partial k}, \quad \dot{\mu} = -\frac{\partial J}{\partial N}. \tag{7.9.6}$$

From the definition of the Hamiltonian, we can express the conditions $\partial J / \partial c = 0$ and $\dot{\upsilon} = -\partial J / \partial k$ in terms of the growth rate of $c(t)$

$$\frac{\dot{c}}{c} = \frac{1}{\theta}\left\{r - \rho - (n - d)[1 - \psi(1 - \theta)] - nb + \frac{\phi(1 - \theta)\dot{n}}{n - d}\right\}.$$

In particular, if we choose $\theta = 1$, the above equation becomes

$$\frac{\dot{c}}{c} = r - \rho - (n - d) - nb. \tag{7.9.7}$$

The conditions $\partial J / \partial c = \partial J / \partial n = 0$ are expressed as

$$\mu = e^{-\rho t} N^{\psi(1-\theta)-1} c^{1-\theta} (n - d)^{\phi(1-\theta)} \Omega, \tag{7.9.8}$$

where

$$\Omega \equiv (1 + b)\frac{k}{c} - \frac{\phi}{n - d}.$$

Differentiating this expression for μ with respect to time and then using the condition $\dot{\mu} = - \partial J / \partial N$ yields

$$\dot{\Omega} = - \psi + \frac{\Omega}{\theta}\left\{\rho - (1 - \theta)\left[r - (1 - \psi)(n - d) - nb + \frac{\phi \dot{n}}{n - d}\right]\right\}.$$

In the case of $\theta = 1$, the above equation becomes

$$\dot{\Omega} = - \psi + \rho\Omega. \tag{7.9.9}$$

This linear differential equation is solved as

$$\Omega(t) = \frac{\psi}{\rho} + \left[\Omega(0) - \frac{\psi}{\rho}\right] e^{\rho t}.$$

Because ψ is constant and ρ is positive, the above equation is unstable. The unstable path violates the transversality condition associated with N. To show this, we note that Eq. (7.9.8) becomes $\mu N = e^{-\rho t}\Omega$ when $\theta = 1$. That is

$$\mu N = \frac{\psi e^{-\rho t}}{\rho} + \Omega(0) - \frac{\psi e^{\rho t}}{\rho}.$$

Therefore, the transversality condition associated with N

$$\lim_{t \to \infty} (\mu N) = 0,$$

is satisfied only when $\Omega(0) = \psi / \rho$. In this case, $\dot{\Omega} = 0$ for all t, and $\Omega(t)$ always equals its initial value.

By $\Omega(t) = \psi / \rho$ and the definition of Ω, we conclude that the fertility rate always satisfies the condition

$$n = d + \frac{\phi\rho(c/k)}{\rho(1+b) - \psi(c/k)}.$$

The fertility rate is proportionally related to the mortality rate. Since higher values of ϕ and ψ raise the marginal utility associated respectively with n and N, an increase in these values increases n.

The dynamics are determined by Eqs. (7.9.5) and (7.9.7). These two equations can be rewritten as

$$\frac{\dot{\chi}}{\chi} = -\rho - (1-\alpha)z + \chi,$$

$$\frac{\dot{z}}{z} = -(1-\alpha)\left[z - \delta - bd - x - \chi - \frac{\phi\rho\chi(1+b)}{\rho(1+b) - \psi\chi}\right],$$

where

$$\chi \equiv \frac{c}{k}, \quad z \equiv A\hat{k}^{\alpha-1}.$$

The dynamics can be analyzed by the methods in Chaps. 5 and 6. The task is left to the reader. Barro and Sala-i-Martin simulated the model with different combinations of the parameter.

7.10 The Ramsey Model with Endogenous Time

We now introduce endogenous time into the Ramsey model. The model reviewed here is based on a model by Barro and Sala-i-Martin.[18] Let us denote $N(t)$ population, which grows at a fixed growth rate n. Let $T(t)$ denote worker's efforts (e.g., working time) at time t. Labor input, denoted by $N^*(t)$, are given by

$$N^*(t) = T(t)N(t).$$

The production sides is described by the Cobb-Douglas production function

$$\hat{y} = A\hat{k}^{\alpha},$$

[18] See Sec. 9.3 in Barro and Sala-i-Martin (1995).

where $0 < \alpha < 1$, $\hat{k} \equiv K / \hat{N}$, $\hat{y} \equiv Y / \hat{N}$, and $\hat{N} \equiv N^* e^{xt}$ is the effective labor input with $x \geq 0$ being the rate of exogenous, labor-augmenting technological progress in the economy. If capital depreciates at the constant rate δ_k, then the marginal conditions are given by

$$r = \alpha A \hat{k}^{\alpha - 1} - \delta_k,$$

$$w = (1 - \alpha) A \hat{k}^{\alpha} e^{xt}.$$

We now introduce the utility which includes a disutility of work effort as

$$U = \int_0^\infty u[c(t), T(t)] e^{-(\rho - n)} dt,$$

where the usual concavity conditions are required

$$u_c > 0, \ u_T < 0, \ u_{cc} \leq 0, \ u_{TT} \leq 0.$$

Let $w(t)$ and $k(t)$ respectively stand for the wage rate paid for per unit of labor input and per capita wealth. Evolution of individual wealth follows

$$\dot{k} = wT + (r - n)k - c. \tag{7.10.1}$$

The Hamiltonian for the problem is

$$J = u(c, T) + \upsilon[wT + (r - n)k - c],$$

where υ is the shadow price associated with the state variable, $k(t)$. The first-order conditions for maximization are

$$\frac{\partial J}{\partial c} = \frac{\partial J}{\partial T} = 0,$$

$$\dot{\upsilon} = -\frac{\partial J}{\partial k}.$$

The first-order condition that reflects the substitution between consumption and leisure at a point in time is

$$-\frac{u_T}{u_c} = w.$$

The first-order condition that provides a relation between the rate of interest rate and growth rate of per capita consumption is

$$r = \rho - \left(\frac{u_{cc}c}{u_c}\right)\frac{\dot{c}}{c} - \left(\frac{u_{cT}T}{u_c}\right)\frac{\dot{T}}{T}. \tag{7.10.2}$$

Since it is so difficult to analyze the behavior of the model, let us consider the steady state at which w and c grow at the same rate, x. We ask whether there is a utility function that the model has a steady state in which c grows at a constant rate and T is constant. According to Eq. (7.10.2), these requirements imply that the elasticity of the marginal utility of consumption must be constant, i.e.

$$\frac{u_{cc}c}{u_c} = -\theta, \text{ a constant}. \tag{7.10.3}$$

We now rewrite $-u_T/u_c = w$ as

$$\frac{w}{c} = -\frac{u_T}{cu_c}.$$

To find a steady state in which w and c grow at the same rate, we take logs of the above equation and differentiate with respect to time

$$\frac{u_{Tc}\dot{c} + u_{TT}\dot{T}}{u_T} - \frac{u_{cc}\dot{c} + u_{cT}\dot{T}}{u_c} - \frac{\dot{c}}{c} = 0.$$

Since $\dot{T} = 0$ and $\dot{c}/c$ is generally non-zero in the steady state, the above equation can be rewritten as

$$\frac{cu_{Tc}}{u_T} = 1 + \frac{cu_{cc}}{u_c} = 1 - \theta.$$

That is

$$\frac{1}{u_T}\frac{\partial u_T}{\partial c} = \frac{1-\theta}{c}.$$

Integration of the above equation with respect to c yields

$$\log(u_T) = (1-\theta)\log(c) + H(T),$$

where $H(T)$ is a function of work time to be determined. Again, integrating the above equation with respect to T, we have

$$u(c,T) = c^{1-\theta}\varphi_1(T) + \varphi_2(c), \tag{7.10.4}$$

where φ_1 and φ_2 are respectively arbitrary functions of T and c. Eqs. (7.10.3) and (7.10.4) imply

$$c\varphi_2'' = -\theta\varphi_2'.$$

This differential equation is solved as, except multiplicative and additive constants

$$\varphi_2 = c^{1-\theta}, \text{ if } \theta \neq 1, \quad \varphi_2 = \log(c), \text{ if } \theta = 1. \tag{7.10.5}$$

Substituting Eq. (7.10.5) into Eq. (7.10.4), we can get the required form of $u(c,T)$. The following function satisfies the requirements

$$u(c,T) = \frac{c^{1-\theta}\exp[(1-\theta)\omega(T)] - 1}{1-\theta}, \tag{7.10.6}$$

where $\omega'(T) < 0$, $\omega''(T) < 0$, and $\theta \geq 1$. With the utility function specified by Eq. (7.10.6) and $-u_T / u_c = w$, we get

$$-\omega'(t) = \frac{w}{c}.$$

For simplicity of analysis, let us consider $\theta = 1$. In this case, the utility function specified by Eq. (7.10.6) is

$$u(c,T) = \log(c) + \log(T). \tag{7.10.7}$$

With this utility function, by Eq. (7.10.2) the growth rate of c is

$$g_c = r - \rho. \tag{7.10.8}$$

Introduce the variables per unit of effective labor to include the effect from variable work effort T

$$\hat{k} \equiv \frac{K}{TNe^{xt}}, \quad \hat{c} \equiv \frac{C}{TNe^{xt}}.$$

Using

$$g_T = 0, \quad r = f'(\hat{k}) - \delta_k,$$

we have

$$g_{\hat{c}} = f'(\hat{k}) - (\delta_k + \rho + x),$$

$$g_{\hat{k}} = \frac{f(\hat{k})}{\hat{k}} - (\delta_k + \rho + x) - \frac{\hat{c}}{\hat{k}}.$$

Through these two equations, the growth rates of variables are examined. Further discussions on the following specified forms

$$f = A\hat{k}^{\alpha}, \quad \omega(t) = -\xi T^{1+\sigma}$$

are referred to Barro and Sala-i-Martin. The dynamics can be analyzed by the methods in Chaps. 5 and 6. The task is left to the reader.

Part III

Higher Dimensions

Chapter 8

Higher-Dimensional Differential Equations

In this chapter we study higher-dimensional differential equations. Section 8.1 gives general solutions to systems of linear differential equations. Section 8.2 examines homogeneous linear systems with constant coefficients. Section 8.3 solves higher-order homogeneous linear differential equations. Section 8.4 introduces diagonalization and introduces concepts of stable and unstable subspaces of the linear systems. Section 8.5 studies the Fundamental Theorem for linear systems and provides a general procedure of solving linear equations.

8.1 Systems of Linear Differential Equations

By a system of first-order linear differential equations, we mean a system that can be written in the form

$$\dot{x}_i(t) = a_{i1}(t)x_1(t) + a_{i2}(t)x_2(t) + \ldots + a_{in}(t)x_n(t) + h_i(t),$$

$$i = 1, 2, \ldots n,$$

where $a_{ij}(t)$ are parameters and $x_i(t)$ are variables. In vector notation, the equations are written as

$$\dot{x}(t) = A(t)x(t) + h(t), \tag{8.1.1}$$

where $x(t)$ and $h(t)$ are $n \times 1$ vectors, and $A_{nxn}(t)$ is an $n \times n$ matrix. We have been concerned with the case of $n = 2$. In this chapter, we are mainly concerned with higher-dimensional problems. As in the case of $n = 2$, if $h(t) = 0$, the system is *homogeneous*; otherwise it is called *nonhomogeneous*.

Without loss of generality, if we add the initial conditions $x(t_0) = b$, where b is a given $n \times 1$ vector, we call the system of linear differential equations an *initial value problem*. A solution to the system is an $n \times 1$ vector $x(t)$ which satisfies (8.1.1).

Theorem 8.1.1 If $a_{ij}(t)$ and $h_i(t)$ are continuous on the interval (a, b) containing t_0 for $1 \le i \le n$ and $1 \le j \le n$, then the initial value problem (8.1.1) has a unique solution on (a, b).[1]

We now investigate the stability of a solution $x^*(t)$ to the problem (8.1.1). Let $x(t)$ represent any other solution and define $\xi(t)$ by

$$\xi(t) = x(t) - x^*(t).$$

The initial condition for $\xi(t)$ is

$$\xi(t_0) = x(t_0) - x^*(t_0).$$

Also $\xi(t)$ satisfies the homogeneous equation derived from (8.1.1)

$$\dot{\xi}(t) = A(t)\xi(t). \tag{8.1.2}$$

It can be seen that the stability property of $x^*(t)$ is the same as the stability of the zero solution of Eq. (8.1.2).

Theorem 8.1.2 All solutions of the regular linear system

$$\dot{x}(t) = A(t)x(t) + h(t),$$

have the same Liapunov stability property (unstable, stable, uniformly stable, asymptotically stable, uniformly and asymptotically stable). This is the same as that of the zero (or any other) solution of the homogeneous equation

$$\dot{\xi}(t) = A(t)\xi(t).$$

Example All the solutions of the system.

$$\dot{x}_1 = x_2,$$

---

[1] See Chap. 6 in Peterson and Sochacki (2002). The other theorems in this section are referred to the same source.

$$\dot{x}_2 = -\omega^2 x_1 + f(t),$$

are uniformly stable, but not asymptotically stable. Equation (8.1.2) becomes $\dot{\xi}_1 = \xi_2$ and $\dot{\xi}_2 = -\omega^2 \xi_1$. The zero solution is a center, which has the specified properties.

We are now concerned with a homogeneous system $\dot{x} = A(t)x$. It is known that the solutions to $\dot{x} = A(t)x$ form a vector space of dimension n. We call a set of n linearly independent solutions $X_1, X_2, ..., X_n$ to $\dot{x} = A(t)x$ a *fundamental set of solutions*. Here, linear independence is defined as follows.

Definition 8.1.1 Let $X_1(t), X_2(t), ..., X_n(t)$ be vector functions (real or complex), continuous on $-\infty < t < \infty$, none being identically zero. If there exist (scalar) constants (real or complex) $k_1, k_2, ..., k_n$, not all zero, such that

$$k_1 X_1(t) + k_2 X_2(t) + ... + k_n X_n(t) = 0,$$

for $-\infty < t < \infty$, the functions are *linearly dependent*. Otherwise they are *linearly independent*.

As a consequence of the definition, note that the vector

$$\begin{bmatrix} 1 \\ 1 \end{bmatrix}, \begin{bmatrix} t \\ t \end{bmatrix},$$

are linearly independent, although the constant vectors

$$\begin{bmatrix} 1 \\ 1 \end{bmatrix}, \begin{bmatrix} t_0 \\ t_0 \end{bmatrix},$$

are linearly dependent for every t_0.

Example $\cos t$ and $\sin t$ are linearly independent on $-\infty < t < \infty$. This is confirmed by observing

$$k_1 \cos t + k_2 \sin t = \alpha \sin(t + \beta)$$

where $\alpha \equiv \sqrt{k_1^2 + k_2^2}$ and β is defined by $k_1 = \alpha \sin \beta$.

Theorem 8.1.2 There exists a set of n linearly independent solutions of $\dot{x} = A(t)x$.

Theorem 8.1.3 Any $n+1$ nonzero solutions of the homogeneous system $\dot{x} = A(t)x$ are linearly dependent.

These two theorems settle the dimension of the solution space: every solution is a linear combination of the solutions $X_1, X_2, ..., X_n$ of Theorem 8.1.2; but since these solutions are themselves linearly independent, we cannot do without any of them. Instead of the special solutions $X_j(t)$ we may take any set of n linearly independent solutions as the basis.

If $X_1, X_2, ..., X_n$ form a fundamental set of solutions, then the *general solution* to $\dot{x} = A(t)x$ is given by

$$X_H = \sum_{i=1}^{n} c_i X_i = X_c C,$$

where c_i are constant, and

$$X_c \equiv [X_1, X_2, ... X_n]_{nxn}, \quad C_{nx1} \equiv [c_1, c_2, ... c_n]^T.$$

Here, the $n \times n$ matrix $X_c(t)$ is called a *fundamental matrix* of the homogeneous system

$$\dot{x} = A(t)x. \qquad (8.1.3)$$

Theorem 8.1.4 The solution of the homogeneous system $\dot{x} = A(t)x$ with the initial conditions $x(t_0) = x_0$ is given by

$$x(t) = X_c(t) X_c^{-1}(t_0) x_0,$$

where $X_c(t)$ is any fundamental matrix of the system.

Example Verify that $\begin{bmatrix} 2 & e^t \end{bmatrix}^T$ and $\begin{bmatrix} e^{-t} & 1 \end{bmatrix}^T$ are solutions of

$$\dot{x} = \begin{bmatrix} 1 & -2e^{-t} \\ e^t & -1 \end{bmatrix} x.$$

Find the solution $x(t)$ such that $x_0 = [3 \quad 1]^T$.

Direct substitution confirms that the two vectors are linearly independent solutions. So we have

$$X_c = \begin{bmatrix} 2 & -e^{-t} \\ e^t & 1 \end{bmatrix}, \; X_c(0) = \begin{bmatrix} 2 & -1 \\ 1 & 1 \end{bmatrix} \Rightarrow X_c^{-1}(0) = \begin{bmatrix} 1 & -1 \\ -1 & 1 \end{bmatrix}.$$

Hence

$$x(t) = X_c(t)X_c^{-1}(t_0)x_0 = \begin{bmatrix} 4 - e^{-t} \\ 2e^t - 1 \end{bmatrix}.$$

Theorem 8.1.5 Suppose that $X_1(t), X_2(t), ..., X_n(t)$ form a fundamental set of solutions to $\dot{x} = A(t)x$ and that $X_p(t)$ is a solution to the nonhomogeneous system of first-order linear differential equations

$$\dot{x}(t) = A(t)x(t) + h(t).$$

Then every solution to $\dot{x} = A(t)x + h(t)$ has the form

$$x = \sum_{i=1}^{n} c_i X_i + X_p = X_c C + X_p.$$

We call $X_p(t)$ a *particular solution* to the nonhomogeneous system.

Example Find all solutions of the system

$$\dot{x}_1 = x_2, \quad \dot{x}_2 = -x_1 + t.$$

The corresponding homogeneous system is $\dot{\phi}_1 = \phi_2$, $\dot{\phi}_2 = -\phi_1$. From $\ddot{\phi}_1 + \phi_1 = 0$ and $\phi_2 = \dot{\phi}_1$, we obtain two linearly independent solutions

$$\begin{bmatrix} \cos t \\ -\sin t \end{bmatrix}, \begin{bmatrix} \sin t \\ \cos t \end{bmatrix}.$$

It can be confirmed that $x_1 = t$ and $x_2 = 1$ is a particular solution of the original system. Therefore, all solutions are given by

$$x = \begin{bmatrix} \cos t & \sin t \\ -\sin t & \cos t \end{bmatrix} \begin{bmatrix} C_1 \\ C_2 \end{bmatrix} + \begin{bmatrix} t \\ 1 \end{bmatrix}.$$

Theorem 8.1.6 If $a_{ij}(t)$ and $h_i(t)$ are continuous and $X_1(t), X_2(t), \ldots, X_n(t)$ form a fundamental set of solutions to $\dot{x} = A(t)x$ on the interval (a, b). If X_c is the matrix of fundamental solutions, then a particular solution to $\dot{x}(t) = A(t)x(t) + h(t)$ on (a, b) is given by

$$X_p = X_c \int X_c^{-1} h(t) dt .$$

The theorems in this section shows how to solve $\dot{x} = A(t)x + h(t)$.

Example Consider a two-dimensional system

$$\dot{x} = \begin{bmatrix} 1 & 2 \\ -1 & 4 \end{bmatrix} x + \begin{bmatrix} 2 \\ t \end{bmatrix}.$$

We leave the reader to show

$$X_1 = e^{2x}\begin{bmatrix} 2 \\ 1 \end{bmatrix}, \ X_2 = e^{3x}\begin{bmatrix} 1 \\ 1 \end{bmatrix}, \ X_c = \begin{bmatrix} 2e^{2x} & e^{3x} \\ e^{2x} & e^{3x} \end{bmatrix},$$

$$X_c^{-1} = \begin{bmatrix} e^{-2x} & -e^{-2x} \\ -e^{-3x} & 2e^{-3x} \end{bmatrix}.$$

A particular solution is given by

$$X_p = \begin{bmatrix} 2e^{2x} & e^{3x} \\ e^{2x} & e^{3x} \end{bmatrix} \int \begin{bmatrix} e^{-2x} & -e^{-2x} \\ -e^{-3x} & 2e^{-3x} \end{bmatrix} \begin{bmatrix} 2 \\ t \end{bmatrix} dt = \begin{bmatrix} -\dfrac{19}{18} + \dfrac{t}{3} \\ -\dfrac{11}{36} - \dfrac{t}{6} \end{bmatrix}.$$

Hence, the general solution to the problem is

$$X = c_1 X_1 + c_2 X_2 + X_c,$$

where c_1 and c_2 are constant.

From Theorems 8.1.4 to 8.16, we see that the following theorem holds.

Corollary 8.1.1 The solution of the system

$$\dot{x} = A(t)x + f(t),$$

with the initial conditions $x(t_0) = x_0$ is given by

$$x(t) = X_c(t)X_c^{-1}(t_0)x_0 + X_c(t)\int_{t_0}^{t} X_c^{-1}(s)f(s)ds ,$$

where $X_c(t)$ is any fundamental matrix of the corresponding homogeneous system $\dot{x} = A(t)x$.

Example Find the solution of

$$\begin{bmatrix} \dot{x}_1 \\ \dot{x}_2 \\ \dot{x}_3 \end{bmatrix} = \begin{bmatrix} 0 & 1 & 0 \\ 1 & 0 & 0 \\ te^{-t} & te^{-t} & 1 \end{bmatrix} \begin{bmatrix} x_1 \\ x_2 \\ x_3 \end{bmatrix} + \begin{bmatrix} e^t \\ 0 \\ 1 \end{bmatrix},$$

which satisfies the initial conditions $x(0) = [0 \ \ 1 \ \ -1]^T$.

We first find a fundamental solution matrix of the associated homogeneous system $\dot{X} = A(t)X$. In component form, this equation separates into

$$\dot{X}_1 = X_2,$$

$$\dot{X}_2 = X_1,$$

$$\dot{X}_3 - X_3 = te^{-t}(X_1 + X_2).$$

From the first two equations

$$X_1 = Ae^t + Be^{-t}, \quad X_2 = Ae^t - Be^{-t},$$

whilst the third equation now becomes

$$\dot{X}_3 - X_3 = 2tA .$$

This equation has the general solution

$$X_3 = -2A(1+t) + Ce^t .$$

Hence we obtain a fundamental solution matrix

$$X_c(t) = \begin{bmatrix} e^t & e^t & 0 \\ e^t & -e^{-t} & 0 \\ -2(1+t) & 0 & e^t \end{bmatrix}.$$

Calculate $\det X_c(t) = -2e^t$ and

$$X_c^{-1}(t) = \frac{1}{2}\begin{bmatrix} e^{-t} & e^{-t} & 0 \\ e^{t} & -e^{t} & 0 \\ 2(1+t)e^{-2t} & (1+t)e^{-2t} & 2e^{-t} \end{bmatrix},$$

$$X_c^{-1}(0) = \frac{1}{2}\begin{bmatrix} 1 & 1 & 0 \\ 1 & -1 & 0 \\ 2 & 1 & 2 \end{bmatrix}.$$

It is easy to check that the solution to

$$\begin{bmatrix} x_1 \\ x_2 \\ x_3 \end{bmatrix} = \begin{bmatrix} \left(\frac{3}{4} + \frac{1}{2}t\right)e^{t} - \frac{3}{4}e^{-t} \\ \left(\frac{3}{4} + \frac{1}{2}t\right)e^{t} + \frac{3}{4}e^{-t} \\ 3e^{t} - t^2 - 3t - 4 \end{bmatrix}.$$

Corollary 8.1.2 If (i) A is a constant $n \times n$ matrix and the eigenvalues of A have negative real parts; and $C(t)$ is continuous for $t > t_0$ and

$$\int_{t_0}^{t} \|C(t)\| dt \text{ is bounded for } t > t_0,$$

then all solutions of

$$\dot{x} = \{A + C(t)\}x,$$

are asymptotically stable.[2]

Example Show that when $a > 0$ and $b > 0$ all solutions of

$$\ddot{y} + a\dot{y} + (b + c\cos t)y = 0,$$

are asymptotically stable for $t \geq t_0$. The system can be rewritten as

$$\dot{x} = \{A + C(t)\}x,$$

where

---

[2] Here, $\|C\| \equiv \left\{\sum_{i,j} |c_{ij}|^2\right\}^{1/2}$.

$$\begin{bmatrix} x_1 \\ x_2 \end{bmatrix} = \begin{bmatrix} y \\ \dot{y} \end{bmatrix}, \quad A = \begin{bmatrix} 0 & 1 \\ -b & -a \end{bmatrix}, \quad C = \begin{bmatrix} 0 & 0 \\ -ce^{-t}\cos t & 0 \end{bmatrix}.$$

As the eigenvalues of A are negative when a and b are positive and

$$\int_{t_0}^{\infty} \|C\| dt = |c| \int_{t_0}^{\infty} e^{-t} |\cos t| dt < \infty ,$$

the conditions in the corollary are satisfied. All solutions are asymptotically stable.

Theorem 8.1.8 (Superposition of solutions for nonhomogeneous equations) If $X_1(t)$ and $X_2(t)$ are solutions of two nonhomogeneous linear differential equations

$$\dot{x}(t) = A(t)x(t) + h_1(t),$$

and

$$\dot{x}(t) = A(t)x(t) + h_2(t),$$

respectively, with the same associated homogeneous equation $\dot{x}(t) = A(t)x(t)$, then

$$X(t) = X_1(t) + X_2(t),$$

is a solution to the equation

$$\dot{x}(t) = A(t)x(t) + h_1(t) + h_2(t).$$

Exercise 8.1

1 All solutions of the equations $\dot{x}_1 = x_2$ and $\dot{x}_2 = -\omega^2 x_1 - kx_2 + f(t)$ are asymptotically stable.

2 Find the general solution to the following two-dimensional systems

(a) $\dot{x} = \begin{bmatrix} 0 & 1 \\ -2 & -3 \end{bmatrix} x + \begin{bmatrix} 0 \\ \sin t \end{bmatrix};$

(b) $\dot{x} = \begin{bmatrix} 0 & 1 \\ -2 & -3 \end{bmatrix} x + \begin{bmatrix} 0 \\ e^t \end{bmatrix}.$

3 Find the general solution to the following two dimensional system

$$\dot{x} = \begin{bmatrix} 0 & 1 \\ -2 & -3 \end{bmatrix} x + \begin{bmatrix} 0 \\ \sin t + e^t \end{bmatrix}.$$

8.2 Homogeneous Linear Systems with Constant Coefficients

We are concerned with systems of homogeneous linear equations with constant coefficients

$$\dot{x}(t) = A_{n\times n} x(t), \tag{8.2.1}$$

where A is a (real) constant $n \times n$ matrix. We assume that A is nonsingular. The origin is the unique (and hence isolated) equilibrium point of this system. We have dealt with the case of $n = 2$. It is actually straightforward to extend to the general case. It should be noted that for the system

$$\dot{x}(t) = A_{n\times n} x(t) + b_{n\times 1},$$

we introduce $y(t) \equiv x(t) + A^{-1}b$. Here, we assume the existence of the inverse of the matrix A. Under the transformation, the above system becomes

$$\dot{y}(t) = A_{n\times n} y(t).$$

Hence, it is sufficient for us to be only concerned with Eq. (8.2.1). Like in Sec. 5.1, we seek solutions of Eq. (8.2.1) of the form

$$x_{n\times 1}(t) = C_{n\times 1} e^{\rho t}$$

where the exponent ρ and the constant vector C are to be determined. Substituting $x_{n\times 1}(t) = C_{n\times 1} e^{\rho t}$ into Eq. (8.2.1) yields

$$(A - \rho I)C = 0,$$

where I is the $n \times n$ identity matrix. Hence, to solve the system of differential equations is to determine the eigenvalues and eigenvectors of A. The eigenvalues ρ are the roots of the nth degree polynomial equation

$$\det(A - \rho I) = 0.$$

There are three possibilities for the eigenvalues of A: (1) all eigenvalues are real and different from each other; (2) some eigenvalues occur in complex conjugate pairs; and (3) some eigenvalues are repeated.

If all eigenvalues, $\rho_1, \rho_2, \cdots, \rho_n$, are real and different from each other, then associated with each ρ_i is a real eigenvector $v^{(i)}$ and the set of n eigenvectors is linearly independent. The corresponding solutions of the differential system (8.2.1) are $x^{(i)}(t) = v^{(i)}e^{r_i t}$, $i = 1, 2, \cdots, n$. Since

$$\begin{vmatrix} v_1^{(1)}e^{\rho_1 t} & v_1^{(2)}e^{\rho_2 t} & \cdots & v_1^{(n)}e^{\rho_n t} \\ v_2^{(1)}e^{\rho_1 t} & v_2^{(2)}e^{\rho_2 t} & \cdots & v_2^{(n)}e^{\rho_n t} \\ \vdots & \vdots & & \vdots \\ v_n^{(1)}e^{\rho_1 t} & v_n^{(2)}e^{\rho_2 t} & \cdots & v_n^{(n)}e^{\rho_n t} \end{vmatrix} = e^{\sum_{i=1}^{n}\rho_i t} \begin{vmatrix} v_1^{(1)} & v_1^{(2)} & \cdots & v_1^{(n)} \\ v_2^{(1)} & v_2^{(2)} & \cdots & v_2^{(n)} \\ \vdots & \vdots & & \vdots \\ v_n^{(1)} & v_n^{(2)} & \cdots & v_n^{(n)} \end{vmatrix} \neq 0,$$

we see that $x^{(1)}(t), x^{(2)}(t), \cdots, x^{(n)}(t)$ form a fundamental set of solutions. Hence the general solution of Eq. (8.2.1) is

$$x(t) = \sum_{i=1}^{n} c_i v^{(i)} e^{\rho_i t}, \tag{8.2.2}$$

where c_i are scalar constants.

Theorem 8.2.1 If A is an $n \times n$ matrix and all the eigenvalues of A have negative real parts, then all solutions $x(t)$ of $\dot{x} = Ax$ satisfy

$$\lim_{t\to\infty} x(t) \to 0.$$

The origin is called a *sink* when Theorem 8.2.1 holds. There is an exactly analogous statement when all eigenvalues have a positive real part, turning time backward; then the origin is called a *source*.[3]

It is known that if A is symmetric and negative definite, then every eigenvalues of A is real and negative. Hence, a sufficient condition for the stability of $x^* = 0$ is that A is symmetric and negative definite. We now provide the conditions for all the eigenvalues of an arbitrary $n \times n$ matrix to have negative real parts. The following Routh-Hurwitz theorem determines this.

[3] When eigenvalues are not all distinct the formal situation is more complicated, and for the theory the reader is referred to, for instance, to Wilson (1971).

Theorem 8.2.2 We consider that the eigenvalues ρ are the roots of the following nth degree polynomial equation

$$a_0\rho^n + a_1\rho^{n-1} + \cdots + a_{n-1}\rho + a_n = 0,$$

with real coefficients having real negative real parts, which in turn holds if and only if

$$a_1 > 0,\quad \begin{vmatrix} a_1 & a_0 \\ a_3 & a_2 \end{vmatrix} > 0,\quad \begin{vmatrix} a_1 & a_0 & 0 \\ a_3 & a_2 & a_1 \\ a_5 & a_4 & a_3 \end{vmatrix} > 0, \cdots$$

$$\begin{vmatrix} a_1 & a_0 & 0 & 0 & \cdots & \cdots \\ a_3 & a_2 & a_1 & a_0 & \cdots & \cdots \\ a_5 & a_4 & a_3 & a_2 & \cdots & \cdots \\ \cdots & \cdots & \cdots & \cdots & \cdots & \cdots \\ \cdots & \cdots & \cdots & \cdots & \cdots & \cdots \\ 0 & 0 & 0 & 0 & 0 & a_n \end{vmatrix} > 0.$$

Here a_0 is taken to be positive (if $a_0 < 0$, then multiply the equation by -1).

The theorem provides a necessary and sufficient condition for stability in a linear system. If A is real and symmetric, then it is known that all the eigenvalues $\rho_1, \rho_2, \cdots, \rho_n$ must be real. Even if some of the eigenvalues are repeated, there is always a full set of n eigenvectors $v_1, v_2, \cdots, v_n$ that are linearly independent. In this case, the general solution is still given by Eq. (8.2.2). The following example demonstrates this.

Example Consider the system

$$\dot{x} = \begin{pmatrix} 0 & 1 & 1 \\ 1 & 0 & 1 \\ 1 & 1 & 0 \end{pmatrix} x.$$

We calculate the three eigenvalues, $\rho_1 = 2$ and $\rho_2 = \rho_3 = -1$ and corresponding three eigenvectors

$$v^{(1)} = \begin{pmatrix} 1 \\ 1 \\ 1 \end{pmatrix}, \quad v^{(2)} = \begin{pmatrix} 1 \\ 0 \\ -1 \end{pmatrix}, \quad v^{(3)} = \begin{pmatrix} 0 \\ 1 \\ -1 \end{pmatrix}.$$

The general solution is

$$x(t) = c_1 \begin{pmatrix} 1 \\ 1 \\ 1 \end{pmatrix} e^{2t} + c_2 \begin{pmatrix} 1 \\ 0 \\ -1 \end{pmatrix} e^{-t} + c_3 \begin{pmatrix} 0 \\ 1 \\ -1 \end{pmatrix} e^{-t}.$$

We now discuss the case that some eigenvalues occur in complex conjugate pairs. As A is real, any complex eigenvalues must occur in conjugate pairs. For instance, if $\rho = \lambda_1 + i\lambda_2$, where λ_1 and λ_2 are real, is an eigenvalue of A, then so is $\bar{\rho} = \lambda_1 - i\lambda_2$. Moreover, if v is an eigenvector associated with ρ, then $\bar{v}$ is an eigenvector associated with $\bar{\rho}$. The corresponding solutions

$$x^{(1)}(t) = ve^{\rho t}, \quad x^{(2)}(t) = \bar{v}e^{\bar{\rho} t},$$

of the differential equations (8.2.1) are complex conjugates of each other. Let $v = u_1 + iu_2$ where u_1 and u_2 are real vectors. We have

$$x^{(1)}(t) = (u_1 + iu_2)e^{(\lambda_1 + i\lambda_2)t} = (u_1 + iu_2)e^{\lambda_1 t}(\cos\lambda_2 t + i\sin\lambda_2 t),$$

where we use

$$e^{i\lambda_2 t} = \cos\lambda_2 t + i\sin\lambda_2 t.$$

Hence, from the above formula we can write $x^{(1)}(t)$ in the form of

$$x^{(1)}(t) = \sigma_1(t) + i\sigma_2(t), \tag{8.2.3}$$

where $\sigma_1(t)$ and $\sigma_2(t)$ are real vectors

$$\sigma_1(t) = e^{\lambda_1 t}(u_1\cos\lambda_2 t - u_2\sin\lambda_2 t),$$

$$\sigma_2(t) = e^{\lambda_1 t}(u_1\sin\lambda_2 t + u_2\cos\lambda_2 t).$$

Here, it can be shown that $\sigma_1(t)$ and $\sigma_2(t)$ are linearly independent solutions. To find general solutions, suppose that

$$\rho_1 = \lambda_1 + i\lambda_2, \quad \rho_2 = \lambda_1 - i\lambda_2,$$

and that $\rho_3, \cdots, \rho_n$ are all real and distinct. Let the corresponding eigenvectors be $v^{(1)} = u_1 + iu_2$, $v^{(2)} = u_1 - iu_2$, and $v_3, \cdots, v_n$. Then the general solution of Eq. (8.2.1) is

$$x(t) = c_1\sigma_1 + c_2\sigma_2 + \sum_{i=3}^{n} c_i v^{(i)} e^{\rho_i t}. \qquad (8.2.4)$$

Theorem 8.2.3 Corresponding to an eigenvalue of $A_{n\times n}$, $\rho = \rho_i$, multiplicity $m \le n$, there are m linearly independent solutions of the system $\dot{x} = Ax$. They are of the form

$$p_1(t)e^{\rho_i t},\ p_2(t)e^{\rho_i t}, \cdots,\ p_m(t)e^{\rho_i t},$$

where the $p_i(t)$ are vector polynomials of degree less than m.

Note that when an eigenvalue is complex, the eigenvectors and the polynomials in the theorem will be complex, and the arrays consist of complex-valued solutions. Here, we will not discuss difficult cases when an eigenvalue is repeated.[4]

Theorem 8.2.4 Let A be constant in the system $\dot{x} = A_{n\times n}x$, with eigenvalues, ρ_i, $i = 1, \cdots, n$.

(i) If the system is stable, then $\text{Re}\{\rho_i\} \le 0$, $i = 1, \cdots, n$.

(ii) If either $\text{Re}\{\rho_i\} < 0$, $i = 1, \cdots, n$; or if $\text{Re}\{\rho_i\} \le 0$, $i = 1, \cdots, n$ and there is no zero repeated eigenvalue, then the system is uniformly stable.

(iii) The system is asymptotically stable if and only if $\text{Re}\{\rho_i\} < 0$, $i = 1, \cdots, n$.

(iv) If $\text{Re}\{\rho_i\} > 0$ for any i, the solution is unstable.

Exercise 8.2

1 Find the general solutions of the following equations and describe the behavior of the solutions as $t \to +\infty$

(i) $\dot{x} = \begin{pmatrix} 3 & 2 & 4 \\ 2 & 0 & 2 \\ 4 & 2 & 3 \end{pmatrix} x$;

---

[4] See Boyce and DiPrima (2001: Chap. 7).

(ii) $\dot{x} = \begin{pmatrix} 1 & -1 & 4 \\ 3 & 2 & -1 \\ 2 & 1 & -1 \end{pmatrix} x$;

(iii) $\dot{x} = \begin{pmatrix} 0 & 1 & 0 \\ 0 & 0 & 1 \\ -5 & 5 & 1 \end{pmatrix} x$.

2 Find the general solution of the following equation, using (8.2.4)

$$\dot{x} = \begin{pmatrix} -\frac{1}{2} & 1 \\ -1 & -\frac{1}{2} \end{pmatrix} x .$$

8.3 Higher-Order Equations

Many differential equations encountered in economics involve higher-order derivatives of unknown functions. We saw in Chap. 2 that a second-order equation is "equivalent" to a system of first-order equations. This is held also for higher-order equations. Since the theory for systems of first-order equations is simple and the intuitive idea of what a differential equation means is clear, it is usually convenient to replace a higher-order equation by system of first order equations. One can show that the nth order homogeneous linear differential equation

$$a_n(t)x^{(n)} + a_{n-1}(t)x^{(n-1)} + ... + a_0(t)x = 0 , \quad a_n(t) \neq 0 ,$$

is equivalent to the system of n homogeneous linear differential equation

$$\dot{X}(t) = A(t)X(t),$$

where

$$X \equiv \begin{bmatrix} x_0 \\ x_1 \\ \vdots \\ x_{n-1} \end{bmatrix}, \quad A(t) \equiv \begin{bmatrix} 0 & 1 & 0 & \cdots & 0 \\ 0 & 0 & 1 & \cdots & 0 \\ & & \vdots & & \\ 0 & & \cdots & 0 & 1 \\ -\frac{a_0(t)}{a_n(t)} & -\frac{a_1(t)}{a_n(t)} & \cdots & & -\frac{a_{n-1}(t)}{a_n(t)} \end{bmatrix},$$

where $x_0 = x,\ x_i = \dot{x}_{i-1},\ i = 1, \cdots, n-1$. It should be noted that if $a_0, a_1, \ldots, a_n$ are constants, then the characteristic equation of the nth order constant coefficient homogeneous linear differential equation and the characteristic equation of the matrix A are the same.

This result can be generalized in the following way. A differential equation of order n in one variable is an equation of the form

$$x^{(n)} = f\left(x, x', \cdots, x^{(n-1)}, t\right). \tag{8.3.2}$$

Obviously, Eq. (8.3.1) is a special case of Eq. (8.3.2). Here, f is a function defined in some region in $R^n \times R$. The key idea to solution of an nth order differential equation is to introduce new variables representing successive derivatives, generalizing what we did before.

Theorem 8.3.1 The differential equation (8.3.2) is equivalent, if we set $x_0 \equiv x$, to the first-order differential equation in R^n

$$\dot{X} = \begin{bmatrix} x' \\ x'' \\ x''' \\ \vdots \\ x^{(n)} \end{bmatrix} = \begin{bmatrix} \dot{x}_0 \\ \dot{x}_1 \\ \dot{x}_2 \\ \vdots \\ \dot{x}_{n-1} \end{bmatrix} = \begin{bmatrix} x_1 \\ x_2 \\ x_3 \\ \vdots \\ f(x_0, x_1, x_2, \cdots, x_{n-1}) \end{bmatrix} \equiv f(X, t), \tag{8.3.3}$$

in the sense that a function $x(t) = u(t)$ is a solution of Eq. (8.3.2) if and only if the n-dimensional vector function

$$U(t) = \begin{bmatrix} u(t) \\ u'(t) \\ u''(t) \\ \vdots \\ u^{n-1}(t) \end{bmatrix},$$

is a solution of the system (3.3.3).

The theorem tells us that for an nth order differential equation, we set up n variables:

$$x_0 = x,\ x_i = \dot{x}_{i-1},\ i = 1, \cdots, n-1;$$

then we solve the system of the first-order differential equations to obtain the solution to the original equation.

Example Consider

$$x''' - 3tx'' + e^{2t}x' - 2x + 3t = 0.$$

The third-order differential equation is equivalent to

$$\begin{bmatrix} \dot{x}_0 \\ \dot{x}_1 \\ \dot{x}_2 \end{bmatrix} = \begin{bmatrix} x_1 \\ x_2 \\ 2tx_2 - e^{2t}x_1 + 2x_0 - 3t \end{bmatrix}.$$

The reduction of a higher-order differential equation to a system enables us to directly have a uniqueness and existence theory for higher-order differential equations, using Theorem 8.3.2 and the uniqueness and existence theorem for system of the first order differential equations in Chap. 3.

Theorem 8.3.1 Let $f(X, t)$ be a function defined on some region Ω in $R^n \times R$ and satisfying a Lipschitz condition with regard to X. Given any $t_0 \in R$ and a vector $V \in R^n$, there exists a unique solution $u(t)$ of

$$x^{(n)} = f\left(x, x', \cdots, x^{(n-1)}, t\right),$$

such that $U(t_0) = V$; that is

$$U(t_0) = \begin{bmatrix} u(t_0) \\ u'(t_0) \\ u''(t_0) \\ \vdots \\ u^{n-1}(t_0) \end{bmatrix} = \begin{bmatrix} v_0 \\ v_1 \\ v_2 \\ \vdots \\ v_{n-1} \end{bmatrix} (= V).$$

It is also possible to carry out the reverse process. For illustration, we examine how to go from a first-order equation in R^2 to transform from a system of first-order equations in R^2 to a second-order equation in one variable.

Suppose

$$\dot{x}_1 = f_1(x_1, x_2),$$

$$\dot{x}_2 = f_2(x_1, x_2),$$

is a first order system in R^2. Differentiate $\dot{x}_1 = f_1(x_1, x_2)$ to find

$$\ddot{x}_1 = \frac{\partial f_1}{\partial x_1}\dot{x}_1 + \frac{\partial f_1}{\partial x_2}\dot{x}_2 \Rightarrow \dot{x}_2 = \left(\ddot{x}_1 - \frac{\partial f_1}{\partial x_1}\dot{x}_1\right) / \frac{\partial f_1}{\partial x_2}.$$

Use $\dot{x}_1 = f_1(x_1, x_2)$ to express x_2 as a function of x_1 and $\dot{x}_1$, say $x_2 = F(x_1, \dot{x}_1)$. Substituting these two new relations into $\dot{x}_2 = f_2(x_1, x_2)$ yields

$$\ddot{x}_1 = \frac{\partial f_1(x_1, F)}{\partial x_2} f_2(x_1, F) + \frac{\partial f_1(x_1, F)}{\partial x_1}\dot{x}_1.$$

The above equation is a second-order equation solely in terms of x_1 and its derivatives.

Example Consider the system of equations

$$\dot{x}_1 = -5x_1 + x_2,$$

$$\dot{x}_2 = \cos(x_1 + x_2).$$

From $\dot{x}_1 = -5x_1 + x_2$, we get

$$x_2 = \dot{x}_1 + 5x_1,$$

$$\dot{x}_2 = \ddot{x}_1 + 5\dot{x}_1.$$

Substituting them into $\dot{x}_2 = \cos(x_1 + x_2)$ yields

$$\ddot{x}_1 = -5\dot{x}_1 + \cos(6x_1 + \dot{x}_1).$$

Exercise 8.3

1 Convert the following linear differential equation to a system of linear equations

$$x^{(3)} + 4x'' - x' - 4x = 0.$$

Also show that the two systems have the same characteristic equation. Finally, determine the general solution to the system of linear systems.

2 Convert the following system into a first-order system of linear differential equations

$$\ddot{x} = x + y,$$

$$\ddot{y} = x + \dot{x} + y + \dot{y}.$$

8.4 Diagonalization

The method of separation of variables can be used to solve a one-dimensional differential equation, $\dot{x} = ax$. The general solution to this linear differential equation is $x(t) = ce^{at}$. Now consider an uncoupled linear system $\dot{x} = A_{n\times n}x$, where A is a diagonal matrix,

$$A = diag[a_1, \cdots, a_n].$$

The general solution of the uncoupled linear system can once again be solved by the method of separation of variables. It is given by $x_i(t) = c_i e^{a_i t}$, $i = 1, \cdots, n$.

The algebraic technique of diagonalizing a square matrix A can be used to reduce the linear system

$$\dot{x} = Ax, \tag{8.4.1}$$

to an uncoupled linear system, which can be easily solved by the method just mentioned.

Lemma 8.4.1 If the eigenvalues $\rho_1, \cdots, \rho_n$ of an $n \times n$ matrix A are real and distinct, then any set of corresponding eigenvectors $\{v_1, \cdots, v_n\}$ forms a basis for R^n, the matrix $P = [v_1, \cdots, v_n]$ is invertible and

$$P^{-1}AP = diag[\rho_1, \cdots, \rho_n].$$

To reduce the system (8.4.1) to an uncoupled system using the above theorem, introduce the linear transformation of coordinates

$$y = P^{-1}x,$$

where P is defined as in Lemma 8.4.1. Substituting $y = P^{-1}x$ into Eq. (8.4.1) yields

$$\dot{y} = diag[\rho_1, \cdots, \rho_n]y.$$

The solution to this uncoupled system is

$$y(t) = diag[e^{\rho_1 t}, \cdots, e^{\rho_n t}]y(0).$$

It follows

$$x(t) = Pdiag[e^{\rho_1 t}, \cdots, e^{\rho_n t}]P^{-1}x(0).$$

This is the general solution to the initial problem (8.4.1).

Example Consider

$$\begin{bmatrix} \dot{x}_1 \\ \dot{x}_2 \end{bmatrix} = \begin{bmatrix} -1 & -3 \\ 0 & 2 \end{bmatrix} \begin{bmatrix} x_1 \\ x_2 \end{bmatrix}.$$

The matrix A has two eigenvalues $\rho_1 = -1$ and $\rho_2 = 2$. A pair of corresponding eigenvectors is given by

$$v_1 = \begin{bmatrix} 1 \\ 0 \end{bmatrix}, \quad v_2 = \begin{bmatrix} -1 \\ 1 \end{bmatrix}.$$

The matrix P and its inverse P^{-1} is given by

$$P = \begin{bmatrix} 1 & -1 \\ 0 & 1 \end{bmatrix}, \quad P^{-1} = \begin{bmatrix} 1 & 1 \\ 0 & 1 \end{bmatrix}.$$

We have

$$PAP^{-1} = \begin{bmatrix} -1 & 0 \\ 0 & 2 \end{bmatrix}.$$

Under $y = P^{-1}x$, we have

$$\dot{y}_1 = -y_1,$$

$$\dot{y}_2 = 2y_2.$$

The general solution to this equation is

$$y_1(t) = c_1 e^{-t},$$

$$y_2(t) = c_2 e^{2t}.$$

We thus have

$$x_1(t) = c_1 e^{-t} + c_2\left(e^{-t} - e^{2t}\right),$$

$$x_2(t) = c_2 e^{2t}.$$

In the above example, the subspaces spanned by the eigenvectors v_1 and v_2 of the matrix A determine the stable and unstable subspaces of the linear system (8.4.1) according to the following definition.

Definition 8.4.1 Suppose that the $n \times n$ matrix A has k negative eigenvalues $\rho_1, \cdots, \rho_k$ and $n - k$ positive eigenvalues $\rho_{k+1}, \cdots, \rho_n$. Let $\{v_1, \cdots, v_n\}$ be a corresponding set of eigenvectors. Then the *stable* and *unstable subspaces* of the linear system (8.4.1), E^s and E^u are the linear subspaces spanned by

$$\{v_1, \cdots, v_k\}, \ \{v_{k+1}, \cdots, v_n\},$$

respectively, i.e.

$$E^s = Span\{v_1, \cdots, v_k\},$$

$$E^u = Span\{v_{k+1}, \cdots, v_n\}.$$

Exercise 8.4

1 Solve the following equations with the technique of diagonalizing

(i) $\begin{bmatrix} \dot{x}_1 \\ \dot{x}_2 \end{bmatrix} = \begin{bmatrix} 3 & 1 \\ 1 & 3 \end{bmatrix} \begin{bmatrix} x_1 \\ x_2 \end{bmatrix}$;

(ii) $\begin{bmatrix} \dot{x}_1 \\ \dot{x}_2 \\ \dot{x}_3 \end{bmatrix} = \begin{bmatrix} 1 & 0 & 0 \\ 1 & 2 & 0 \\ 1 & 0 & -1 \end{bmatrix} \begin{bmatrix} x_1 \\ x_2 \\ x_3 \end{bmatrix}$.

8.5 The Fundamental Theorem for Linear Systems

This section establishes the fundamental fact that for $x_0 \in R^n$ the initial value problem

$$\dot{x} = A_{nxn}x, \quad x(0) = x_0, \tag{8.5.1}$$

has a unique solution for all $t \in R$ which is given by

$$x(t) = e^{At}x_0. \tag{8.5.2}$$

Theorem 8.5.1 (The fundamental theorem for linear systems) Let A be an $n \times n$ matrix. Then for a given $x_0 \in R^n$, the initial value problem (8.5.1) has a unique solution for all $t \in R$ given by the solution (8.5.2).

Example Solve

$$\begin{bmatrix} \dot{x}_1 \\ \dot{x}_2 \end{bmatrix} = \begin{bmatrix} -2 & -1 \\ 1 & -2 \end{bmatrix} \begin{bmatrix} x_1 \\ x_2 \end{bmatrix}, \quad x_0 = \begin{bmatrix} 1 \\ 0 \end{bmatrix}.$$

Using the results in Sec. 5.4, we have

$$x(t) = e^{At}x_0 = e^{-2t}\begin{bmatrix} \cos t & -\sin t \\ \sin t & \cos t \end{bmatrix}\begin{bmatrix} 1 \\ 0 \end{bmatrix} = e^{-2t}\begin{bmatrix} \cos t \\ \sin t \end{bmatrix}.$$

Lemma 8.5.1[5] If the $2n \times 2n$ A has $2n$ distinct complex eigenvalues, $\rho_j = a_j + ib_j$ and $\overline{\rho}_j = a_j - ib_j$ and corresponding complex eigenvectors $w_j = u_j + iv_j$ and $\overline{w}_j = u_j - iv_j$, then

$$\{u_1, v_1, \cdots, u_n, v_n\},$$

[5] The lemma is proved in Hirsch and Smale (1974).

is a basis for R^{2n}, the matrix

$$P = [u_1 \quad v_1 \quad \cdots \quad u_n \quad v_n],$$

is invertible and

$$P^{-1}AP = diag\begin{bmatrix} a_j & -b_j \\ b_j & a_j \end{bmatrix},$$

is a real $2n \times 2n$ matrix with 2×2 blocks along the diagonal.

We immediately have that under the hypotheses of the above lemma, the solution of the initial problem (8.5.1) is given by

$$x(t) = Pdiage^{a_j t}\begin{bmatrix} \cos b_j t & -\sin b_j t \\ \sin b_j t & \cos b_j t \end{bmatrix} P^{-1}x_0 .$$

Example Solve $\dot{x} = Ax$ where

$$A = \begin{bmatrix} 1 & -1 & 0 & 0 \\ 1 & 1 & 0 & 0 \\ 0 & 0 & 3 & -2 \\ 0 & 0 & 1 & 1 \end{bmatrix}.$$

The matrix has the eigenvalues $1 \pm i$ and $2 \pm i$. The corresponding complex eigenvectors

$$u_1 \pm iv_1 = \begin{bmatrix} \pm i \\ 1 \\ 0 \\ 0 \end{bmatrix}, \quad u_2 \pm iv_2 = \begin{bmatrix} 0 \\ 0 \\ 1 \pm i \\ 1 \end{bmatrix}.$$

It is straightforward to calculate

$$P = \begin{bmatrix} 1 & 0 & 0 & 0 \\ 0 & 1 & 0 & 0 \\ 0 & 0 & 1 & 1 \\ 0 & 0 & 1 & 1 \end{bmatrix}, \quad P^{-1} = \begin{bmatrix} 1 & 0 & 0 & 0 \\ 0 & 1 & 0 & 0 \\ 0 & 0 & 1 & -1 \\ 0 & 0 & 1 & 1 \end{bmatrix},$$

$$P^{-1}AP = \begin{bmatrix} 1 & -1 & 0 & 0 \\ 1 & 1 & 0 & 0 \\ 0 & 0 & 2 & -2 \\ 0 & 0 & 1 & 2 \end{bmatrix}.$$

Hence the solution is

$$x(t) = \begin{bmatrix} e^t \cos t & -e^t \sin t & 0 & 0 \\ e^t \sin t & e^t \cos t & 0 & 0 \\ 0 & 0 & e^{2t}(\cos t + \sin t) & -e^{2t} \sin t \\ 0 & 0 & e^{2t} \sin t & e^{2t}(\cos t - \sin t) \end{bmatrix} x_0 .$$

In the case A has both real and complex eigenvalues and they are distinct, we have the following result: If A has distinct real eigenvalues ρ_j and corresponding eigenvectors v_j, $j = 1, \cdots, k$ and distinct eigenvalues $\rho_j = a_j + ib_j$ and $\overline{\rho}_j = a_j - ib_j$ and corresponding complex eigenvectors $w_j = u_j + iv_j$ and $\overline{w}_j = u_j - iv_j$, $j = k+1, \cdots, n$, then the matrix

$$P = [v_1 \quad \cdots \quad v_k \quad v_{k+1} \quad u_{k+1} \quad \cdots \quad v_n \quad u_n],$$

is invertible and

$$P^{-1}AP = diag[\rho_1 \quad \cdots \quad \rho_k \quad B_{k+1} \quad \cdots \quad B_n],$$

were the 2×2 blocks

$$B_j = \begin{bmatrix} a_j & -b_j \\ b_j & a_j \end{bmatrix}, \quad j = k+1, \cdots, n .$$

Example Solve $\dot{x} = Ax$ where

$$A = \begin{bmatrix} -3 & 0 & 0 \\ 0 & 3 & -2 \\ 0 & 1 & 1 \end{bmatrix}.$$

Its eigenvalues are $\rho_1 = -3$ and $\rho_{2,3} = 2 \pm i$ and the corresponding eigenvectors are

$$[1 \quad 0 \quad 0], \ [0 \quad \pm 1 \quad 1].$$

We calculate

$$P = \begin{bmatrix} 1 & 0 & 0 \\ 0 & 1 & 1 \\ 0 & 0 & 1 \end{bmatrix}, \quad P^{-1} = \begin{bmatrix} 1 & 0 & 0 \\ 0 & 1 & -1 \\ 0 & 0 & 1 \end{bmatrix}, \quad P^{-1}AP = \begin{bmatrix} 3 & 0 & 0 \\ 0 & 2 & -1 \\ 0 & 1 & 2 \end{bmatrix}.$$

The solution is

$$x(t) = \begin{bmatrix} e^{-3t} & 0 & 0 \\ 0 & e^{2t}(\cos t + \sin t) & -2e^{2t}\sin t \\ 0 & 2e^{2t}\sin t & e^{2t}(\cos t - \sin t) \end{bmatrix} x_0 .$$

Here, we don't discuss the case when the matrix A has multiple eigenvalues.[6]

Exercise 8.5

1 Solve $\dot{x} = Ax$ for

(i) $A = \begin{bmatrix} 1 & 0 & 0 \\ 0 & 2 & -3 \\ 0 & 3 & 2 \end{bmatrix}$;

(ii) $A = \begin{bmatrix} -1 & -1 & 0 & 0 \\ 1 & -1 & 0 & 0 \\ 0 & 0 & 0 & -2 \\ 0 & 0 & 1 & 2 \end{bmatrix}$.

---

[6] General solutions to $\dot{x} = Ax$ when A has multiple eigenvalues can be found in Sec. 1.7 of Perko (2000).

Chapter 9

Higher-Dimensional Nonlinear Differential Equations

The chapter examples higher dimensional nonlinear differential equations. Section 9.1 studies local stability and validity of linearization. Section 9.2 introduces the Liapunov methods and studies Hamiltonian systems. In Sec. 9.3, we examine differences between conservative and dissipative systems. We examine the Goodwin model in detail. Section 9.4 defines the Poincaré maps. In Sec. 9.5, we introduce center manifold theorems. Section 9.6 applies the center manifold theorem and Liapunov theorem to a simple planar system. In Sec. 9.7, we introduce the Hopf bifurcation theorem in higher dimensional cases and apply it to a predator-prey model. Section 9.8 simulates the Lorenz equations, demonstrating chaotic motion of deterministic dynamical systems.

9.1 Local Stability and Linearization

We now consider a general autonomous system of the form

$$\dot{x}(t) = f(x), \tag{9.1.1}$$

where

$$x = (x_1, \cdots, x_n)^T, \quad f = (f_1, \cdots, f_n)^T .$$

Suppose x^* is an equilibrium point of Eq. (9.1.1). Introduce $X(t) = x(t) - x^*$. From the Taylor theorem for functions,[1] we know

[1] The Taylor series is referred to App. A4.

$$f_j(x) = f_j(x^*) + \sum_k \frac{\partial f_j}{\partial x_k}(x^*) X_k + g_j(X),$$

$$j = 1, \cdots, n, \tag{9.1.2}$$

where $g_j(X)$ are higher order terms and

$$\frac{g_j(X)}{\|X\|} \to 0 \text{ as } \|X\| \to 0,$$

where

$$\|X\| \equiv \sqrt{\sum_j X_j^2}.$$

Using $\dot{x} = \dot{X}$ and $f(x^*) = 0$, Eq. (9.1.1) can be expressed in vector form as

$$\dot{X} = AX + g(X), \tag{9.1.3}$$

where the matrix

$$A = \left(\frac{\partial f_i}{\partial x_j} \right)_{n \times n},$$

is the *Jacobian matrix* of f at x^*. The linear system

$$\dot{X} = AX,$$

is called the *linearized system* of Eq. (9.1.1).

Example Consider

$$\begin{bmatrix} \dot{x}_1 \\ \dot{x}_2 \\ \dot{x}_3 \end{bmatrix} = \begin{bmatrix} x_1^2 - x_2 + x_3 \\ x_1 - x_2 \\ 2x_2^2 + x_3 - 2 \end{bmatrix} \begin{bmatrix} x_1 \\ x_2 \\ x_3 \end{bmatrix}.$$

The Jacobian at equilibrium is

$$J = \begin{bmatrix} 2x_1 & -1 & 1 \\ 1 & -1 & 0 \\ 0 & 2x_2 & 0 \end{bmatrix}.$$

The system contains two equilibrium points $[-2 \quad -2 \quad -6]^T$ and $[1 \quad 1 \quad 0]^T$. Hence,

$$J = \begin{bmatrix} -4 & -1 & 1 \\ 1 & -1 & 0 \\ 0 & -4 & 0 \end{bmatrix}_{[-2\ \ -2\ \ -6]}, \quad J = \begin{bmatrix} 2 & -1 & 1 \\ 1 & -1 & 0 \\ 0 & 2 & 0 \end{bmatrix}_{[1\ \ 1\ \ 0]}.$$

The following theorem includes the case that g explicitly depends on t.

Theorem 9.1.1 If (i) $g(0,t) = 0$ and A is an $n \times n$ constant matrix, (ii) the solutions of

$$\dot{x} = Ax,$$

are asymptotically stable; and (iii)

$$\lim_{\|x\| \to 0} \left\{ \frac{h(x,t)}{\|x\|} \right\} = 0,$$

uniformly in t, $0 \le t < \infty$, then the zero solution,

$$x(t) = 0 \text{ for } t \ge 0,$$

is an asymptotically stable solution of the regular system

$$\dot{x} = Ax + g(x,t).^2$$

Example Consider the Van der Pol's equation.

$$\ddot{x} + e(x^2 - 1)\dot{x} + x = 0, \quad e < 0.$$

We replace the system by

$$\begin{bmatrix} \dot{x}_1 \\ \dot{x}_2 \end{bmatrix} = \begin{bmatrix} 0 & 1 \\ -1 & e \end{bmatrix} \begin{bmatrix} x_1 \\ x_2 \end{bmatrix} + \begin{bmatrix} 0 \\ -x_1^2 x_2 \end{bmatrix}.$$

The eigenvalues of A are negative when $e < 0$; therefore all solutions of $A[x_1 \ \ x_2]^T$ are stable. It is straightforward to demonstrate that the conditions of the theorem are satisfied. Hence, the system has an asymptotically stable zero solution.

---

[2] Here, by regular system we mean that $g(x,t)$ is continuous and $\partial g_i / \partial x_j$ for all i, j, are continuous on $-\infty < x_j < \infty$ and $-\infty < t < \infty$. The proof of the theorem is referred, for instance, to Cesari (1971).

Definition 9.1.1 An equilibrium point x^* of the system (9.1.1) is called a *sink* if all of the eigenvalues of the Jacobian matrix A at x^* have negative real part; it is called a *source* if all of the eigenvalues of A have positive real part. It is called a *saddle* if it is a hyperbolic equilibrium point and A has at least one eigenvalue with a positive real part and at least one with a negative part.

If x^* is sink, there is a neighborhood U of x^* such that any solution $u(t)$ with $u(t_0) \in U$ remains in U for $t \geq t_0$, and $\lim_{t\to\infty} u(t) = x^*$; if x^* is a source, there is a neighborhood U of x^* such that any solution $u(t)$ with $u(t_0) \in U$ remains in U for $t \leq t_0$, and $\lim_{t\to-\infty} u(t) = x^*$.

Theorem 9.1.2 If at a zero of an autonomous differential equation the linearization is a sink or a source, then zero is itself a sink or a source. Furthermore all solutions sufficiently close to the zero tend to it exponentially fast as $t \to \infty$ for a sink or as $t \to -\infty$ for a source.

The following example shows that it is possible for a zero to be a sink without the linearization being a sink.

Example Consider

$$\dot{x}_1 = -x_2 - x_1\left(x_1^2 + x_2^2\right),$$

$$\dot{x}_2 = x_1 - x_2\left(x_1^2 + x_2^2\right).$$

The linearization at $(0, 0)$ is a center. Introduce

$$d(t) = \left(x_1^2 + x_2^2\right).$$

Then $\dot{d} = -2d^2$ for the original system. The general solution to this new system is $u = 1/(2t - C)$. As d is the distance from the origin of a solution to the original equation, the distance goes to zero as $t \to \infty$.

Theorem 9.1.3 Let f be a C^1 function. If all the eigenvalues of the Jacobian matrix A have negative real parts, then the equilibrium point x^* of the differential equation $\dot{x} = f(x)$ is asymptotically stable.

Theorem 9.1.4 Let f be a C^1 function. If at least one of the eigenvalues of the Jacobian matrix A has positive real parts, then the equilibrium point x^* of the differential equation $\dot{x} = f(x)$ is unstable.

In the absence of eigenvalues with zero real parts, linearization captures many of the local qualitative features, such as stability type and local stable and unstable manifolds of nonlinear systems near equilibria. The following theorem demonstrates that linearization determines the full orbit structure locally under certain conditions.

Theorem 9.1.5 (Grobman-Hartman theorem) If x^* is a hyperbolic equilibrium point of $\dot{x} = f(x)$ (that is, all the eigenvalues of the Jacobian matrix A have nonzero real parts), then there is a neighborhood of x^* in which f is topologically equivalent to the linear vector field $\dot{x} = Ax$.

The Grobman-Hartman Theorem shows that the stability type of a hyperbolic equilibrium point is preserved under arbitrarily but small nonlinear perturbations.

$$\dot{X} = AX + g(X). \tag{9.1.3}$$

Exercise 9.1

1 Show that the zero solution of the equation

$$\ddot{x} + k\dot{x} + \sin x = 0, \quad k > 0$$

is asymptotically stable.

2 For the differential equation

$$\dot{x}_1 = x_2,$$

$$\dot{x}_2 = -x_1 - x_2^3.$$

(i) Show that at the origin the linearization gives a center, so we need further analysis to determine the nonlinear behavior; (ii) Show that $V = x_1^2 + x_2^2$ serves as a Liapunov function.

9.2 Liapunov Functions

Section 6.2 introduces the Liapunov methods for stability analysis. This section introduces the analysis for higher dimensions.

Definition 9.2.1 Consider the function: $V: U \times R_+ \to R$, where $U \subset R^n$ is open and connected, $0 \in U$, and assume that $V \in C^1$; we say that V is *positive semidefinite* if $V(x, t) \geq 0$ for all $(x, t) \in U \times R_+$; V is positive definite if there is a function $W \in C^0(U, R)$ such that for all $(x, t) \in U \times R_+$, $x \neq 0$:

$$V(x, t) > W(0) > 0,$$

and $V(0, t) = W(0) = 0$; V is *indefinite* if for every neighborhood B of the origin it assumes positive as well as negative values in $B \times R_+$.

We can similarly define *negative semidefinite* and *negative definite* functions. We say that the function $h: R^+ \to R^+$ *belongs to the function class* H if $h(0) = 0$, and it is strictly increasing and continuous. Let $X \subset R^n$ be open and connected

$$x \in X,\ f \in C^0(X \times R_+),\ f_x^{'} \in C^0(X \times R_+),\ f(0, t) = 0,$$

and consider the system

$$\dot{x} = f(x, t). \tag{9.2.1}$$

The derivative of V with respect to Eq. (9.2.1) at $(x, t) \in C^0(X \times R_+)$ is

$$\dot{V}(x, t) = V_t^{'}(x, t) + \sum_{i=1}^{n} V_{x_k}^{'}(x, t) f_i =$$

$$V_t^{'}(t, x) + \langle gardV(x, t), f(x, t) \rangle . \tag{9.2.2}$$

In the following theorems,[3] U denotes an open and connected subset of X which contains the origin.

Theorem 9.2.1 (Liapunov's first theorem) If there exists a function $V: U \times R_+ \to R$, where $0 \in U \subset X$, and a function $h \in H$ such that for

$$(x, t) \in U \times R_+ : V(x, t) \geq h(|x|),\ V(t, 0) = 0,$$

[3] The proofs of these theorems can be found in Rouche, Habets and Laloy (1977).

and $\dot{V}$ with regard to the problem (9.2.1) is negative semidefinite, then the origin is stable in the Liapunov sense.

The conditions in the theorem imply that V is positive definite. Inversely, one may prove that if V is positive definite, then such $h \in H$ exists.

Theorem 9.2.2 (Liapunov's second theorem) If there exists a function $V : U \times R \to R$, where $0 \in U \subset X$, and functions $h_1, h_2, h_2 \in H$ such that: $h_1(|x|) \le V(x, t) \le h_2(|x|)$, and $\dot{V} \le -h_3(|x|)$, then the origin is uniformly asymptotically stable.

Theorem 9.2.3 (Liapunov's third theorem) If there exist a function $V : U \times R \to R$, where $0 \in U \subset X$, functions $h_2, h_2 \in H$ such that: $V(x, t) \le h_2(|x|)$, and $\dot{V} \le -h_3(|x|)$, $(x, t) \in U \times R_+$, and a $t_0 \in R_+$ such that in every neighborhood $B(0, \delta)$ of the origin, there is an $x \in B(0, \delta)$ for which $V(x, t_0) < 0$, then the origin is unstable.

The functions V in Theorems 9.2.1-9.2.3 are called loosely *Liapunov function*. The method sketched in these theorems is often called the *method of Liapunov functions*.

Example Consider

$$\dot{x}_1 = x_2,$$

$$\dot{x}_2 = -a \sin x_1 - b x_2, \quad a > 0,$$

where a and b are parameters. Introduce

$$V(x_1, x_2) = a(1 - \cos x_1) + x_2^2 > 0,$$

$$(x_1, x_2) \ne (0, 0), \ |x_1| < \pi, \ x_2 \in R. \tag{9.2.3}$$

This is a positive definite function on the indicated domain satisfying the conditions of Liapunov's First and Second Theorems. Its derivative with respect to the system is

$$\dot{V} = -b x_2^2,$$

which is negative semedefinite for positive b. From Eq. (9.2.3), we conclude that the origin is stable in the Liapunov sense. Nevertheless, if we modify V by

$$V_1(x_1, x_2) = 4a(1 - \cos x_1) + x_2^2 + (bx_1 + x_2)^2 > 0,$$

$$(x_1, x_2) \neq (0, 0),\ |x_1| < \pi,\ x_2 \in R.$$

This function is positive definite and its derivative with respect to Eq. (9.2.1) is

$$\dot{V}_1(x_1, x_2) = -2b(x_2^2 + ax_1 \sin x_1) < 0.$$

Thus V satisfies the conditions of Liapunov's Second Theorem for positive b. We see that the origin is uniformly asymptotically stable. If $b < 0$, then the Liapunov function $-V_1$ and its derivative $-\dot{V}_1$ satisfy the conditions of Liapunov's Third Theorem, so the origin is unstable.

Example Consider

$$\dot{x}_1 = -2x_2 + x_2 x_3,$$

$$\dot{x}_2 = x_1 - x_1 x_3,$$

$$\dot{x}_3 = x_1 x_2.$$

The origin is an equilibrium point for this system and the Jacobian matrix at the origin is

$$J = \begin{bmatrix} 0 & -2 & 0 \\ 1 & 0 & 0 \\ 0 & 0 & 0 \end{bmatrix}.$$

Thus J has eigenvalues $\rho_1 = 0$, $\rho_{2,3} = \pm 2i$, i.e., $x = 0$ is a nonhyperbolic equilibrium point. Choose

$$V(t) = \frac{c}{2}(x_1^2 + 2x_2^2 + x_3^2),$$

where $c > 0$ is a constant. We have $V(t) > 0$ for $x \neq 0$ and $\dot{V} = 0$ for all $x \in R^3$. Therefore, the origin is stable.

Liapunov's theorems are valid for autonomous systems as well. But there are results about the asymptotic stability of the equilibrium of an autonomous system that relaxes the requirements in Liapunov's Second Theorem.[4]

Example Hamiltonian problems in dynamics.[5]
Conservative problems can be expressed in the form

$$\dot{p}_i = -\frac{\partial H}{\partial q_i},$$

$$\dot{q}_i = -\frac{\partial H}{\partial p_i}, \quad i = 1, \cdots, n,$$

where H is a given function called the Hamiltonian of the system, q_i is a generalized coordinate, and p_i is a generalized momentum. The Hamiltonian is defined by

$$H(p, q) = T(p, q) + V(q),$$

where T is the kinetic energy and V is the potential energy. Assume $V(0) = 0$. T is a positive definite quadratic form in p_i, so $T(0, 0) = 0$. Suppose that $q = 0$ is a minimum of V so that V, and hence H, is positive definite in a neighborhood of the origin. Then

$$\dot{H} = \sum_i \frac{\partial H}{\partial p_i} \dot{p}_i + \sum_i \frac{\partial H}{\partial q_i} \dot{q}_i = 0.$$

Hence, H is a weak Liapunov function for the dynamical system. The zero solution is stable when it is at a minimum of V.

Consider an n-dimensional dynamic system

$$\dot{x} = A_{n \times n} x + g(x), \quad x \in R^n, \tag{9.2.4}$$

where A is real and $g(x)$ is of smaller order of magnitude than Ax. We now construct explicit Liapunov functions for the linearized system $\dot{x} = Ax$ and show that they also work for the original system.

---

[4] See Sec. 1.5 in Farkas (1994).
[5] See Perko (2001) for further examination of Hamiltonian systems.

Theorem 9.2.4 Let the n- dimensional system (9.2.4) be regular and the origin be an equilibrium point of the system (9.2.4). If (i) the zero solution of $\dot{x} = Ax$ is asymptotically stable; (ii) $g(0) = 0$ and $O(\|x\|)/\|x\| \to 0$ as $\|x\| \to 0$, then $x(t) = 0$ for any t_0 is an asymptotically stable solution of the system (9.2.4).

Proof: From the conditions, we know all the real parts of eigenvalues are negative, i.e., $\operatorname{Re}\{\rho_j\} < 0$ for any j. Define $V(x) = x^T Kx$ (where K is to be determined) and differentiate V with regard to t for the system (9.2.4)

$$\dot{V} = \dot{x}^T Kx + x^T K\dot{x} = x^T \left(A^T K + KA\right)x + g^T Kx + x^T Kg . \quad (9.2.5)$$

We now want to determine K such that $A^T K + KA = -I$ and V is positive definite. We choose

$$K = \int_0^\infty e^{A^T t} e^{At}\, dt .$$

We see that K is symmetry. As $\dot{x} = Ax$ is asymptotically stable, there are $c > 0$ and $b < 0$ such that $\|e^{At}\|, \|e^{A^T t}\| < ce^{bt}$. This ensures the convergence of the integrals below

$$\dot{K} = \frac{1}{dt}\int_0^\infty e^{A^T t} e^{At}\, dt = A^T \int_0^\infty e^{A^T t} e^{At}\, dt + \int_0^\infty e^{A^T t} e^{At}\, dt A = -I , \quad (9.2.6)$$

where we use

$$\frac{1}{dt}\int_0^\infty e^{A^T t} e^{At}\, dt = e^{A^T t} e^{At}\Big|_0^\infty = -I_{n\times n} .$$

From the right-hand sides of Eq. (9.2.6) and the definition of K, we have

$$A^T K + KA = -I . \quad (9.2.7)$$

To show that V is positive definite, we first note

$$V = \int_0^\infty \left(x^T e^{A^T t}\right)\left(e^{At} x\right) dt = \int_0^\infty x^T e^{A^T t} e^{At} x\, dt .$$

The integrand is simply the sum of certain squares, and is therefore positive definite. Using Eq. (9.2.7) and the symmetry of K, we reduce Eq. (9.2.5) into

$$\dot{V} = -x^T x + 2g^T Kx. \tag{9.2.8}$$

It is possible to find a neighborhood of the origin in which the first term in Eq. (9.2.8) dominates. Now

$$\left|2g^T Kx\right| \le 2\|g(x)\|\|K\|\|x\|.$$

By (ii), given any $\varepsilon > 0$ there exists $\delta > 0$ such that

$$\|x\| < \delta \Rightarrow \|g(x)\| < \varepsilon\|x\|.$$

Suppose ε to be chosen so that $\varepsilon < 1/4\|K\|$. Then we have

$$\left|2g^T Kx\right| \le 2\|g(x)\|\|K\|\|x\| \le 2\varepsilon\|K\|\|x\|^2 < \frac{\|x\|}{2}.$$

From this inequality and Eq. (9.2.8), we conclude that $\dot{V}$ is negative definite on $\|x\| < \delta$. Therefore, the zero solution is asymptotically stable.

Theorem 9.2.5 Let the n-dimensional system (9.2.4) be regular and the origin be an equilibrium point of the system (9.2.4). If (i) the eigenvalues of A are distinct, none are zero, and at least one has positive real part; (ii) $g(0) = 0$ and $O(\|x\|)/\|x\| \to 0$ as $\|x\| \to 0$, then the zero solution $x(t) = 0$ for any t_0 to the system (9.2.4) is unstable.

Proof: Here, we shall only prove the case that all the eigenvalues ρ_j are real and different. As the origin is unstable for the linearized system, at least one eigenvalue is positive. We know that a real $n \times n$ matrix invertible C can be chosen such that $C^{-1}AC = diag[\rho_j] \equiv D$, where at least one ρ_j is positive. Introduce $x = CX$. Now the system (9.2.4) is transformed into

$$\dot{X} = DX + C^{-1}g(CX). \tag{9.2.9}$$

Introduce

$$V(X) = X^T D^{-1} X = \sum_{j=1}^{n} \frac{X_j^2}{\rho_j}.$$

If $\rho_1 > 0$, then $V(X) > 0$ when $X_1 \neq 0$ and all $X_j = 0,\ j \neq 1$. Using the system (9.2.9), we calculate

$$\dot{V} = 2\sum_{j=1}^{n} X_j^2 + 2X^T D^{-1} g(CX). \tag{9.2.10}$$

By (ii), given any $\varepsilon > 0$ there exists $\delta > 0$ such that

$$\|C\|\|X\| < \delta \Rightarrow \|CX\| < \delta \Rightarrow \|g(CX)\| < \varepsilon|CX|.$$

Therefore, $\|X\| < \delta / \|C\|$ implies

$$|2X^T D^{-1} g(CX)| \leq 2\|X^T\|\|D^{-1}\|\|g(CX)\| \leq 2\varepsilon\|D^{-1}\|\|C\|\|X\|.$$

If we choose $\varepsilon < 1/2\|D^{-1}\|\|C\|$, then $|2X^T D^{-1} g(CX)| \leq \|X\|^2$. Under this condition, $\dot{V}$ is positive definite. Hence, the zero solution is unstable.

Exercise 9.2

1 Show the origin is an asymptotically stable equilibrium of the system

$$\dot{x}_1 = 2x_1^3 x_2^2 - x_1,\ \ \dot{x}_2 = -x_2,$$

and that the ball $B(0, 1)$ is a subset of the basin.

2 Show that for the system

$$\dot{x}_1 = -2x_2 + x_2 x_3 - x_1^3,$$

$$\dot{x}_2 = x_1 - x_1 x_3 - x_2^3,$$

$$\dot{x}_3 = x_1 x_2 - x_3^3,$$

the origin is asymptotically stable, but it is not a sink.

9.3 Conservative Systems

We are now concerned with the concept of conservative systems and examine the properties of such systems. We also discuss the relation between conservative systems and structural stability.

Consider a dynamic system $\dot{x} = f(x)$. The fundamental property of a conservative system is the existence of a function for the dependent variables which is a constant of the motion equation and plays the role of "energy". Accurately, the system is called *conservative* if there exists a function $G(x)$, known as a first integral, or simply an integral, of the system, such that

$$\frac{dG(x)}{dt} = \sum_{i=1}^{n} \frac{\partial G}{\partial x_i} \frac{dx_i}{dt} = 0 .$$

In physical terms, conservative systems are characterized by the fact that during evolutionary processes a "volume" element in phase space changes only its shape but retains its volume in the course of time. This difference is illustrated in Fig. 9.3.1. In dissipative systems, trajectories are attracted to a fixed point, and volume shrinks, but in conservative systems the points rotate around an elliptic fixed point and volume is conserved.

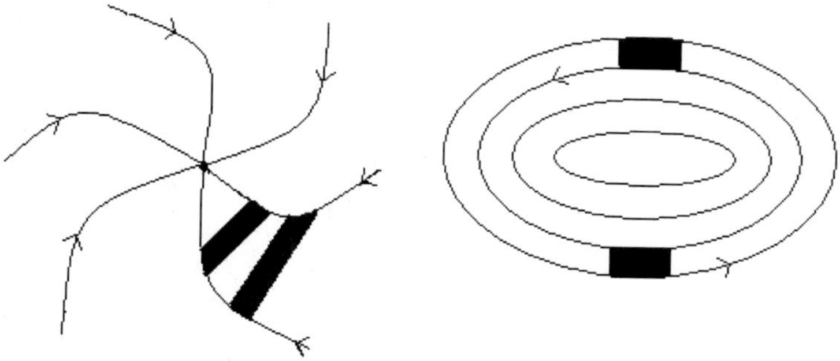

Fig. 9.3.1 Dissipative systems and conservative systems.

Definition 9.3.1 If a differential equation in R^n implies for some function $F(x)$ that $F'(x) = 0$, then

$$F(x) = A ,$$

where A is a constant, along the trajectories of the solutions, and the equation $F(x) = A$ is called a *conservation law*.

The system of ordinary differential systems $\dot{x} = f(x)$ is called *dissipative* if there are numbers $R > 0$ and $t_1 > 0$ such that for all solutions $x(t)$ of the system it is the case that $|x(0)| \le R$ always implies that $|x(t)| < R$ for all times $t > t_1$. About dissipative systems, the following theorem holds.

Theorem 9.3.1 The dissipative system $\dot{x} = f(x, t), x \in R^n$ has a solution of period $p > 0$ if (i) the function f is p-periodic with respect to t; and (ii) for every initial value $x_0 \in R^n$ there is a unique solution $x(t)$ with $x(0) = x_0$ which exists for all times $t \in [0, +\infty]$. Here, $x(t)$ depends continuously on x_0.

Proof: To prove the theorem, we use the following Brouwer fixed point theorem. Let $A: X \to X$ be a compact operator on R^n. Suppose that for some fixed natural number m the set $A^m(X)$ is bounded. Then A has a fixed point.

Construct the shift operator $A: R^n \to R^n$ by $Ax_0 = x(p)$. Here $x(.)$ is the solution of the system. Then

$$A^m x_0 = x(mp).$$

Set $G = \{x \in R^n : |x| < R\}$. Hence $A^m x_0 \in G$ for all x_0 belongs to the closure of G and sufficiently large m. Thus A has a fixed point, to which the desired periodic solution corresponds.

Example Find the trajectories of the system

$$\dot{x}_1 = 4 - 2x_2, \quad \dot{x}_2 = 12 - 3x_1^2 .$$

The equilibrium points of the system are $(-2, 2)$ and $(2, 2)$. For this system, we have

$$\frac{dx_2}{dx_1} = \frac{12 - 3x_1^2}{4 - 2x_2} .$$

Separating the variables in the above equation and integrating, we find that solutions satisfy

$$H(x_1, x_2) = 4x_2 - x_2^2 - 12x_1 + x_1^3 = c ,$$

where c is an arbitrary constant. It is straightforward to check that $(-2, 2)$ is a center and $(2, 2)$ is a saddle point.

Example A well-known example of differential equations possessing first integrals is the existence of a first integral in the so-called Hamiltonian formations. For a given C^1 function

$$H: R^2 \to R ,$$

a planar system of differential equations of the form

$$\dot{x}_1 = \frac{\partial H}{\partial x_2},$$

$$\dot{x}_2 = -\frac{\partial H}{\partial x_1},$$

is called a *Hamiltonian system with the Hamiltonian* H. The total energy of a mechanical system, up to a multiplicative or additive constant, can often be taken as the Hamiltonian of the system. The Hamiltonian function is obviously a first integral – *conservation of energy*, as $\dot{H} = 0$. It is straightforward to check that

$$H(x_1, x_2) = \frac{x_2^2}{2} + \int_0^{x_1} g(s)ds,$$

is the Hamiltonian of the equivalent first-order system

$$\dot{x}_1 = x_2,$$

$$\dot{x}_2 = -g(x_1).$$

Conservative systems often have oscillatory solutions and have therefore been widely used to model phenomena such as oscillations in prey and predator populations, urban land rent and land use density interactions, unemployment and economic growth dynamics, and so on.

We consider the predator-prey system, which has been applied in economic dynamics. The system consists of two differential equations

$$\dot{x}(t) = \alpha(y_0 - y(t))x(t),$$

$$\dot{y}(t) = \beta(x(t) - x_0)y(t), \tag{9.3.1}$$

where $x(t)$ and $y(t)$ are respectively the population of preys and predators, and α, β, x_0 and y_0 are parameters.[6]

[6] This is called the *Lotka-Volterra predator-prey* system. It was suggested by Volterra (1931) to explain the change in the composition of catch observed by fishermen on the Adriatic Sea after World War One. The same model occurred in Lotka (1924). See also Freedman (1980) and Sec. 3.4 in Farkas (1994) for analysis of other (generalized) Lotka-Volterra models.

In the dynamic urban literature, this model has been employed to describe the dynamics of a small urban area. We interpret $x(t)$ and $y(t)$ respectively as the land use density and the land rent. This is a simple demand-supply model of speculative land rent under foresight, with particularly congruent expectations from demanders and suppliers.[7]

We now provide another application of this type of model to economics. The *Goodwin* model reviewed below is built to describe the class struggle in labor market.[8] Consider two kinds of households: workers and capitalists. It is assumed that workers spend all their income $w(t)L(t)$ on consumption, where $w(t)$ is the wage rate at time t and $L(t)$ is the labor force. Capitalists save all their income, which is equal to $Y - wL$, where $Y(t)$ is production. The goods price is normalized to unity. Let $K(t)$ denote capital stocks and $a(t)$ $(= Y(t)/L(t))$ denote labor productivity. Assume that labor productivity grows at the constant rate g, that is, $a(t) = a_0 \exp(nt)$, where a_0 is the initial level of labor productivity. The wage income share of national income is $wL/Y = w/a$. Hence, the profit share is equal to $1 - w/a$. As the savings are determined by

$$S(t) = Y - wL = \left(1 - \frac{w}{a}\right)Y,$$

the investment is

$$\dot{K} = S = Y - wL = \left(1 - \frac{w}{a}\right)Y, \tag{9.3.2}$$

where we neglect any depreciation. Introducing the capital output ratio $k = K/Y$, we rewrite Eq. (9.3.2) as

$$\frac{\dot{K}}{K} = \left(1 - \frac{w}{a}\right)\frac{1}{k}. \tag{9.3.3}$$

We assume k to be constant. Hence, by $k = K/Y$, we have $\dot{K}/K = \dot{Y}/Y$. From $a = Y/L$, we have

---

[7] See Dendrinos and Mullally (1985) and Zhang (1988) for further explanations of the urban model.

[8] See Gabisch and Lorenz (1989) for the explanation of the model.

$$\frac{\dot{Y}}{Y} - \frac{\dot{L}}{L} = g .$$

From this equation, $\dot{K} / K = \dot{Y} / Y$, and Eq. (9.3.3), we have

$$\frac{\dot{L}}{L} = \left(1 - \frac{w}{a}\right)\frac{1}{k} - g . \tag{9.3.4}$$

Introducing the labor bill share, $y = w / a$, and the employment rate, $x = L / N$, we can show that the dynamics of the class struggle are described by

$$\dot{x} = x\left(\frac{1 - y}{k} - (g + n)\right),$$

$$\dot{y} = y\left(\frac{\dot{w}}{w} - g\right), \tag{9.3.5}$$

where we use $\dot{a} / a = g$. The wage rate is assumed to be a fast variable and be determined by a Phillips curve relation as follows

$$\dot{w} = wf(x)x, \ \lim_{x \to 1} f(x) = +\infty, \ \lim_{x \to 0} f(x) < 0, \ f' > 0 .$$

Approximating this relation linearly by

$$\frac{\dot{w}}{w} = -r + bx .$$

From this equation and Eqs. (9.3.5), we have

$$\dot{x} = \frac{x}{k}(y_0 - y),$$

$$\dot{y} = by(x - x_0), \tag{9.3.6}$$

where

$$y_0 \equiv 1 - (g + n)k , \ \ x_0 \equiv \frac{r + g}{b} .$$

We see that the Goodwin model (9.3.6) is dynamically identical to the predator-prey model. The general discussion about properties of the system (9.3.1) should be applicable to the Goodwin model. The formal identity of the Goodwin model with the Lotka-Volterra predator-prey system establishes an analogy between the class struggle and the struggle of competitive species. The Goodwin model, with its interaction of the employment rate and the wage bill share, is strongly reminiscent of the models of classical political economics. The model, sometimes referred to as a neo-Marxian model, has stimulated modern attention to the classical economists such as Ricardo, Smith, and Marx. There are different extensions of the model.[9] The model is simple and may exhibit oscillations. However, the property of structural instability limits its applications. It is known that even small perturbations in the functional forms will change the qualitative properties of the system. It can seen that the model can hardly be transferred to the real process under consideration because when we construct a model, the real situation is simplified and idealized. The parameters would be determined only approximately. The question then arises of how to choose those properties of the model of a process, which are not very sensitive to small changes in the model. The concept of structural stability answers the question.[10]

To show that the system (9.3.6) is conservative, we make the following transformation

$$u \equiv \frac{x}{x_0}, \quad v \equiv \frac{y}{y_0}, \quad \sigma \equiv \frac{bkx_0}{y_0}, \quad t^* = \frac{t}{k}.$$

Under this transformation, the system (9.3.6) becomes

$$\dot{u} = u(1 - v),$$

$$\dot{v} = \sigma y(u - 1), \tag{9.3.7}$$

where the derivatives are with respect to t^*. The following first integral can easily be identified as

---

[9] See Desai (1973), Velupillai (1978), Shah and Desai (1981), van der Ploeg (1983, 1987), and Zhang (1988).

[10] The fundamental ideas for the concept of structural stability were introduced by Poincaré. The model development of the concept was initiated by Andronov and Pontrjagin in 1973. Smale (1967) made significant progresses for phase spaces with small dimension. He showed that for phase space of large dimension, systems exist in the neighborhood of which there is no structural stable system. This result means that the problem of complete topological classification of differential equations with high-dimensional phase space is hopeless, even when restricted to generic equations and nondegenerate cases. See App. A.5 for the definition of of structural stability.

$$G(u, v) = \sigma(u - \ln u) + v - \ln v = A,$$

where A is a constant. Since $G(u, v)$ does not change as we move along a trajectory or solution curve of the equations, these trajectories are defined by the curves $G(u, v) = A$ for different values of the constant A. It follows from this that the equilibrium point $(1, 1)$ cannot be a stable focus. For if it were, then all curves in a neighborhood of it would tend to it, and hence would have

$$G(u, v) = G(1, 1),$$

since G is a continuous function. But this implies that G is constant in the neighborhood of $(1, 1)$, which contradicts its definition. It also follows by similar arguments that there are no stable or unstable limit cycles surrounding the equilibrium point. All trajectories starting in the positive quadrant are bounded, so the only possibility is that the phase plane consists of closed trajectories around the equilibrium point, each with a different value of the "energy" $G(u, v)$. The model is thus orbitally stable but not stable.

In order to show how the behavior of the system (9.3.7) can be affected by small perturbations, let us add a term $-ru^2$ to the first of the system (9.3.7) to obtain

$$\dot{u} = u(1 - v) - ru^2,$$

$$\dot{v} = \sigma y(u - 1). \tag{9.3.8}$$

If the parameter r is extremely small, it is reasonable to require that the new term will not have a significant effect on the solution of the original system. However, an eigenvalue analysis shows that the equilibrium point $(1, 1 - r)$ is a stable focus in the linearized system and is therefore a stable focus in the nonlinear system, however small r may be. It is straightforward to demonstrate that the following function

$$V(u, v) = \sigma(u - \ln u) + v - (1 - r)\ln v$$

is a global (in the positive quadrant) Liapunov function for the system. The solutions now spiral into the equilibrium point, and system can no longer be put forward as a model for oscillations.

In fact, for the general perturbation problem to the system (9.3.7)

$$\dot{u} = u(1 - v) + \varepsilon f_1(u, v),$$

$$\dot{v} = \sigma y(u - 1) - \varepsilon f_2(u, v), \tag{9.3.9}$$

where f_1 and f_2 are perturbation functions, even when ε is extremely small, it is possible to observe stable or unstable cycles for some specified functions f, which are qualitatively different from the original periodic solution. It is known that conservative systems tend to be amenable to analysis, but they have some major disadvantages as models for real systems. As all conservative systems are structurally unstable,[11] they should be used with great care.

9.4 Poincaré Maps

Poincaré maps transform continuous-time dynamical systems defined by differential equations to discrete-time dynamical systems (maps). The introduction of such maps allows us to apply the results concerning maps to differential equations. This is particularly efficient if the resulting map is defined in a lower-dimensional space than the origin system.

Consider a dynamic system

$$\dot{x} = f(x), \quad x \in R^n, \tag{9.4.1}$$

with smooth $f(x)$. Assume that the system (9.4.1) has a periodic orbit L_0. Take a point $x_0 \in L_0$ and introduce a cross-section Σ at this point, see Fig. 9.4.1.

The cross-section Σ is a smooth hypersurface of dimension $n-1$, intersecting L_0 at a nonzero angle. Since the dimension of Σ is one less than the dimension of the state space, we say that Σ is of "*codimension*" one, $\operatorname{codim}\Sigma = 1$. Suppose that Σ is defined near the point x_0 by the zero-level set of a smooth scalar function

$$g: R^n \to R, \; g(x_0) = 0$$

as

$$\Sigma = \{x \in R^n : g(x) = 0\}.$$

[11] Britton (1986).

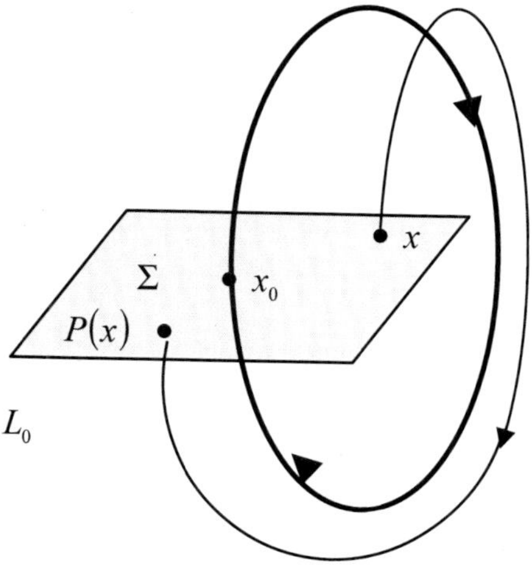

Fig 9.4.1 The Poincaré map associated with a cycle.

A nonzero intersection angle ("*transversality*") means that the gradient

$$\nabla g(x) = \left(\frac{\partial g(x)}{\partial x_1}, \frac{\partial g(x)}{\partial x_2}, \ \cdots, \frac{\partial g(x)}{\partial x_n} \right)^T,$$

is not orthogonal to L_0 at x_0, that is

$$\langle \nabla g(x_0), f(x_0) \rangle \neq 0,$$

where $\langle \cdot, \cdot \rangle \neq 0$ is the standard scalar product in R^n. A possible choice of Σ is a hyperplane orthogonal to the cycle L_0 at x_0 given by

$$\langle f(x_0), x - x_0 \rangle = 0.$$

Consider now the orbits of the system (9.4.1) near the cycle. The cycle starts at the point x_0 on Σ and returns to Σ at the same point. Since the solutions of the system depend smoothly on their initial points, an orbit starting at a point $x \in \Sigma$ sufficiently close to x_0 also returns to Σ at some point $\widetilde{x} \in \Sigma$ near x_0. Moreover, nearby orbits will also intersect Σ transversally. Thus, we have constructed a map

$$P : \Sigma \to \Sigma, \ \ x \mapsto \widetilde{x} = P(x).$$

This map P is called a *Poincaré map associated with the cycle* L_0. The point $\tilde{x}$ is the *first return* or *Poincaré map of the point* x. We are not implying that such a point must exist, but if it does then it is called a first return. If we continue on, then the first return of x_1 $(=\tilde{x})$ is x_2. We can represent this process as a mapping by an operator P_Σ, $\tilde{x} = P_\Sigma(x)$. For successive returns starting from x_0, we use the notation $x_k = P_\Sigma(x_{k-1})$, $k = 1, \cdots, n$. Note that the "time" lapse between returns is not in general constant.

The Poincaré map is invertible near x_0 because of the invertibility of the dynamical system defined by the system (9.4.1). From the construction process, we see that the intersection point x_0 is a fixed point of the Poincaré map, that is, $P(x_0) = x_0$.

Let us introduce local coordinates $\xi = (\xi_1, \cdots, \xi_{n-1})$ on Σ such that $\xi = 0$ corresponds to x_0. Then the Poincaré map will be characterized by a locally defined map

$$P : R^{n-1} \to R^{n-1},$$

which transforms ξ corresponding to x into $\tilde{\xi}$ corresponding to $\tilde{x}$

$$P(\xi) = \tilde{\xi}.$$

The origin of $\xi = 0$ is a fixed point of the map

$$P: \; P(0) = 0.$$

The stability of the cycle L_0 is equivalent to the stability of the fixed point $\xi_0 = 0$ of the Poincaré map. Thus, the cycle is stable if all eigenvalues (multipliers) $\rho_1, \cdots, \rho_{n-1}$ of the $(n-1)\times(n-1)$ Jacobian matrix of P

$$A = \left.\frac{dP}{d\xi}\right|_{\xi=0},$$

are located inside the unit circle. The following lemma guarantees that the multipliers are not dependent on the choice of the point x_0 on Σ or the coordinates ξ on it.

Lemma 9.4.1[12] The multipliers $\rho_1, \cdots, \rho_{n-1}$ of the Jacobian matrix A of the Poincaré map associated with a cycle L_0 are independent of the point x_0 on L_0, the cross-section Σ, and local coordinates on it.

We now state the relationship between the multipliers of a cycle and the differential equations (9.4.1) defining the dynamical system that has this cycle. Let $x^*(t)$ denote a periodic solution of Eqs. (9.4.1),

$$x^*(t + T_0) = x^*(t)$$

corresponding to a cycle L_0. Represent a solution of Eqs. (9.4.1) in the form

$$x(t) = x^*(t) + u(t),$$

where $u(t) \in R^n$ is a deviation from the periodic solution. Then

$$\dot{u} = A(t)u + O(\|u\|^2),$$

where

$$A(t) = f_x(x^*(t)), \quad A(t + T_0) = A(t).$$

Truncating $O(\|u\|^2)$ terms results in the linear T_0-periodic system

$$\dot{u} = A(t)u, \quad u \in R^n. \tag{9.4.2}$$

Definition 9.4.1 The system (9.4.2) is called the *variational equation* about the cycle L_0.

The stability of the cycle depends on the properties of the variational equation.

Definition 9.4.2 The time-dependent matrix $M(t)$ is called the *fundamental matrix* solution of Eqs. (9.4.1) if it satisfies

$$\dot{M} = A(t)M,$$

with the initial condition $M(0) = I_{n \times n}$. Moreover, the matrix $M(T_0)$ is called a *monodromy matrix of the cycle* L_0.

---

[12] The proofs of the lemma and the next theorems are referred to Kuznetsov (1998). See also Hartman (1964) and Farkas (1994) on the relation between Poincaré maps, multipliers, and stability of limit cycles.

The following Liouville formula expresses the determinant of the monodromy matrix in terms of the matrix $A(t)$

$$\det M(T_0) = \exp\left\{ \int_0^{T_0} trA(t)dt \right\}.$$

Theorem 9.4.1 The monodromy matrix $M(T_0)$ has eigenvalues

$$1,\ \rho_1, \cdots, \rho_{n-1},$$

where ρ_i are the multipliers of Poincaré map associated with the cycle L_0.

Example Obtain the map of first returns P_Σ for the differential equations

$$\dot{x}_1 = \mu x_1 + x_2 - x_1\sqrt{x_1^2 + x_2^2},$$

$$\dot{x}_2 = -x_1 + \mu x_2 - x_2\sqrt{x_1^2 + x_2^2},$$

for the cross-section Σ given by $x_2 = 0$, $x_1 > 0$ with $t_0 = 0$. In polar coordinates the system becomes

$$\dot{r} = r(\mu - r),$$

$$\dot{\theta} = -1.$$

We solve

$$r = \frac{\mu r_0}{r_0 + (\mu - r_0)e^{-\mu t}},$$

$$\theta = -t + \theta_0.$$

We see that $r = \mu$ is a limit cycle. Eliminating t in the first equation, we obtain

$$r = \frac{\mu r_0}{r_0 + (\mu - r_0)e^{(\theta-\theta_0)\mu}}.$$

The section given corresponds to $\theta_0 = 0$, and required successive returns occur for $\theta = -2\pi, -4\pi$, ... with initial point $(r_0, 0)$. Hence

$$r_n = \frac{\mu r_0}{r_0 + (\mu - r_0)e^{-2\pi n\mu}}, \quad \theta_n = 0, \quad n = 1, 2, \cdots. \tag{9.4.3}$$

As $n \to \infty$ the sequence of point approaches the fixed point $(\mu, 0)$, as shown in Fig. 9.4.2, corresponding to the intersection with the limit cycle.

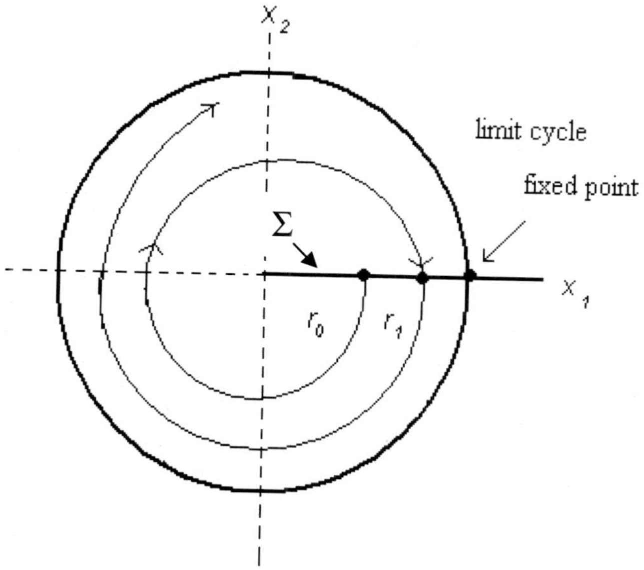

Fig. 9.4.2 First returns approaching the fixed point on Σ.

We can find the difference equation of which (9.4.3) is the solution. From (9.4.3), we solve r_0 as a function of r_n and substitute it into (9.4.3) for $n+1$

$$r_{n+1} = \frac{\mu r_0}{r_0 + (\mu - r_0)e^{-2\pi\mu(n+1)}} = \frac{\mu r_n}{r_n\left(1 - e^{-2\pi\mu}\right) + \mu e^{-2\pi\mu}}$$
$$\equiv f(r_n). \tag{9.4.4}$$

This is a first-order difference equation for r_n.

9.5 Center Manifold Theorems

We studied bifurcations of equilibria and fixed points in one- and two-dimensional systems. We will show that the corresponding bifurcations occur in "essentially" the same way for "generic" n-dimensional systems. We shall see that there are certain parameter-dependent one- or two-dimensional invariant "manifolds" on which the system exhibits the corresponding bifurcations, while the behavior off the manifolds is somehow "trivial". This section states the main theorems that allow us to reduce the dimension of a given system near a local bifurcation.

First, consider a dynamical system

$$\dot{x} = f(x), \quad x \in R^n, \tag{9.5.1}$$

where f is sufficiently smooth,

$$f(0) = 0.$$

Let the eigenvalues of the Jacobian matrix A evaluated at the origin be $\rho_1, \cdots, \rho_n$. Suppose that the equilibrium is not hyperbolic and that there are thus eigenvalues with zero real part. Assume that there are n_+ eigenvalues (counting multiplicities) with $\mathrm{Re}\,\rho > 0$, n_0 eigenvalues with $\mathrm{Re}\,\rho = 0$, and n_- eigenvalues (counting multiplicities) with $\mathrm{Re}\,\rho < 0$, see Fig. 9.5.1.

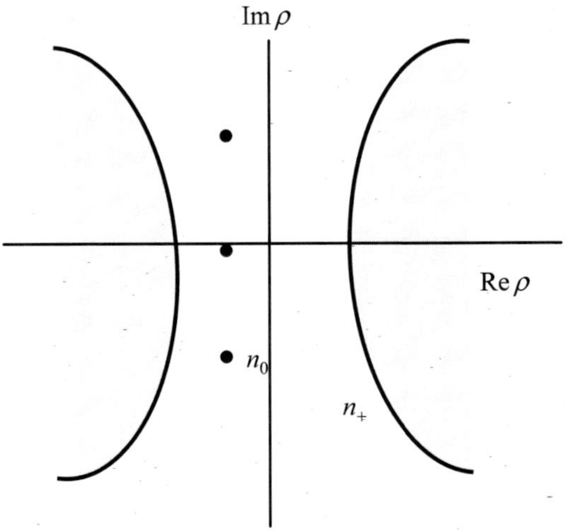

Fig. 9.5.1 Critical eigenvalues of an equilibrium point.

Let E^c denote the linear eigenspace of A corresponding to the union of the n_0 eigenvalues on the imaginary axis. The eigenvalues with $\operatorname{Re}\rho = 0$ are called *critical*, as is the eigenspace E^c. Let $\phi(t)$ denote the flow associated with the system (9.5.1). Before stating the center manifold theorem, we introduce the concept of manifold. Technically a *manifold* is a subspace of dimension of $m \le n$ in R^n usually satisfying continuity and differentiability conditions. For our purpose, it is sufficient to consider the manifold $M \in R^n$ as a set of points in R^n that satisfy a system of m scalar equations, $F(x) = 0$, where $F : R^n \to R^m$ for some $m \le n$. The manifold M is smooth (differentiable) if F is smooth and the rank of the Jacobian matrix F_x is equal to m at each point $x \in M$. Thus the sphere surface

$$x_1^2 + x_2^2 + x_3^2 = 1,$$

is a manifold of dimension 2, the solid sphere

$$x_1^2 + x_2^2 + x_3^2 \le 1,$$

is a manifold of dimension 3 in R^3; and the parabola $y = x^2$ is a manifold of dimension 1 in R^2. If a solution of a differential equation starts on a given space, surface or curve (manifold) and remains within it for all time, the manifold is said to be *invariant*. For instance, the Kaldor model has a limit cycle. Any solution which starts on the limit cycle will remain on it for all time. Hence, this closed curve in the phase plane is an invariant manifold. Equilibrium points are invariant manifolds.

At each point x of a smooth manifold, an $n-m$-dimensional *tangent space* T_xM is defined. This space consists of all vectors $v \in R^n$ that can be represented as $v = \dot{h}(t)$, where $h : R^1 \to M$ is a smooth curve on the manifold satisfying $h(0) = 0$. Alternatively, T_xM can be characterized as the orthogonal complements to

$$span\{\nabla F_1, \cdots, \nabla F_m\},$$

where

$$\nabla F_k = \left(\frac{\partial F_k}{\partial x_1}, \cdots, \frac{\partial F_k}{\partial x_n} \right)^T, \quad k = 1, \cdots, m,$$

are linear independent gradient vectors at point x. One can introduce $n-m$ coordinates near each point $x \in M$ by projecting to $T_x M$, so that a smooth manifold M is locally equivalent to R^{n-m}.[13]

Before introducing the center manifold theorem, we introduce the stable manifold theorem, which is one of the most important results in the local qualitative theory of ordinary differential equations. It shows that near a hyperbolic equilibrium point x^*, the nonlinear system (9.5.1) has stable and unstable manifolds S and U tangent at x^* to the stable and unstable subspaces E^s and E^u of the linearized system

$$\dot{x} = Ax,$$

where $A = Df(x^*)$. To explain the theorem, consider

$$\dot{x}_1 = -x_1,$$

$$\dot{x}_2 = -x_2 + x_1^2,$$

$$\dot{x}_3 = x_3 + x_1^2.$$

The origin is an equilibrium point. The Jacobian matrix is

$$J = Diag[-1 \quad -1 \quad 1].$$

The stable and unstable subspaces E^s and E^u of $\dot{x} = Ax$ are the x_1-x_2-plane and the x_3-axis respectively. It is straightforward to solve the original nonlinear equation as

$$x_1(t) = c_1 e^{-t},$$

$$x_2(t) = c_2 e^{-t} + c_1^2\left(e^{-t} - e^{-2t}\right),$$

$$x_3(t) = c_3 e^{-t} + \frac{c_1^2}{3}\left(e^{-t} - e^{-2t}\right),$$

where $c = x(0)$. Clearly $\phi(t, x_0) = 0$ as $t \to \infty$ if and only if

$$c_3 + c_1^3/3 = 0,$$

and $\phi(t, x_0) = 0$ as $t \to -\infty$ if and only if

$$c_1 = c_2 = 0.$$

Thus we have

$$S = \{c \in R^3 \mid 3c_3 + c_1^3 = 0\}, \quad U = \{c \in R^3 \mid c_1 = c_2 = 0\}.$$

[13] See Arrowsmith and Place (1990) and Jordan and Smith (1999).

Theorem 9.5.1 (The stable manifold theorem) Let E be an open subset of R^r containing the origin, let $f \in C^1(E)$, and let $\phi(t)$ be the flow of the nonlinear system (9.5.1). Suppose that $f(0)=0$ and $Df(0)$ has n_- eigenvalues with negative real part and $n-n_-$ eigenvalues with positive real part. Then there exists an n_--dimensional differentiable manifold S tangent to the stable subspace E^s of the linear system

$$\dot{x} = Ax,$$

at 0 such that for all $t \geq 0$, $\phi(t, S) \subset S$ and for all $x_0 \in S$, $\phi(t, x_0) = 0$ as $t \to \infty$; and there exists an $(n - n_-)$-dimensional differentiable manifold U tangent to the unstable subspace E^u of the linear system $\dot{x} = Ax$ at 0 such that for all $t \leq 0$, $\phi(t, U) \subset U$ and for all $x_0 \in U$, $\phi(t, x_0) = 0$ as $t \to -\infty$.

Let $\phi(t)$ be the flow of the nonlinear system (9.5.1). The *global stable* and *global unstable manifolds* of the nonlinear system (9.5.1) at 0 are defined respectively by

$$W^s(0) = \underset{t \leq 0}{U} \phi(t, S), \quad W^u(0) = \underset{t \geq 0}{U} \phi(t, S).$$

It can be shown that the global stable and unstable manifolds are unique and that they are invariant with respect to the flow $\phi(t)$.

Theorem 9.5.2 (Center manifold theorem) There is a locally defined smooth n_0-dimensional invariant manifold $W^c(0)$, called the *center manifold*, of the nonlinear system (9.5.1) that is tangent to E^c at $x = 0$. Moreover, there is a neighborhood U of $x_0 = 0$ such that if $\phi(x, t) \in U$ for all $t \geq 0$ ($t \leq 0$), then

$$\phi(x, t) \in W^c(0),$$

for $t \to \infty$ ($t \to -\infty$).

It should be mentioned that W^c need not be unique. For instance, the system

$$\dot{x}_1 = x_1^2,$$

$$\dot{x}_2 = -x_2,$$

has an equilibrium point $(x_1, x_2) = (0, 0)$ with $\rho_1 = 0, \rho_2 = -1$. It possesses a family of one-dimensional center manifolds

$$W_\beta^c = \{(x_1, x_2) : x_2 = \psi_\beta(x_1)\},$$

where

$$\psi_\beta(x_1) = \begin{cases} \beta \exp(-1/x), & \text{for } x_1 < 0, \\ 0, & \text{for } x_1 \geq 0. \end{cases}$$

In its eigenbasis which is a basis formed by all (generalized) eigenvectors of A (or their linear combinations if the corresponding eigenvalues are complex), the system (9.5.1) can be rewritten as

$$\dot{u} = Cu + g(u, v),$$

$$\dot{v} = Pv + h(u, v), \tag{9.5.2}$$

where $u \in R^{n_0}$ and $v \in R^{n_+ + n_-}$, C is an $n_0 \times n_0$ matrix with all its n_0 eigenvalues on the imaginary axis, while P is an $(n_+ + n_-) \times (n_+ + n_-)$ matrix with no eigenvalues on the imaginary axis. Functions g and h have Taylor expansions starting with at least quadratic terms. The center manifold W^c of the system (9.5.2) can be locally represented as a graph of a smooth function

$$W^c = \{(u, v) : v = V(u)\}.$$

Here

$$V : R^{n_0} \to R^{n_+ + n_-},$$

and due to the tangent property of W^c, $V(u) = O(\|u\|^2)$.

Theorem 9.5.3 (Reduction principle) The system (9.5.2) is logically topologically equivalent near the origin to the system

$$\dot{u} = Cu + g(u, V(u)),$$

$$\dot{v} = Pv. \tag{9.5.3}$$

The equations for u and v are uncoupled in the system (9.5.3). The first equation is the restriction of (9.5.2) to its center manifold. Thus, the dynamics of the structurally unstable system (9.5.2) are essentially determined by this restriction, since the second equations for v in the system (9.5.3) is linear and has exponentially decaying/growing solutions. The following local center manifold theorem often helps us to find center manifolds in applications.

Theorem 9.5.4 (The local center manifold theorem)[14] Let $f \in C^k(U)$, where U is an open subset of R^n containing the origin and $k \geq 1$. Suppose that $f(0) = 0$ and that $Df(0)$ has an n_0 eigenvalues with zero real parts and n_- eigenvalues with negative real parts, where $n_- + n_0 = n$. The system (9.5.1) can then be written in diagonal form

$$\dot{x} = Cx + F(x, y),$$

$$\dot{y} = Py + G(x, y), \tag{9.5.4}$$

where $(x, y) \in R^{n_0} \times R^{n_-}$, C is a square matrix with n_0 eigenvalues having zero real parts, P is a square matrix with n_- eigenvalues with negative real parts, and

$$F(0) = G(0) = 0, \; DF(0) = DG(0) = 0;$$

furthermore, there exists a $\delta > 0$ and a function $h \in C^k(N_\delta(0))$ that defines the local center manifold

$$W_{loc}^c = \{(x, y) \in R^{n_0} \times R^{n_-} \mid y = h(x) \text{ for } |x| < \delta\},$$

and satisfies

$$Dh(x)[Cx + F(x, h(x))] - Ph(x) - G(x, h(x)) = 0, \tag{9.5.5}$$

for all $|x| < \delta$; and the flow on the center manifold is defined by the system of differential equations

$$\dot{x} = Cx + F(x, h(x)), \tag{9.5.6}$$

for all $x \in R^{n_0}$ with $|x| < \delta$.

Equation (9.5.5) for the function h follows from the fact that the center manifold is invariant under the flow defined by the system (9.5.1) by substituting $\dot{x}$ and $\dot{y}$ from the differential equations (9.5.4) into the equation

$$\dot{y} = Dh(x)\dot{x},$$

---

[14] The theorem is proved in Carr (1981).

which follows from the chain rule applied to the equation $y = h(x)$ defining the center manifold. Equation (9.5.5) gives us a method for approximating the function h to any degree of accuracy that we wish, provided that the degree k is sufficiently large.

Although there may be many different functions h which determine different center manifolds, the flows on the various center manifolds are determined by Eq. (9.5.6) and they are all topologically equivalent in a neighborhood of the origin.

Example Consider the following system with $n_- = n_0 = 1$

$$\dot{x}_1 = x_1^2 x_2 - x_1^5,$$

$$\dot{x}_2 = -x_2 + x_1^2.$$

In this case, we have

$$C = 0,\ P = [-1],\ F(x_1, x_2) = x_1^2 x_2 - x_1^5,\ G(x_1, x_2) = x_1^2.$$

We substitute the expressions

$$h(x_1) = ax_1^2 + bx_1^3 + O(x_1^4),\ Dh(x_1) = 2ax_1 + 3bx_1^2 + O(x_1^3),$$

into Eq. (9.5.5) to obtain

$$(2ax_1 + 3bx_1^2 + \cdots)(ax_1^4 + bx_1^5 + \cdots - x_1^5)$$

$$+ ax_1^2 + bx_1^3 + \cdots - x_1^2 = 0.$$

Setting the coefficients of like powers of x_1 equal to zero yields $a = 1$, $b = 0$. Thus

$$h(x_1) = x_1^2 + O(x_1^5).$$

Substituting this result into Eq. (9.5.6) yields

$$\dot{x}_1 = x_1^4 + O(x_1^5),$$

on the center manifold near the origin. We see that the origin is a saddle-node and it is unstable. The local phase portrait of the system and the center manifold $W^c(0)$ are depicted by Fig. 9.5.2.

Example Consider the following example with $n_0 = 2$ and $n_- = 1$

$$\dot{x}_1 = x_1 x_3 - x_1 x_2^2,$$

$$\dot{x}_2 = x_2 x_3 - x_2 x_1^2,$$

$$\dot{x}_3 = -x_3 + x_1^2 + x_2^2.$$

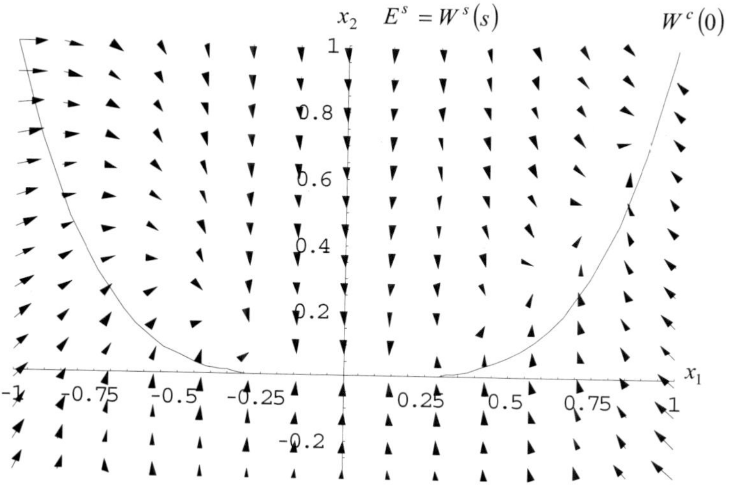

Fig. 9.5.2 The phase portrait.

We have $C = 0,\ P = [-1]$

$$F = \begin{bmatrix} x_1 x_3 - x_1 x_2^2 \\ x_2 x_3 - x_2 x_1^2 \end{bmatrix},\quad G = x_1^2 + x_2^2 .$$

We substitute the expression

$$h(x_1,\ x_2) = ax_1^2 + bx_1 x_2 + cx_2^2 + O(|x|^2),$$

into Eq. (9.5.5) and set the coefficients of like powers of x_1 and x_2 equal to zero. We obtain

$$a = 1,\quad b = 0,\quad c = 1.$$

Thus

$$h(x_1,\ x_2) = x_1^2 + x_2^2 + O(|x|^3).$$

Using this result, we obtain

$$\dot{x}_1 = x_1^3 + O(|x|^4),$$

$$\dot{x}_2 = x_2^3 + O(|x|^4),$$

on the center manifold near the origin. As

$$r\dot{r} = x_1^4 + x_2^4 + O(|x|^5) > 0,$$

we see that the origin is unstable.

Theorem 9.5.3 can be generalized to the case when the dimension of the unstable manifold is not equal to zero.[15] In its eigenbasis which is a basis formed by all (generalized) eigenvectors of A (or their linear combinations if the corresponding eigenvalues are complex), the system (9.5.1) can be rewritten as

$$\dot{x} = Cx + F(x, y, z),$$

$$\dot{y} = Py + G(x, y, z),$$

$$\dot{z} = Qz + H(x, y, z), \tag{9.5.7}$$

where

$$x \in R^{n_0}, \ y \in R^{n_-}, \ z \in R^{n_+},$$

C is an $n_0 \times n_0$ matrix with all its n_0 eigenvalues on the imaginary axis, P is an $n_- \times n_-$ matrix with negative real parts, Q is an $n_+ \times n_+$ matrix with positive real parts,

$$F(0) = DF(0) = 0, \ G(0) = DG(0) = 0, \quad H(0) = DH(0) = 0.$$

The local center manifold is now given by

$$W_{loc}^c(0) = \{(x, y, z) \in R^{n_0} \times R^{n_-} \times R^{n_+} \mid y = h_1(x), z = h_2(x) \text{ for } |x| < \delta\},$$

for some $\delta > 0$ where

$$h_j \in C^k(N_\delta(0)), \ h_j(0) = 0, \ Dh_j(0) = 0, \ j = 1, 2.$$

The functions h_j can be approximated to any desired degree of accuracy (provided that k is sufficiently large) by substituting their power series expansions into the following equations

$$Dh_j(x)[Cx + F(x, h_1(x), h_2(x))] - Ph_j(x) - G(x, h_1(x), h_2(x)) = 0. \tag{9.5.8}$$

[15] The following theorem is proved in Carr (1981). See also Wiggins (1990).

Theorem 9.5.5 Let $f \in C^1(U)$, where U is an open subset of R^n containing the origin. Suppose that $f(0) = 0$ and that $Df(0)$ has an n_0 eigenvalues with zero real parts, n_- eigenvalues with negative real parts, n_+ eigenvalues with positive real parts, where

$$n_- + n_+ + n_0 = n.$$

Then there exist C^1 functions $h_1(x)$ and $h_2(x)$ satisfying Eqs. (9.5.8) in a neighborhood of the origin such that the original system (9.5.1), which can be written in the form (9.5.7), is topologically conjugate to the C^1 system[16]

$$\dot{x} = Cx + F(x, h_1(x), h_2(x)),$$

$$\dot{y} = Py,$$

$$\dot{z} = Qz$$

in a neighborhood of the origin.

Exercise 9.5

1 Apply Theorem 9.5.3 to examine the dynamics near a unique equilibrium of the following system

$$\dot{x}_1 = x_2 + x_3,$$

$$\dot{x}_2 = x_3 + x_1^2,$$

$$\dot{x}_3 = -x_3 + x_2^2 + x_1 x_3.$$

9.6 Applying the Center Manifold Theorem and the Liapunov Theorem to a Simple Planar System

This section applies the center manifold theorem and the Liapunov theorem to examine dynamic properties of a simple dynamical system.[17]

We show how to determine the stability of a nonhyperbolic equilibrium point of a planar vector field with one zero and one negative

---

[16] Two maps, f and g, satisfying $f = h^{-1} \circ g \circ h$ for some homeomorphism h are called *conjugate*.

[17] The example is from Sec. 10.1 in Hale and Koçak (1991).

eigenvalue. Since linearization cannot help us to solve the problem, we have to examine how a particular term of a vector field affects the flow near such a nonhyperbolic equilibrium point. To illustrate the essence of the problem, we begin with an example easy to examine.

Let $k \geq 1$ be an integer, $a \neq 0$ be a real number. Consider

$$\dot{x}_1 = a x_1^k,$$

$$\dot{x}_2 = -x_2.$$

The eigenvalues of the linearized system at the origin are always 0 and -1, independent of a and k. Linearization cannot determine the stability type. We need to study the effect of the nonlinear term of the vector field. Since $x_2(t) \to 0$ as $t \to \infty$, the stability properties of the equilibrium point $x = 0$ are determined by the first scalar equation

$$\dot{x}_1 = a x_1^k .$$

The origin is asymptotically stable if a is negative and k is odd, and unstable otherwise. Figure 9.6.1 depicts a case of stability when $k = 3$ and $a < 0$.

We now consider a general system

$$\dot{x}_1 = f_1(x_1, x_2),$$

$$\dot{x}_2 = -x_2 + f_2(x_1, x_2), \tag{9.6.1}$$

where f is a given C^k function with $k \geq 1$,

$$f(0) = 0, \; Df(0) = 0 .$$

We may also write the system in vector form

$$\begin{bmatrix} \dot{x}_1 \\ \dot{x}_2 \end{bmatrix} = \begin{bmatrix} 0 & 0 \\ 0 & -1 \end{bmatrix} \begin{bmatrix} x_1 \\ x_2 \end{bmatrix} + \begin{bmatrix} f_1 \\ f_2 \end{bmatrix}. \tag{9.6.2}$$

The linear part of the vector field about the equilibrium point at the origin is in Jordan form with eigenvalues 0 and -1. We may consider x_2 in the first equation as the zero $h(x_1)$ of

$$-x_2 + f_2(x_1, x_2).$$

The dynamics in a neighborhood of the origin should be determined by the scalar differential equation

$$\dot{x}_1 = f_1(x_1, h(x_1)).$$

To find $h(x_1)$, consider

$$Z(x_1, x_2) \equiv -x_2 + f_2(x_1, x_2) = 0.$$

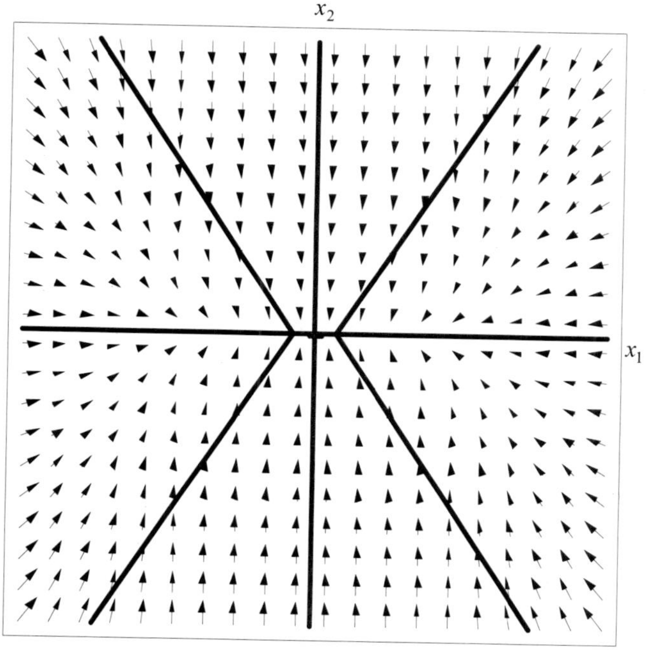

Fig. 9.6.1 Phase portrait for $k = 3$ and $a < 0$.

Since $Z(0, 0) = 0$ and $Z_{x_2}(0, 0) = 0$, the Implicit Function Theorem implies that there is a constant $\delta > 0$ and a unique C^1 function

$$h : \{x_1 : |x_1| < \delta\} \to \{x_2 : |x_2| < \delta\}$$

such that

$$-h(x_1) + f_2(x_1, h(x_1)) = 0,$$

$$h(0) = 0, \quad h'(0) = 0. \tag{9.6.3}$$

The relation $h'(0) = 0$ follows from differentiating the first equation in Eqs. (9.6.3) with respect to x_1 and setting $x_1 = 0$.

Theorem 9.6.1 Suppose that $f = (f_1, f_2)$ is a C^{k+1} function with

$$f_1(x_1, h(x_1)) = ax_1^k + O(|x_1^{k+1}|), \text{ as } x_1 \to 0, \tag{9.6.4}$$

where $a \neq 0$ is a real number, k is a positive integer, and $h(x_1)$ is as by

$$-h(x_1) + f_2(x_1, h(x_1)) = 0.$$

Then the equilibrium point at the origin of Eqs. (9.6.1) is asymptotically stable if $a < 0$ and k is an odd integer; otherwise, it is unstable.

Proof: Introduce the new variables $y = (y_1, y_2)$

$$x_1 = y_1,$$

$$x_2 = y_2 + h(x_1).$$

In these variables the original system (9.6.1) becomes

$$\dot{y}_1 = g_1(y_1, y_2),$$

$$\dot{y}_2 = -y_2 + g_2(y_1, y_2), \tag{9.6.5}$$

where

$$g_1 = f_1(y_1, h(y_1) + y_2),$$

$$g_2 = f_2(y_1, h(y_1) + y_2) - f_2(y_1, h(y_1)) - h'(y_1) f_1(y_1, h(y_1) + y_2).$$

The stability properties of the equilibrium point of Eq. (9.6.5) are the same as those of the equilibrium point $x = 0$ of Eqs. (9.6.1). As $\|y\| \to 0$, from Eqs. (9.6.3) and (9.6.4), the first several terms of the Taylor series of these functions about the origin are given

$$g_1(y_1, y_2) = ay_1^k (1 + O(\|y\|)) + y_2 O(\|y\|),$$

$$g_2(y_1, y_2) = O(\|y_1^{k+1}\|) + y_2 O(\|y\|). \tag{9.6.6}$$

Let us now consider the function

$$V(y_1, y_2) = -\frac{1}{a(k+1)} y_1^{k+1} + \frac{1}{2} y_2^2, \tag{9.6.7}$$

and compute its derivative along the solution of Eqs. (9.6.5). Utilizing Eq. (9.6.7), we observe that there is a $\delta > 0$ such that for $\|y\| < \delta$

$$\dot{V}(y_1, y_2) = -y_1^{2k}(1 + O(\|y\|)) + y_1^k y_2 O(\|y\|) - y_2^2 (1 + O(\|y\|))$$

$$\leq -\frac{1}{2}\left(y_1^{2k} - y_1^k y_2 + y_2^2\right).$$

The function $-\dot{V}(y_1, y_2)$ can easily be seen to be positive definite by treating it as a quadratic function in y_1^k and y_2. We conclude that if $a < 0$ and k is odd, then $V(y_1, y_2)$ is positive definite, thus from the Liapunov Theorem, $y = 0$ is asymptotically stable. If $a > 0$ and k is odd, then we apply Theorem 6.2.4 of Četaev to the function $-V$ to conclude the instability of $y = 0$. The remaining case $a \neq 0$ and k follows from the same theorem.

Example Consider

$$\dot{x}_1 = ax_1^3 + x_1 x_2,$$

$$\dot{x}_2 = -x_2 + x_2^2 + x_1 x_2 - x_1^3, \tag{9.6.8}$$

where a is a real constant. The function $h(x_1)$ is a solution of

$$-h(x_1) + h^2(x_1) + x_1 h(x_1) - x_1^3 = 0.$$

Substituting a Taylor series for $h(x_1)$ and equating the coefficients of like powers x_1, we obtain

$$h(x_1) = -x_1^3 + O(|x_1^4|),$$

$$f_1(x_1, h(x_1)) = ax_1^3 - x_1^4 + O(|x_1^5|).$$

Theorem 9.6.1 implies that the equilibrium $x = 0$ of Eqs. (9.6.8) is asymptotically stable if $a < 0$ and unstable if $a \geq 0$.

9.7 The Hopf Bifurcation Theorem and Its Applications

We studied the Hopf bifurcation in planar dynamical systems. We now examine the Hopf bifurcation in higher dimensions.

Theorem 9.7.1 (Andronov-Hopf bifurcation theorem) Consider the system

$$\dot{x} = f(x, \lambda),$$

where

$$f \in C^{k+1}(R^n \times R),\ k \geq 4,\ \ f(0, \lambda) \equiv 0.$$

Suppose that for small $|\lambda|$ the matrix $f_x'(0, \lambda)$ has a pair of complex conjugate eigenvalues

$$\alpha(\lambda) \pm i\omega(\lambda),\ \omega(\lambda) > 0,\ \alpha(0) = 0,\ \alpha'(0) > 0,$$

and the other $n-2$ eigenvalues have negative real part; then

(i) there is a $\delta > 0$ and a function $\lambda \in C^{k-2}((-\delta, \delta), R)$ such that for $\varepsilon \in (-\delta, \delta)$ the system

$$\dot{x} = f(x, \lambda(\varepsilon)),$$

has a periodic solution $p(t, \varepsilon)$ with period $T(\varepsilon) > 0$, also

$$T \in C^{k-2},\ \lambda(0) = 0,\ T(0) = 2\pi / \omega(0),\ \ p(t, 0) \equiv 0,$$

and the amplitude of this periodic solution (the approximate distance of the corresponding periodic orbit from the origin) is proportional to $\sqrt{|\lambda(\varepsilon)|}$;

(ii) the origin $(x, \lambda) = (0, 0)$ of the space $R^n \times R$ has a neighborhood $U \subset R^n \times R$ that does not contain any periodic orbit of Eqs. (9.7.1) but those of the family $p(t, \varepsilon)$, $\varepsilon \in (-\delta, \delta)$;

(iii) if the origin $x = 0$ is a 3-asymptotically stable (resp. 3-unstable) 3-unstable equilibrium of the system $\dot{x} = f(x, 0)$, then $\lambda(\varepsilon) > 0$ (resp. $\lambda(\varepsilon) < 0$) for $\varepsilon \neq 0$, and the periodic solution $p(t, \varepsilon)$ is asymptotically orbitally stable (resp. unstable).[18]

Example We examined a predator-prey model in Sec. 9.3. This section applies the Andronov-Hopf bifurcation theorem to a *predator-prey model with memory*.[19] The model is as follows

$$\dot{N}(t) = \varepsilon N(t)\left(1 - \frac{N(t)}{K}\right) - \alpha P(t)N(t),$$

---

[18] The concepts of h- asymptotical stability and instability are referred to Chap. 7 in Farkas (1994).

[19] This example is from Sec. 7.3 in Farkas (1994). We illustrate the analysis, leaving some terms and results unexplained. The analysis in detail is referred to the reference. See also MacDonald (1978), Wörz-Busekros (1978), Dai (1981), Farkas (1984), Farkas, Farkas and Szabó (1988).

$$\dot{P}(t) = -\gamma P(t) + \beta P\int_{-\infty}^{t} N(\tau)G(t-\tau)d\tau,$$

where $N(t)$ and $P(t)$ are respectively the quantity of prey and predator, respectively, $\varepsilon > 0$, $\alpha > 0$, $\beta > 0$, and $\gamma > 0$ are respectively the growth rate of prey, the predation rate, the mortality of predator and the conversion rate, $K > 0$ is the carrying capacity of the environment for the prey, and $G : [0, \infty) \to R_+$ be a C^1 *density function* satisfying

$$\int_0^{\infty} G(s)ds = 1.$$

The term $-N/K$ takes account of the intraspecific competition in the prey species: this has a saturation effect, and as a consequence that the prey is not growing exponentially in the absence of predation but tends to a finite limit. The term $\beta\int N(\tau)G(t-\tau)d\tau$ reflects that the present growth rate of a predator depends not only on the present quantity of food but also on past quantities (in the period of gestation, say). Assume the density function is exponentially decaying

$$G(s) = a\,e^{-as},\ \ a > 0.$$

Introduce

$$Q(t) \equiv \int_{-\infty}^{t} N(\tau)G(\tau)d\tau = a\int_{-\infty}^{t} N(\tau)e^{-a(t-\tau)}d\tau.$$

Under this transformation, the original system becomes

$$\dot{N} = \varepsilon N\left(1 - \frac{N}{K}\right) - \alpha PN,$$

$$\dot{P} = -\gamma P + \beta PQ,$$

$$\dot{Q} = a(N - Q), \tag{9.7.1}$$

where the last equation is obtained by differentiation with respect to t. Instead of the delay equation with exponentially fading memory, we will study the system (9.7.1) of ordinary differential equations on $t \in [0, \infty)$.

First, we introduce

$$N(t) = Kn(t),\ \ P(t) = Kp(t),\ \ Q(t) = Kq(t),\ \ t = \frac{s}{\varepsilon}.$$

To transform the system (9.7.1) into

$$\dot{n} = n(1-n) - \frac{\alpha_0}{\varepsilon} np,$$

$$\dot{p} = -\frac{\gamma}{\varepsilon}\gamma p + \frac{1}{\varepsilon b} pq,$$

$$\dot{q} = \frac{a}{\varepsilon}(n-q), \tag{9.7.2}$$

where $\alpha_0 \equiv \alpha K$, $b \equiv 1/\beta K$, and the dot now denotes differentiation with respect to the variable s.

The system (9.7.2) has three equilibria: the origin which is unstable and of no interest; the point $(1, 0, 1)$ which is asymptotically stable if $\gamma b > 1$ and unstable if $\gamma b < 1$. The third equilibrium is

$$\left(n^*, p^*, q^*\right) = \left(\gamma b, \frac{1-\gamma b}{\alpha_0}\varepsilon, \gamma b\right),$$

where $\gamma_0 \equiv \gamma b$. The equilibrium point is in the positive octant of the three variables if and only if $\gamma_0 < 1$. This condition also implies that the system has no asymptotically stable equilibrium except, possibly $\left(n^*, p^*, q^*\right)$. The Jacobian matrix at this point is

$$\begin{bmatrix} -\gamma b & -\alpha\gamma/\varepsilon\beta & 0 \\ 0 & 0 & (1-\gamma b)\beta/\alpha \\ \frac{a}{\varepsilon} & 0 & -\frac{a}{\varepsilon} \end{bmatrix}.$$

The characteristic equation is

$$\rho^3 + \left(\gamma b + \frac{a}{\varepsilon}\right)\rho^2 + \frac{\gamma b a \rho}{\varepsilon} + (1-\gamma b)\frac{a\gamma}{\varepsilon^2} = 0. \tag{9.7.3}$$

From the Routh-Hurwitz criteria we know that this is a stable polynominal if and only if

$$\gamma b < 1, \quad \left(\gamma b + \frac{a}{\varepsilon}\right)\gamma \frac{ba}{\varepsilon} > (1 - \gamma b)\frac{a\gamma}{\varepsilon^2}.$$

That is

$$\gamma b < 1, \quad a > \frac{1}{b} - \gamma - \gamma b\varepsilon .$$

If $1/b - \gamma - \gamma b\varepsilon$ is negative or equal to zero (which can be guaranteed if ε is sufficiently large), then the equilibrium point (n^*, p^*, q^*) is asymptotically stable for all positive a. Here, we are concerned with

$$1/b - \gamma - \gamma b\varepsilon > 0 .$$

The equilibrium is losing its stability at the positive value

$$a_0 = \frac{1 - \gamma b - {}^2\gamma b\varepsilon}{b}.$$

At $a = a^*$, the characteristic polynomial (9.7.3) assumes

$$\left(\rho^2 + \frac{1 - \gamma b - \gamma b^2\varepsilon}{\varepsilon}\gamma\right)\left(\rho + \frac{1 - b\gamma}{b\varepsilon}\right) = 0 .$$

The three eigenvalues are

$$\rho_0(a_0) = \frac{b\gamma - 1}{b\varepsilon},$$

$$\rho_{1,2} = \pm i\omega,$$

where

$$\omega \equiv \sqrt{\frac{1 - \gamma b - {}^2\gamma b\varepsilon}{\varepsilon}\gamma} > 0 .$$

Denote $\rho_1(a)$ the root of the polynomial (9.7.3) as a function of a that assumes the value $i\omega$ at a_0 and by

$$F(\rho, a) = \rho^3 + \left(\gamma b + \frac{a}{\varepsilon}\right)\rho^2 + \frac{\gamma b a\rho}{\varepsilon} + (1 - \gamma b)\frac{a\gamma}{\varepsilon^2}, \qquad (9.7.4)$$

the characteristic polynomial in Eq. (9.7.3) as a function of a. Since

$$F(\rho(a_0), a_0) = F(i\omega, a_0) = 0 ,$$

and $i\omega$ is a simple root of the polynomial $F(\rho, a_0)$, the smooth function ρ_1 is uniquely determined by

$$F(\rho(a), a) = 0,\ \rho_1(i\omega) = i\omega .$$

The derivative of the implicit function ρ_1 at a_0 is

$$\rho_1' = -\frac{F_a'(i\omega, a_0)}{F_\rho'(i\omega, a_0)} =$$

$$\frac{(\gamma b + i\omega)\{\gamma b(1 - \gamma b - \gamma b^2\varepsilon) + \omega(1 - \gamma b)\}}{\gamma^2 b^2(1 - \gamma b - \gamma b^2\varepsilon)^2 + \omega^2(1 - \gamma b)^2}\frac{\gamma b^2}{2} .$$

Hence, we have

$$\frac{d}{da}\mathrm{Re}(\rho_1(a)) = \mathrm{Re}\left(\frac{d\rho_1(a)}{da}\right) =$$

$$-\frac{\gamma b^2}{2}\frac{1 - \gamma b - \gamma b^2\varepsilon}{\varepsilon\gamma b^2(1 - \gamma b - \gamma b^2\varepsilon) + (1 - \gamma b)^2} < 0 .$$

We see that the equilibrium (n^*, p^*, q^*) is asymptotically stable for $a > a_0$ and is losing its stability at $a = a_0$. Hence, we established the conditions for occurrence of the Hopf bifurcation. With some more complicated calculations, Farkas proved the following theorem.[20]

Theorem 9.7.2 If

$$a_0 = \frac{1 - \gamma b - {}^2\gamma b\varepsilon}{b} > 0 ,$$

$$a_0 - \frac{\varepsilon^2\gamma b}{2\gamma + \varepsilon} > 0,\ \ (\text{resp.} < 0) ,$$

then there exists a $\delta > 0$ such that for each

$$a \in (a_0 - \delta, a_0)\ \ (\text{resp.}\ a \in (a_0, a_0 + \delta)) ,$$

the system (9.7.1) has a unique periodic orbit in a neighborhood of the equilibrium point (n^*, p^*, q^*) and the corresponding periodic solution is asymptotically orbitally stable (resp. unstable).

---

[20] See Farkas (1994: 447–8).

Exercise 9.7

1 We consider the following augmented Kaldor model[21]

$$\dot{Y} = \alpha\{I(Y, K, r) - S(Y, r)\},$$

$$\dot{r} = \beta\{L(r, Y) - \overline{M}\},$$

$$\dot{K} = I(Y, K) - \delta K,$$

where variables and parameters are define as

Y	= output level;
K	= capital stock;
r	= interest rate;
$I(Y, K, r)$	= investment function ($I_Y > 0$, $I_K < 0$);
$S(Y, r)$	= savings function ($0 < S_Y < 1$, $S_r > 0$);
$L(Y, r)$	= money demand;
$\overline{M}$	= money supply
α and β	= positive adjustment parameters;
δ	= capital depreciation rate.

In Sec. 6.4, we demonstrated that when $\beta \to \infty$, the system exhibits limit cycles under certain conditions. With the similar arguments, try to establish conditions for the existence of limit cycles in the model.

2 We consider the so-called *Oreganor equations*[22]

$$\varepsilon\dot{x} = x + y - qx^2 - xy,$$

$$\dot{y} = -y + 2hz - xy,$$

$$p\dot{z} = x - z,$$

where the parameters ε, q, h and p are all positive. We simulate the model with the following specified parameters

[21] The model is examined by Lorenz (1993: 105–107). As mentioned by Lorenz, the adjustment equation for the interest rate is not unproblematic. The interest rate is determined on the bonds market, and assumed form of its adjustment equation implies that the excess supply of bonds excess demand for money. However, it remains unclear how possible excess demand in the goods market are financed.

[22] Enns and McGuire (2001: 76–79).

$$\varepsilon = 0.03, \ q = 0.006, \ h = 0.75, \ p = 2, x_0 = 100,$$

$$y_0 = 1, z_0 = 10.$$

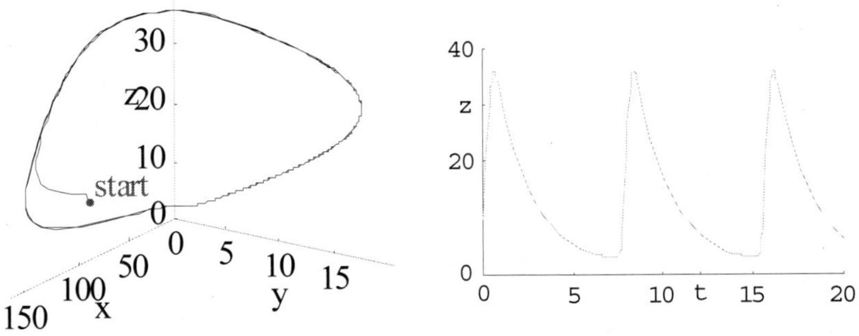

Fig. 9.7.1 The motion of the Oreganor equations.

The system exhibits oscillations at the specified parameter values. Try to apply the Hopf bifurcation theorem to identify existence of limit cycles in the model.

9.8 The Lorenz Equations and Chaos

It has recently become clear that there are different and very complex phenomena that can occur in systems of third and higher order that are not present in second order systems. We now introduce the Lorenz equations to illustrate phenomena of chaos.

The Lorenz equations are a quadratic system of autonomous differential equations in three dimensions modeling a three-mode approximation to the motion of a layer of fluid heated from below. The system of equations that Edward N. Lorenz proposed in 1963 is

$$\dot{x} = \sigma(-x + y), \ \dot{y} = rx - y - xz,$$

$$\dot{z} = -bz + xy, \tag{9.8.1}$$

where x, y, and, z are time-dependent variables.[23]

[23] A thorough treatment of the Lorenz equations is given by Sparrow (1982).

The Lorenz equations are nonlinear as the second and third equations involve quadratic nonlinearities. The system involves three positive, real parameters, σ, r, and b. For the earth's atmosphere reasonable values of these parameters are $\sigma = 10$ and $b = 8/3$. The parameter r is proportional to the vertical temperature difference

To analyze the behavior of the system, first we study critical points by solving

$$\sigma(-x+y)=0,$$

$$rx - y - xz = 0,$$

$$-bz + xy = 0.$$

From $\sigma(-x+y)=0$, we have $x = y$. Substituting $x = y$ into the other two equations yields

$$x(r-1-z)=0, \quad -bz + x^2 = 0.$$

In the case of $x = 0$, we have $y = z = 0$. In the case of $x \neq 0$, the other critical points are

$$z = r-1, \quad x = y = \pm\sqrt{b(r-1)}.$$

The expressions of x and y are real only when $r \geq 1$. Thus $(0, 0, 0)$, denoted by P_1, is a critical point for all values of r, and it is the only critical point for $r < 1$. When $r > 1$, the other two critical points, denoted respected by P_2 and P_3, are

$$\left(\sqrt{b(r-1)}, \sqrt{b(r-1)}, r-1\right), \quad \left(-\sqrt{b(r-1)}, -\sqrt{b(r-1)}, r-1\right).$$

All three critical points coincide at $r = 1$. As r increases through the value 1, the critical point P_1 at the origin bifurcates and the critical points P_2 and P_3 come into existence. We specify $\sigma = 10$ and $b = 8/3$ in the remainder of this section.

Near $P_1\,(0, 0, 0)$, the linearized system is

$$\begin{pmatrix} \dot{X} \\ \dot{Y} \\ \dot{Z} \end{pmatrix} = \begin{pmatrix} -10 & 10 & 0 \\ r & -1 & 0 \\ 0 & 0 & -8/3 \end{pmatrix} \begin{pmatrix} X \\ Y \\ Z \end{pmatrix}. \tag{9.8.2}$$

The eigenvalues are determined by

$$\begin{vmatrix} -10-\rho & 10 & 0 \\ r & -1-\rho & 0 \\ 0 & 0 & -8/3-\rho \end{vmatrix}$$

$$= -\left(\rho + \frac{8}{3}\right)\left\{\rho^2 + 11\rho - 10(r-1)\right\} = 0 .$$

Therefore,

$$\rho_1 = -\frac{8}{3}, \quad \rho_{2,3} = \frac{-11 \pm \sqrt{81+40r}}{2} . \qquad (9.8.3)$$

If $r < 1$, then all three eigenvalues are negative. Hence, the origin is asymptotically stable for this range of r. However, ρ_3 changes sign when r passes from 1 to $r > 1$. The origin is unstable for $r > 1$. Near

$$P_2\left(\sqrt{b(r-1)}, \sqrt{b(r-1)}, r-1\right),$$

the linearized system is

$$\begin{pmatrix} \dot{X} \\ \dot{Y} \\ \dot{Z} \end{pmatrix} = \begin{pmatrix} -10 & 10 & 0 \\ 1 & -1 & -\sqrt{8(r-1)/3} \\ \sqrt{8(r-1)/3} & \sqrt{8(r-1)/3} & -8/3 \end{pmatrix} \begin{pmatrix} X \\ Y \\ Z \end{pmatrix} .$$

The eigenvalues are determined from the equation

$$3\rho^3 + 41\rho^2 + 8(r+10)\rho + 160(r-1) = 0 .$$

The solutions of the above equation depend on r as follows[24]

For $1 < r < r_1 \cong 1.3456$, there are three negative real eigenvalues.
For $r_1 < r < r_2 \cong 24.737$, there are one negative real eigenvalue and two complex eigenvalues with negative real part.
For $r_2 < r$, there are one negative real eigenvalue and two complex eigenvalues with positive real part.

The same results are obtained for the critical point P_3. Summarizing these discussions, we conclude that: (1) For $0 < r < 1$, the only critical point P_1 is asymptotically stable; (2) For $1 < r < r_1$, P_2 and P_3 are asymptotically stable and P_1 is unstable. All nearby solutions approach

[24] See Boyce and Diprima (2001: 534) or Sec. 5.4 in Tabor (1989).

one or the other of the points P_2 and P_3 exponentially; (3) For $r_1 < r < r_2$, P_2 and P_3 are asymptotically stable and P_1 is unstable. All nearby solutions approach one or the other of the points P_2 and P_3; most of them spiral inward to the critical point; and (4) For $r_2 < r$, all three critical points are unstable. Most solutions near P_2 and P_3 spiral away from the critical point.

We consider solutions for $r_2 < r$. All three critical points are unstable. A trajectory can approach any one of the critical points only on certain highly restricted paths. The slightest deviation from these paths causes the trajectory to depart from the critical point. What is surprising with the system is that no trajectory will approach infinity as $t \to \infty$, even though all critical points are unstable. In fact, it can be shown that all solutions remain bounded as $t \to \infty$. It can be shown that all for positive values of r, all solutions ultimately approach a certain limiting set of points that has zero volume.

We first simulate the model, we compute values of $x(t)$ versus t. Figure 9.8.1 describes the motion of $x(t)$ with initial condition $(6, 6, 6)$. The solution changes "randomly" from positive values to negative ones. The Lorenz equations are deterministic and its motion is completely determined by the initial conditions. The attracting set in this case has a complicated structure and is called a *strange attractor*. The term *chaotic* is used to describe solutions such as those shown in Fig. 9.8.1.

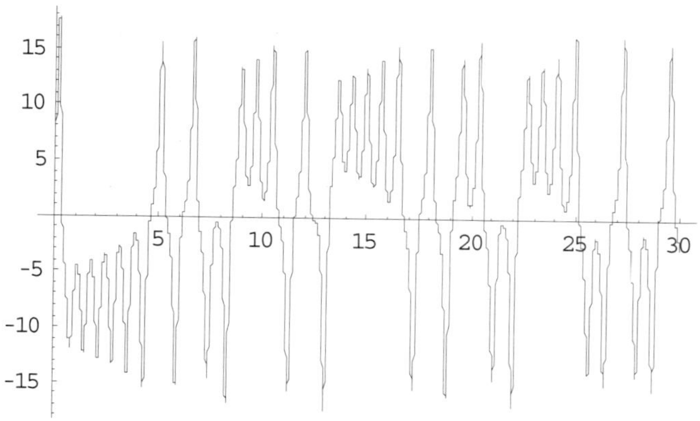

Fig. 9.8.1 A plot of $x(t)$ for $r = 28$ for initial condition $(6, 6, 6)$.

A significant feature of the system is that its solutions are extremely sensitive to small changes in the initial conditions. Figure 9.8.2 shows the solutions $x(t)$ with different initial conditions. The initial conditions of y and z are the same in the two cases; the only difference is for $x(0)$. The two solutions remain close until t is near 10, after which they diverge rapidly. This property of the Lorenz equations caused him to conclude that detailed long-range weather predictions are probably no possible as any small difference in initial condition may lead to great differences in the future.

Figure 9.8.3 shows the corresponding Lorenz attractor. There are two sheets in which trajectories spiral outwards. When the distance from the center of such a spiral becomes larger than some particular threshold, the motion is ejected from the spiral and is attracted by the other spiral, where it again begins to spiral out, and the process is repeated. The motion is not regular. The number of turns that a trajectory spends in one spiral before it jumps to the other is not specified. It may wind around one spiral twice, and then three times around the other, then ten times around the first and so on.

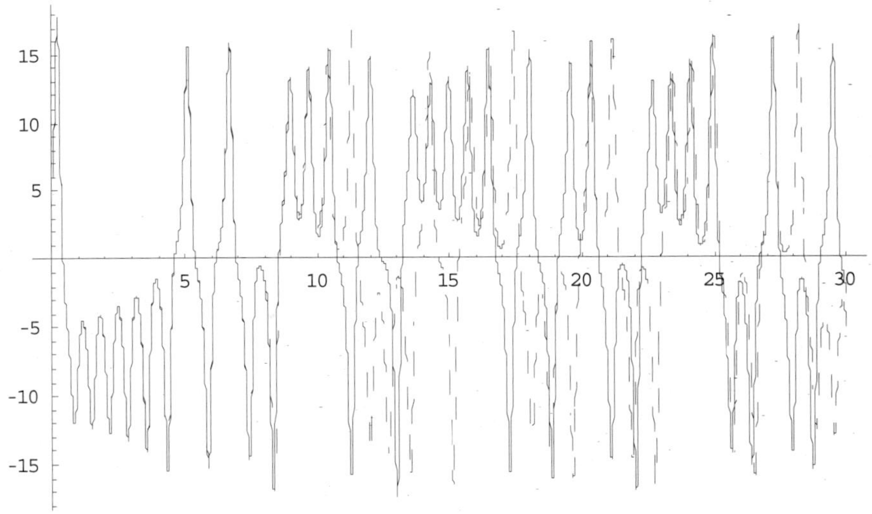

Fig. 9.8.2 A plot of $x(t)$ for $r = 28$; $(x_0, y_0, z_0) = (6, 6, 6)$ for solid curve and $(x_0, y_0, z_0,) = (6.01, 6, 6)$ for dashed curve.

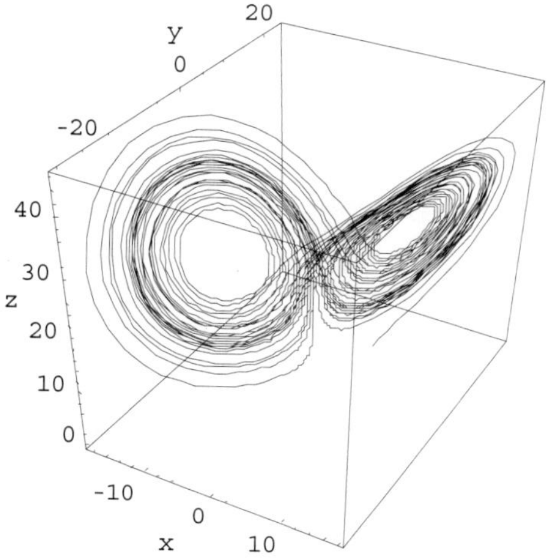

Fig. 9.8.3 The dynamics of the Lorenz equations $r = 28$ and $(x_0, y_0, z_0) = (6, 6, 6)$.

The Lorenz attractor is dubbed a strange attractor because there are no asymptotically stable equilibria or period orbits in a global attractor that is a compact, connected invariant set. The geometry of the attractor is exceedingly complicated. We may also visualize projects in xz - plane. Figure 9.8.4 illustrates the case with $r = 28$ and $(x_0, y_0, z_0) = (6, 6, 6)$. The graphs appear to cross over themselves repeatedly, but this cannot be true for the actual trajectories in three-dimensional space because of the unique theorem. The apparent crossings are due wholly to the two-dimensional character of the figure.

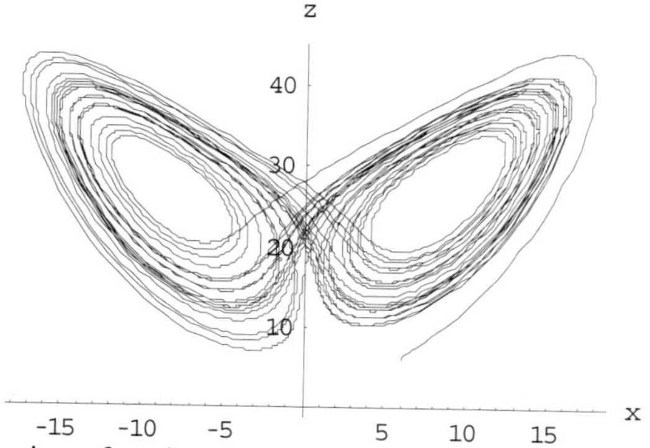

Fig. 9.8.4 Projections of a trajectory of the Lorenz equations in the xz -Plane; $r = 28$.

Example An interpretation of the Lorenz equations for urban dynamics. Zhang found that some urban systems can also be described by the Lorenz equations.[25] The urban model describes an urban system within a metropolitan area. The system under consideration is small, so that its dynamics will have almost no significant impact on the metropolitan area. Businesses and residents are free to choose their location sites either in the urban area or in the outside world. Locational characteristics of the urban system are described by the following three variables:

$x(t)$ = the output of the urban system at time t;
$y(t)$ = the number of residents;
$z(t)$ = the land rent.

With some proper assumptions about the behavior of businesses and consumers, Zhang constructed a dynamic urban system that can be reduced to the Lorenz equations. Since the urban system exhibits the same behavior given by Fig. 9.8.3, the urban system exhibits the following properties: (1) the temporary path of the three urban variables are time-dependent but are not periodic (or "regular"); (2) the motion does not appear to show a transient phenomenon since, regardless of how long the numerical integration is continued, the trajectory is going to continue to wind around and around without settling down to either periodic or stationary behavior; (3) the topology of the figure is not dependent on the choice of initial conditions or integrating route; and (4) it is impossible to predict the details of how the trajectory will develop over any period other than a very short time interval. Considering the above urban interpretation of the Lorenz equations, we see that even if the government is well informed and is composed of well-educated officials and experts, it is impossible for the government to predict the impact of its own actions, such as tax policy, land policy, and infrastructure policy.

---

[25] See Zhang (1991, Chap. 6).

Exercise 9.8

1 (a) Show that the eigenvalues of the linear system (9.8.2) are given by (9.8.3); (b) Determine the corresponding eigenvectors; and (c) Determine the eigenvalues and eigenvectors of the system (9.8.2) when $r = 28$.

2 Using the Liapunov function

$$V(x, y, z) = x^2 + \sigma y^2 + \sigma z^2,$$

to show that the origin is a globally asymptotically stable critical point for the Lorenz equations if $r < 1$.

Chapter 10

Higher-Dimensional Economic Evolution

This chapter applies the mathematical concepts and theorems of higher differential equations introduced in the previous two chapters to differential economic models. Section 10.1 examines some *tâtonnement* price adjustment processes, mainly applying the Liapunov methods. Section 10.2 studies a three-country international trade model with endogenous global economic growth. Section 10.3 extends the trade model of the previous section for examining impacts of global economic group on different groups of people not only among countries but also within countries. We provide insights into complexity of international trade upon different people. Section 10.4 examines a two-region growth model with endogenous capital and knowledge. Different from the trade model where international migration is not allowed, people freely move among regions within the interregional modeling framework. Section 10.5 introduces money into the growth model. We demonstrate the existence of business cycles in the model, applying the Hopf bifurcation theorem. Section 10.6 guarantees the existence of limit cycles and aperiodic behavior in the traditional multi-sector optimal growth model, an extension of the Ramsey growth model. Section 10.7 proposes a dynamic model with interactions among economic growth, human capital accumulation, and opening policy to provide insights into the historical processes of Chinese modernization. Analysis of behavior of this model requires almost all techniques introduced in this book.

10.1 Tâtonnement Price Adjustment Processes

We now use the Liapunov direct method to prove stability of a Tâtonnement price adjustment process of the Arrow-Debreu system. The following example is based on Hahn.[1]

There are N goods in the economy, H households, and F firms in the economy. Let $x^h \in R^N$ stand for the net trade vector of households h; introduce

$$x \equiv \sum_{h=1}^{H} x^h .$$

Define $y^f \in R^N$ as an activity of firm f where positive components of y^f denote outputs and negative ones denote inputs. Also introduce

$$y \equiv \sum_{f=1}^{F} y^f .$$

Let z be the aggregate excess demand vector and s the aggregate excess supply vector defined by

$$z \equiv x - y = -s .$$

Let $p \in R_+^N$ be a price vector. Assume that the price of goods 1 is positive, $p_1 > 0$. Let P stand for the vector $(1/p_1)p$ with its first component deleted. The endowment of household h is denoted by $w^h \in R_+^N$. Define

$$w \equiv \sum_{h=1}^{H} w^h , \quad w^* = \left(w^1, \cdots, w^H\right).$$

We are concerned here with economies, which have continuously differentiable excess supply and demand functions. It is known that as a result of the rational behavior of the households and firms, we can determine the excess supply and demand as functions of p and w^*, i.e.,

$$s\left(p, w^*\right) \text{ and } z\left(p, w^*\right).$$

They are homogeneous of degree one in p and obey Walras' law,[2] respectively

[1] See Hahn (1982).

$$s(p, w^*) = s(1, P, w^*), \quad ps(p, w^*) = 0, \quad \text{for all } p \in R_+^N .$$

Let

$$D = \{p \,|p > 0, \textstyle\sum_i p_i = 1\}.$$

Also let G be the boundary of D.

Definition 10.1.1 $p^* \in D/G$ is an *equilibrium* of the economic system if for each i, (a) $s_i(p^*, w^*)$ is non-negative, and (b)

$$p_i^* s_i(p^*, w^*) = 0 .$$

Under appropriate conditions the existence of a unique equilibrium is guaranteed. We consider the following *Walrasian tâtonnement*

$$\dot{p}_i = 0, \quad \text{if } p_i = 0 \text{ and } s_i(P) > 0, \quad i = 2, \cdots, N ,$$

$$\dot{p}_i = -\, k_i s_i(p), \quad k_i > 0 \text{ otherwise}. \tag{10.1.1}$$

Introduce

$$V(p) = \sum_{i=1}^{N} \frac{(p_i - p_i^*)^2}{k_i} . \tag{10.1.2}$$

Under Walras' law, we have

$$\dot{V} = 2p^* s(p).$$

It can be shown that if all goods are gross substitutes, then

$$p^* s(p) < 0$$

if p is not equal to kp^* for $k > 0$.[3] The following theorem is thus held.

Theorem 10.1.2 If all goods are gross substitutes, then the unique equilibrium is globally asymptotically stable under (10.1.1).

---

[2] Positive homogeneity of degree zero means that if all prices are multiplied by the same positive constant, the excess demands do not vary, and this is a well-known consequence of the utility maximization postulate.

[3] Gross substitutability means that $\partial z_i / \partial p_j > 0$ for all $i, j, i \neq j$. The conclusion of $p^* s(p) < 0$ can be found in Arrow, Block, and Hurwicz (1959: 90). Note that the negativity of the expression means that the aggregate excess demand function satisfies the weak axiom of revealed preference.

The price dynamic process described by the dynamics (10.1.1) can be generalized. Moreover, different possible adjustment processes have been suggested in the literature.[4] For illustration of applying the Liapunov function, we introduce the case of the *dominant negative diagonal*.[5] The assumption of the dominant negative diagonal means that the aggregate excess demand functions satisfy

$$z_{ii} < 0, \ |z_{ii}| > \sum_{j=1, j \neq i}^{n} |z_{ij}| \ \text{ for all } i, \tag{10.1.3}$$

where $z_{ik} \equiv \partial z_i / \partial p_k$. Consider the following Walrasian *tâtonnement*

$$\dot{p}_i = k_i z_i(p), \ k_i > 0 \ \text{ for all } i,$$

where k_i are constant. For the Liapunov function, we choose

$$V = \max_i |k_i z_i|.$$

Let $|k_s z_s| \geq |k_i z_i|$ for all i. Then

$$V = |k_s z_s|.$$

Assume that $\dot{V}$ exists everywhere.[6] Using the notation sgn ,[7] we have

$$\dot{V} = k_s \operatorname{sgn}(z_s) \sum_i z_{si} \dot{p}_i = k_s \operatorname{sgn}(z_s) \sum_i z_{si} k_i z_i . \tag{10.1.4}$$

At the equilibrium, $\dot{V} = 0$. If the system is out of equilibrium, $z_k \neq 0$ at least for some k. By the definition of z_s, $|z_s| > 0$ and

$$|z_{ss}| > \sum_{i=1, j \neq s}^{n} |z_{si}|.$$

Hence, we have

$$|z_s||z_{ss}| > |z_s| \sum_{i=1, i \neq s}^{n} |z_{si}| = |z_s| \sum_{i=1, i \neq s}^{n} |z_{si}| \geq \frac{1}{k_s} \sum_{i=1, i \neq s}^{n} k_i |z_{si}||z_i|,$$

where we use $|z_s| \geq k_i |z_i| / k_s$. Therefore

---

[4] See Hahn (1982) and Arrow and Hahn (1971).

[5] This case is referred to Gandolfo (1996).

[6] The situation when $\dot{V}$ does not exist is referred to Arrow, Block, and Hurwicz (1959: 106).

[7] The notation , $syn(x)$, means the sign of x , i.e., $syn(x) = +1$ if $x > 0$, $syn(x) = -1$ if $x < 0$, $syn(x) = 0$ if $x = 0$.

$$k_s |z_s| |z_{ss}| > \sum_{i=1, i \neq s}^{n} k_i |z_{si}| |z_i| . \tag{10.1.5}$$

As $z_{ss} < 0$, we have

$$k_s |z_s| |z_{ss}| = -k_s z_{ss} z_s syn(z_s) . \tag{10.1.6}$$

We now show

$$k_s |z_s| |z_{ss}| > \sum_{i=1, i \neq s}^{n} k_i |z_{si}| |z_i| \operatorname{sgn}(z_s) = \operatorname{sgn}(z_s) \sum_{i=1, i \neq s}^{n} k_i |z_{si}| |z_i| . \tag{10.1.7}$$

We exclude

$$\operatorname{sgn}(z_s) = 0$$

since the system is out of the equilibrium. There are three possible cases. The first case is that for one or more subscripts i, z_{si} and z_i have the same sign. We thus have $|z_{si}| |z_i| = z_{si} z_i$. The second case is that for one or more subscripts i, z_{si} or/and z_i are equal to zero. In this case,

$$|z_{si}| |z_i| = z_{si} z_i .$$

The third case is when one or more subscripts i, z_{si} and z_i are of opposite sign. If

$$\operatorname{sgn}(z_s) = -1,$$

we have:

$$|z_{si}| |z_i| = z_{si} z_i \operatorname{sgn}(z_s) .$$

If $\operatorname{sgn}(z_s) = 1$,

$$|z_{si}| |z_i| > z_{si} z_i \operatorname{sgn}(z_s) .$$

Summarizing the above discussions, we conclude Eqs. (10.1.6). From Eqs. (10.1.5) to (10.1.7), we obtain

$$-k_s z_{ss} z_s syn(z_s) > \operatorname{sgn}(z_s) \sum_{i \neq s} z_{si} z_i k_i .$$

Therefore

$$0 > \operatorname{sgn}(z_s) \sum_{i=1}^{n} z_{si} z_i k_i . \tag{10.1.8}$$

This inequality holds in any non-equilibrium point. From Eq. (10.1.4), we conclude that the equilibrium point is globally stable.

It should be noted that the proof for the case of the dominant negative diagonal can be extended to the case of a *quasi-dominant negative diagonal*, in which

$$z_{ii} < 0, \quad c_i |z_{ii}| > \sum_{j=1, j \neq i}^{n} c_j |z_{ij}| \quad \text{for all } i,$$

where the c 's are positive constants. If we choose

$$V = \max_i \left| \frac{k_i}{c_i} z_i \right|,$$

then as in the previous case we prove the global stability.

Example An unstable competitive equilibrium[8]
Consider an economy with three consumers and three goods. Each consumer desires only two commodities (as specified in the utility functions below) in a fixed ratio (i.e., the two commodities are perfectly complementary), which is taken to be one to one, and has no desire for the remaining commodity. Formally, the utility functions of the tree consumers are written as

$$U_1(x_{11}, x_{12}, x_{13}) = \min(x_{11}, x_{12}),$$

$$U_2(x_{21}, x_{22}, x_{23}) = \min(x_{22}, x_{23}),$$

$$U_3(x_{31}, x_{32}, x_{33}) = \min(x_{33}, x_{31}),$$

where x_{ij} are the quantity of good j by consumer i. It is assumed that the initial endowments are

$$\bar{x}_{ij} = \begin{cases} 1 & \text{for } i = j, \\ 0 & \text{for } i \neq j. \end{cases}$$

That is, the first consumer possesses initially one unit of goods 1 and zero units of goods 2 and 3, and so on.

Consider consumer 1. The first consumer's income is

$$M_1 = \sum_j p_j \bar{x}_{1j} = p_1,$$

---

[8] This example is from Gandolfo (1996: 417–20). The model was proposed by Scarf (1960).

where p_j is the price of goods j. His budget constraint is

$$\sum_j p_j x_{1j} = M_1 .$$

For any income he will demand the same quantity of goods 1 and 2. His demand functions are given by

$$x_{11}(p_1,\ p_2,\ p_3,\ M_1) = \frac{M_1}{p_1 + p_2} = \frac{p_1}{p_1 + p_2},$$

$$x_{12}(p_1,\ p_2,\ p_3,\ M_1) = \frac{p_1}{p_1 + p_2},$$

$$x_{13}(p_1,\ p_2,\ p_3,\ M_1) = 0.$$

The excess demands of the first consumer are

$$z_{11}(p_1,\ p_2,\ p_3) = x_{11} - 1 = \frac{-p_2}{p_1 + p_2},$$

$$z_{12}(p_1,\ p_2,\ p_3) = x_{12} - 0 = \frac{p_1}{p_1 + p_2},$$

$$z_{11}(p_1,\ p_2,\ p_3) = x_{13} - 0 = 0.$$

In a similar way we can derive the excess demand functions of consumers 2 and 3. Adding the three excess demand functions for each commodity we obtain the following aggregate excess demand functions

$$z_1(p_1,\ p_2,\ p_3) = \sum_i z_{i1} = \frac{-p_2}{p_1 + p_2} + \frac{p_3}{p_1 + p_3},$$

$$z_2(p_1,\ p_2,\ p_3) = \sum_i z_{i2} = \frac{-p_3}{p_2 + p_3} + \frac{p_1}{p_1 + p_2},$$

$$z_3(p_1,\ p_2,\ p_3) = \sum_i z_{i3} = \frac{-p_1}{p_1 + p_3} + \frac{p_2}{p_2 + p_3}.$$

The only equilibrium situation is[9]

$$p_1 = p_2 = p_3 .$$

---

[9] At equilibrium, $z_i = 0$ for all i.

To determine absolute prices, we need a normalization condition, for instance

$$\sum_i p_i^2 = 3.$$

The equilibrium point is thus $(1, 1, 1)$.

We consider the following Walrasian *tâtonnement*

$$\dot{p}_i = z_i(p_1, p_2, p_3). \quad (10.1.9)$$

It is straightforward to show that

$$p_1(t)p_2(t)p_3(t) = \text{constant},$$

for any solution of Eqs. (10.1.9). It follows that equilibrium is not asymptotically stable. In fact, the value of $p_1 p_2 p_3$ at equilibrium is 1, and if

$$p_1(0)p_2(0)p_3(0) \neq 1,$$

equilibrium will never be reached.

10.2 The Three-Country Trade Model with Capital Accumulation

Most aspects of our model are similar to the OSG model defined in Chap. 4, except that the system consists of three countries, indexed by $j = 1, 2, 3$.[10] Only one goods is produced in the system. Perfect competition is assumed to prevail in goods markets both within each country and between the countries, and commodities are traded without any barriers such as transport costs or tariffs. We assume that there is no migration between the countries and the labor markets are perfectly competitive within each country. Each country has a fixed labor force, N_j, ($j = 1, 2, 3$). Let prices be measured in terms of the commodity and the price of the commodity be unity. We denote wage and interest rates by $w_j(t)$ and $r_j(t)$, respectively, in the jth country. In the free trade system, the interest rate is identical throughout the world economy, i.e.,

$$r(t) = r_j(t).$$

---

[10] This model is proposed by Zhang (1992a, 1992b, 1993a, 1994). This section only provides some simulation illustrations. Varied extensions of this model are proposed in Zhang (2000).

Behavior of producers

We specify the production functions as follows

$$F_j = Z_j (K_j + E_j)^\alpha N_j^\beta ,$$

$$\alpha + \beta = 1, \ \ \alpha, \beta \geq 0, \ \ j = 1, 2, 3, \tag{10.2.1}$$

where Z_j are the technological level of country j, K_j is the level of capital stocks owned by country j, and $E_j > (<) \, 0$ are the level of foreign capital stocks (the home capital stocks located abroad).[11]

According to the definition of E_j, we have the following accounting equation

$$\sum_{j=1}^{3} E_j = 0 . \tag{10.2.2}$$

The marginal conditions are given by

$$r = \frac{\alpha F_j}{K_j + E_j}, \ \ w_j = \frac{\beta F_j}{N_j} . \tag{10.2.3}$$

Behavior of consumers

Consumers obtain the current income Y_j

$$Y_j(t) = r(t) K_j(t) + w_j(t) N_j , \tag{10.2.4}$$

from the interest payment rK_j and the wage payment $w_j N$.[12] The total value of wealth that consumers can sell to purchase goods and to save is equal to $K(t)$. Here, we assume that selling and buying wealth can be conducted instantaneously without any transaction cost. This is obviously a strict assumption. The *gross disposable income* of country j is equal to

$$Y_j^*(t) = Y_j(t) + K_j(t) . \tag{10.2.5}$$

---

[11] Chapter 2 in Zhang (2000) examines cases when α 's are different between the two countries.

[12] It should be noted that the current income of the world is equal to its total output. This can be shown by

$$\sum_j Y_j = \sum_j (rK_j + w_j N) = \sum_j (rK_j + rE_j + w_j N) - \sum_j rE_j = \sum_j F_j ,$$

where we use Eqs. (10.2.2) and (10.2.3).

The gross disposable income is used for saving and consumption and for paying the depreciation of the wealth. We assume that consumers pay the depreciation of capital goods which they own. The total amount is equal to $\delta_k K_j(t)$ where δ_k $(0 \le \delta_k < 1)$ is the depreciation rate of physical capital. We assume that the depreciation rate is equal across countries. At each point of time, consumers would distribute the total available budget among saving $S_j(t)$, consumption of goods $C_j(t)$, and payment for depreciation $\delta_k K(t)$. The budget constraint is given by

$$C_j + \delta_k K_j + S_j = Y_j^* = Y_j + K_j .$$

The disposable income is

$$\hat{Y}_j \equiv Y_j + K_j - \delta_k K_j = rK_j + w_j N + \delta K_j, \quad \delta \equiv 1 - \delta_k .$$

The budget constraint is now given by

$$C_j + S_j = \hat{Y}_j . \tag{10.2.6}$$

We use the Cobb-Douglas utility function to describe consumers' preferences

$$U_j(t) = C^{\xi_j} S^{\lambda_j}, \quad \xi_j, \lambda_j > 0, \tag{10.2.7}$$

in which ξ_j and λ_j are respectively country j's propensities to consume goods and to own wealth. Without loss of generality, we require

$$\xi_j + \lambda_j = 1 .$$

Maximizing U_j subject to (10.2.7) yields

$$C_j = \xi_j \hat{Y}_j, \quad S_j = \lambda_j \hat{Y}_j . \tag{10.2.8}$$

Accumulation of capital

The change in the households' wealth is equal to the net saving minus the wealth sold at time t, i.e.

$$\dot{K}_j(t) = S_j(t) - K_j(t) .$$

Substituting Eqs. (10.2.8) and $\hat{Y}_j = rK_j + w_j N + \delta K_j$ into the above equation yields

$$\dot{K}_j = \lambda_j Y_j - \delta_j K_j, \quad j = 1, 2, 3, \tag{10.2.9}$$

where

$$\delta_j \equiv (1 - \delta_k)\xi_j + \delta_k .$$

We have thus built the model. Through the conditions of equalization of interest rates in Eqs. (10.2.3) and (10.2.1), we obtain

$$r = \frac{\alpha A_1 N_1^\beta}{(K_1 + E_1)^\beta} = \frac{\alpha A_2 N_2^\beta}{(K_2 + E_2)^\beta} = \frac{\alpha A_3 N_3^\beta}{(K_3 + E_3)^\beta}.$$

Hence,

$$E_i = a_i(K_1 + E_1) - K_i, \quad i = 2, 3, \tag{10.2.10}$$

where

$$a_i \equiv \left(\frac{Z_i}{Z_1}\right)^{1/\beta} \frac{N_i}{N_1}.$$

We see that $E_2(t)$ and $E_3(t)$ can be considered as functions of $K_j(t)$ ($j = 1, 2, 3$) and $E_1(t)$. From Eqs. (10.2.2) and (10.2.10), we solve $E_1(t)$ as a function of $K_j(t)$ ($j = 1, 2, 3$)

$$E_1(t) = \frac{(K_2 + K_3) - (a_2 + a_3)K_1}{a_2 + a_3 + 1}. \tag{10.2.11}$$

Lemma 10.2.1 For given levels of $K_j(t)$ ($j = 1, 2, 3$) at any time t, all the other variables in the dynamic system are uniquely determined by the following procedure: E_1 by Eq. (10.2.11) $\rightarrow$ E_2 and E_3 by Eqs. (10.2.10) $\rightarrow$ F_j by Eqs. (10.2.1) $\rightarrow$ r and w_j by Eqs. (10.2.3) $\rightarrow$ Y_j Eqs. (10.2.4) $\rightarrow$ $\hat{Y}_j = rK_j + w_jN + \delta K_j$ $\rightarrow$ C_j and S_j by Eqs. (10.2.8) $\rightarrow$ U_j by (10.27). Moreover, the dynamics of $K_j(t)$ ($j = 1, 2, 3$) over time as given as follows

$$\dot{K}_j = \lambda_j Y_j(K_1, K_2, K_3) - \delta_j K_j, \quad j = 1, 2, 3. \tag{10.2.12}$$

This lemma shows how to simulate the model with given initial conditions of $K_j(0)$ ($j = 1, 2, 3$) and the parameters. The simple model reveals the motion of the global economy with different (exogenous) preferences and technology under economic freedom (not freedom of migration). To simulate the model, we specify the parameters as follows

$$\alpha = 0.25, \ \delta_k = 0.05, \ \lambda_1 = 0.7, \ \lambda_2 = 0.65. \ \lambda_3 = 0.5, \ N_1 = 1,$$

$$N_2 = 5, \ N_3 = 10, \ Z_1 = 3, \ Z_2 = 1, \ Z_1 = 0.5.$$

As shown in Fig. 10.2.1, the total capital, output and consumption levels of the world economy increase; but tend to approach long-term equilibrium.

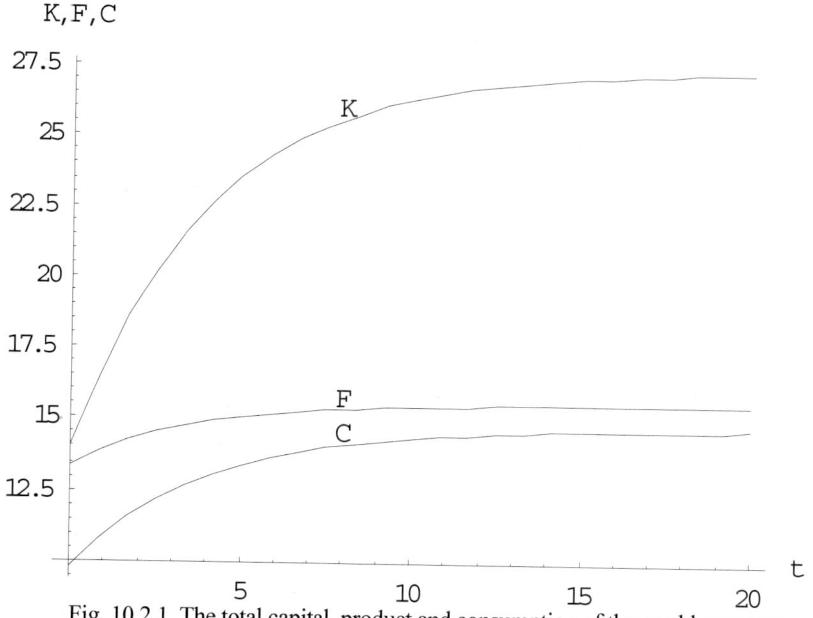

Fig. 10.2.1 The total capital, product and consumption of the world economy.

Figure 10.2.2 shows the dynamics of three countries' wage rates and the rate of interest. Differences in wage rates between countries are great. For instance, the ratio between country 1 and country 3 's human capital, Z_1 / Z_3, is 6; the wage ratio between the two countries w_1 / w_3 is 14 near the long-term equilibrium.

Figure 10.2.3 shows the per-capita consumption levels in the three economies. In free competitive world, living conditions are greatly different among the countries due to differences in human capital and preferences.

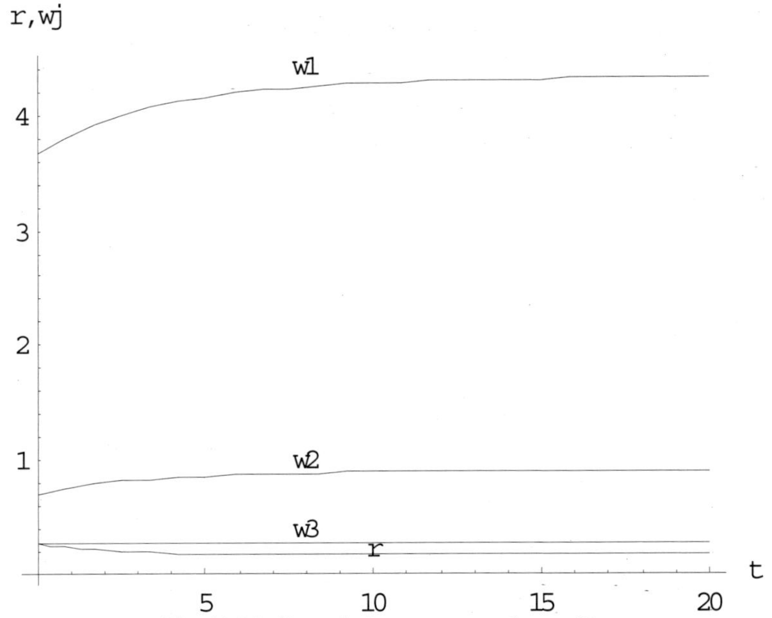

Fig. 10.2.2 Countries' wage rates and rate of interest.

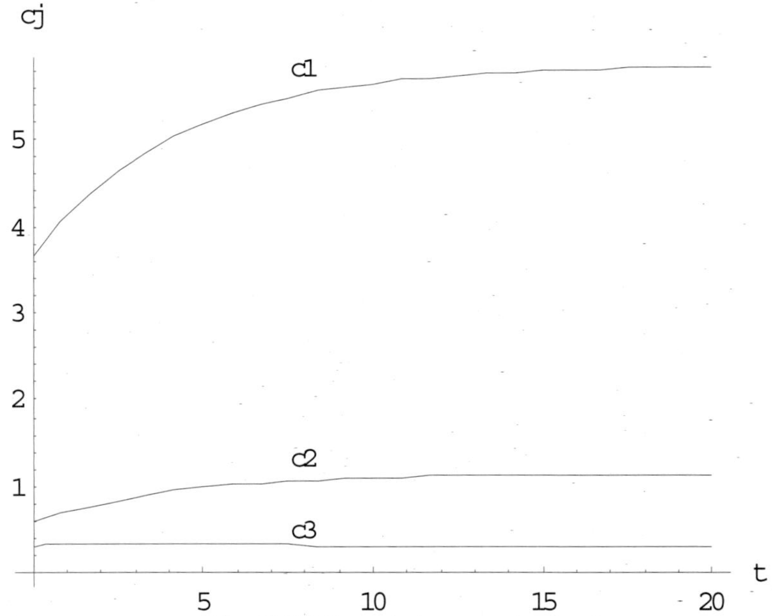

Fig. 10.2.3 Countries' consumption levels.

10.3 Growth, Trade, and Wealth Distribution Among Groups[13]

The model examines how free trade may affect different people from the same national economy. This chapter classifies the population of each country into two groups. The two groups are assumed to have different human capital and utility functions. We are interested in how changes in the preferences and human capital of one group may affect the living conditions of all the groups in the world economy. Most parts of the model are the same as the trade model in Sec. 10.2. The system has two countries, indexed by $j = 1, 2$, and produces one good. The population of each country is classified into two groups, indexed by group 1 and group 2, respectively.

N_{jk} = the fixed population of group k in country $j, k = 1, 2$;

$K_{jk}(t)$ = the capital stocks owned by group k in country j, at time t;

$E(t) > (<) \ 0$ = country 2's (1's) capital stocks employed by country 1 (2);

$F_j(t)$ = country j's output;

$C_{jk}(t)$ = the consumption level of group k in country j;

$w_{jk}(t)$ = the wage rate of group k in country j; and

$r(t)$ = the rate of interest.

Country j's total capital stock $K_j(t)$, the world's capital stocks $K(t)$, country j's qualified labor force N_j, and the world's qualified labor force N are given by

$$K_j = K_{j1} + K_{j2}, \ K = K_1 + K_2, \ N_j = z_{j1}N_{j1} + z_{j2}N_{j2},$$

$$N = N_1 + N_2, \tag{10.3.1}$$

where z_{jk} is the human capital of group k in country $j, k = 1, 2$. The parameter z_{jk} measures the productivity of group k in country j. The production functions of the two countries are

---

[13] This model is explained and analyzed in Zhang (2000).

$$F_1(t) = (K_1 + E)^{\alpha} N_1^{\beta}, \quad F_2(t) = (K_2 - E)^{\alpha} N_2^{\beta},$$

$$\alpha + \beta = 1, \ \alpha, \beta > 0, \ j = 1, 2, \tag{10.3.2}$$

where $K_1 + E$ and $K_2 - E$ are the capital stocks employed by countries 1 and 2, respectively. The marginal conditions are given by

$$r = \frac{\alpha F_1}{K_1 + E} - \frac{\alpha F_2}{K_2 - E}, \quad w_{jk} = \frac{\beta z_{jk} F_j}{N_j}. \tag{10.3.3}$$

It is assumed that the utility level $U_{jk}(t)$ of group k in country j depends on its temporary consumption level $C_{jk}(t)$ and the net saving, $S_{jk}(t)$. The utility functions $U_{jk}(t)$ are specified as

$$U_{jk}(t) = C_{jk}^{\xi_{jk}} S_{jk}^{\lambda_{jk}},$$

$$\xi_{jk} + \lambda_{jk} = 1, \ \xi_{jk}, \lambda_{jk} > 0, \ j, k = 1, 2. \tag{10.3.4}$$

Here, we call ξ_{jk} and λ_{jk} group jk 's propensities to consume goods and to hold wealth, respectively. We interpret U_{jk} as in Chap. 2.

The current income Y_{jk} of group k in country j is given by

$$Y_{jk} = rK_{jk} + w_{jk} N_{jk}. \tag{10.3.5}$$

Households in each country have two decision variables, C_{jk} and S_{jk}. The budget constraints are given by

$$C_{jk} + S_{jk} = \hat{Y}_{jk}, \ j, k = 1, 2, \tag{10.3.6}$$

where

$$\hat{Y}_{jk} = Y_{jk} + K_{jk} - \delta_k K_{jk}$$

and δ_k is the fixed depreciation rate of capital. The consumers' optimal decisions are

$$C_{jk} = \xi_{jk} \hat{Y}_{jk}, \ S_{jk} = \lambda_{jk} \hat{Y}_{jk}, \ j, k = 1, 2. \tag{10.3.7}$$

The wealth accumulation of group k in country j is given by

$$\dot{K}_{jk} = S_{jk} - K_{jk}.$$

Substituting S_{jk} from Eqs. (10.3.7) into the above equations yields

$$\dot{K}_{jk} = \lambda_{jk} Y_{jk} - \delta_{jk} K_{jk}, \tag{10.3.8}$$

where

$$\delta_{jk} \equiv \xi_{jk} + \delta_k \lambda_{jk}.$$

We have thus built the model. The system consists of 30 endogenous variables, K_{jk}, C_{jk}, S_{jk}, Y_{jk}, w_{jk}, U_{jk} ($j, k = 1, 2$), K_1, K_2, F_1, F_2, E and r. It also contains the same number of independent equations. We now show that the system has solutions.

To express the dynamics in terms of $K_{jk}(t)$, it is sufficient to represent $Y_{jk}(t)$ as functions of $K_{jk}(t)$ at any point of time. First, by Eqs. (10.3.2) and

$$r = \frac{\alpha F_1}{K_1 + E} = \frac{\alpha F_2}{K_2 - E},$$

in Eqs. (10.3.3), E is solved as a function of K_1 and K_2 as follows:

$$E(t) = \frac{N_1 K_2 - N_2 K_1}{N}. \tag{10.3.9}$$

Substituting Eqs. (10.3.3), (10.3.2) and (10.3.9) into Eqs. (10.3.5), we get

$$Y_{jk} = \frac{\alpha N K_{jk} / K + \beta z_{jk} N_{jk}}{N^{\alpha}} K^{\alpha}. \tag{10.3.10}$$

The above four equations determine Y_{jk} as functions of K_{jk}. By Eqs. (10.3.10) and (10.3.8), the dynamics of the four variables $K_{jk}(t)$ are determined by the following four-dimensional autonomous differential equations

$$\dot{K}_{jk} = \lambda_{jk} Y_{jk}(\{K_{jk}\}) - \delta_{jk} K_{jk}. \tag{10.3.11}$$

Summarizing the above discussion, we obtain the following lemma.

Lemma 10.3.1 The dynamics of the economic system are determined by the four-dimensional differential equations (10.3.11). The values of all other variables at any point of time are directly given by the following procedure: K_{jk} by Eqs. (10.3.11) $\rightarrow$ $K_j = K_{j1} + K_{j2}$ $\rightarrow$ E by Eq. (10.3.9) $\rightarrow$ Y_{jk} by Eqs. (10.3.10) $\rightarrow$ F_j by Eqs. (10.3.2) $\rightarrow$ r and w_{jk} by Eqs. (10.3.3) $\rightarrow$ S_{jk} and C_{jk} by Eqs. (10.3.7) $\rightarrow$ U_{jk} by (10.3.4).

The dynamic structure of the economic system is explicitly determined by the above lemma. By Eqs. (10.3.10) and (10.3.11), an equilibrium point is given as a solution of the following equations

$$\frac{\alpha N K_{jk} / K + \beta z_{jk} N_{jk}}{N^{\alpha}} K^{\alpha} = \frac{\delta_{jk} K_{jk}}{\lambda_{jk}}.$$

From this equation, we solve K_{jk} as functions of K as follows

$$K_{jk} = \frac{\beta z_{jk} N_{jk} K}{(\delta_{jk} K^{\beta} / \lambda_{jk} N^{\beta} - \alpha) N}. \tag{10.3.12}$$

Since $K_{jk} \geq 0$ have to be satisfied, by Eqs. (10.3.12) it is necessary to require

$$K \geq K_0 \equiv \min\left\{ \left(\frac{\alpha \lambda_{jk}}{\delta_{jk}} \right)^{1/\beta} N, \ j, k = 1, 2 \right\} > 0. \tag{10.3.13}$$

Adding the above four equations and using

$$\sum_{jk} K_{jk} = K,$$

we have

$$\Phi(K) \equiv \frac{N}{\beta} \sum_{jk} \frac{z_{jk} N_{jk}}{\delta_{jk} K^{\beta} / \lambda_{jk} N^{\beta} - \alpha}. \tag{10.3.14}$$

Since $\Phi(K_0) < 0, \Phi(+\infty) > 0$ and $\Phi' > 0$ for $K_0 < K < +\infty$, the equation,

$$\Phi(K) = 0, \quad K_0 < K < +\infty,$$

has a unique solution. The world capital stocks K are thus uniquely determined. By Eqs. (10.3.12), the capital stocks K_{jk} of the two groups in the two countries are uniquely determined. By the procedure in lemma 10.3.1, the equilibrium values of the other variables are uniquely determined. The following proposition is held.

Proposition 10.3.1 The dynamic system has a unique equilibrium.

We now examine the equilibrium trade pattern. By Eqs. (10.3.9), (10.3.1) and (10.3.12), we have

$$E = \frac{\beta N_1 N_2 K}{N^2}\left\{\frac{z_{2k}N_{2k}}{(\delta_{2k}K^\beta / \lambda_{2k}N^\beta - \alpha)N_2} - \sum_k \frac{z_{1k}N_{1k}}{(\delta_{1k}K^\beta / \lambda_{1k}N^\beta - \alpha)N_1}\right\}.$$

As it is not easy to explicitly interpret the above condition, we examine some special cases. It can be seen that E is positive (negative) if

$$\frac{z_{2k}N_{2k}}{(\delta_{2k}K^\beta / \lambda_{2k}N^\beta - \alpha)N_2} > (<) \frac{z_{1k}N_{1k}}{(\delta_{1k}K^\beta / \lambda_{1k}N^\beta - \alpha)N_1},$$

$$k = 1, 2.$$

In the case of $z_{2k}N_{2k} = z_{1k}N_{1k}$, E is positive (negative) if $\delta_{2k} / \lambda_{2k} < (>) \, \delta_{1k} / \lambda_{1k}$, i.e., $1/\lambda_{2k} < (>) \, 1/\lambda_{1k}$. We may thus conclude that in the case in which the qualified labor force of group k ($k = 1$ and 2) in country 2 is equal to that of group k in country 1, then if group k in country 2 has higher (lower) propensities to hold wealth than in country 1, country 1 (2) will utilize some of country 2's (1's) capital stocks.

In the case of $\lambda_{2k} = \lambda_{1k}$, $k = 1, 2,$ E is positive (negative) if $z_{2k}N_{2k} > (<) \, z_{1k}N_{1k}$. In the case in which the preferences of group k ($k = 1$ and 2) in country 2 are identical to those of group k in country 1, then if group k in country 2 has more (less) qualified labor force than in country 1, country 1 (2) will utilize some of country 2's (1's) capital stocks. It is difficult to interpret other cases. From the above discussion, it can be seen that the trade pattern is determined by differences in human capital and propensities to hold wealth between the two countries.

For convenience of discussion, in the remainder of this section it is assumed that the populations of the four groups are identical, i.e., $N_{jh} = N_{jh}$. By Eqs. (10.3.12), the ratios of the capital stocks per capita between group jk and group ih are given by

$$\frac{K_{jk}}{K_{ih}} = \frac{\delta_{ih}K^\beta / \lambda_{ih}N^\beta - \alpha}{\delta_{jk}K^\beta / \lambda_{jk}N^\beta - \alpha}\frac{z_{jk}}{z_{ih}}, \quad k = 1, 2. \tag{10.3.15}$$

We have that $K_{jk} > (<) K_{ih}$ if $z_{jk} > (<) z_{ih}$ and $\lambda_{jk} > (<) \lambda_{ih}$. That is, in the free trade world economy the level of capital stocks owned by group jk is higher (lower) than that owned by group ih if group jk has a higher level of human capital and a higher propensity to hold wealth than group ih.

By $\lambda_{jk} Y_{jk} = \delta_{jk} K_{jk}$ and Eqs. (10.3.7), we obtain

$$\frac{Y_{jk}}{Y_{ih}} = \frac{\delta_{jk}\lambda_{ih} K_{jk}}{\lambda_{jk}\delta_{ih} K_{ih}}, \quad \frac{C_{jk}}{C_{ih}} = \frac{\xi_{jk}\lambda_{ih} K_{jk}}{\lambda_{jk}\xi_{ih} K_{ih}}. \tag{10.3.16}$$

By Eqs. (10.3.15) and (10.3.16), we see directly that $Y_{jk} > (<) Y_{ih}$ if $z_{jk} > (<) z_{ih}$ and $\lambda_{jk} > (<) \lambda_{ih}$. In the case of $\delta_k = 0$, we have: $C_{jk} > (<) C_{ih}$ if $z_{jk} > (<) z_{ih}$ and $\lambda_{jk} > (<) \lambda_{ih}$. In other cases, it is difficult to explicitly judge the signs of $Y_{jk} - Y_{ih}$ and $C_{jk} - C_{ih}$.

We now examine effects of changes in group 11's human capital z_{11} on the world economy. By Eq. (10.3.14), we get

$$\Phi_0 \frac{dK}{dz_{11}} = \frac{(\lambda_{11} N^\beta - \delta_{11} K^\beta) N_{11}}{(\delta_{11} K^\beta - \alpha\lambda_{11} N^\beta)\beta} + \frac{N_{11}}{N},$$

where

$$\Phi_0 \equiv \sum_{jk} \frac{\beta\delta_{jk} z_{jk} N_{jk}}{(\delta_{jk} K^\beta / \lambda_{jk} N^\beta - \alpha)^2 \lambda_{jk} K^\alpha N^\beta} > 0.$$

In the case of $\lambda_{11} N^\beta \ge \delta_{11} K^\beta$, an improvement in group 11's human capital increases the world capital stocks. If $\lambda_{11} N^\beta < \delta_{11} K^\beta$, the impact on K is ambiguous. As the equilibrium value K is not explicitly solved, it is not easy to interpret the condition $\lambda_{11} N^\beta \ge \delta_{11} K^\beta$. Taking derivatives of Eqs. (10.3.12) with respect to z_{11}, we obtain

$$\frac{\delta_{11} K^\beta - \alpha\lambda_{11} N^\beta}{K_{11}} \frac{dK_{11}}{dz_{11}} = \frac{\delta_{11} K^\beta - \alpha\lambda_{11} N^\beta}{z_{11}}$$

$$+ \alpha(\delta_{11} K^\beta - \lambda_{11} N^\beta)\left(\frac{1}{K}\frac{dK}{dz_{11}} - \frac{\delta_{11} K^\beta N_{11}}{N}\right),$$

$$\frac{(\delta_{jk} K^\beta - \alpha\lambda_{jk} N^\beta) K}{\alpha K_{jk}} \frac{dK_{jk}}{dz_{11}}$$

$$= (\delta_{jk} K^{\beta} - \lambda_{jk} N^{\beta}) \left(\frac{dK}{dz_{11}} - \frac{\delta_{jk} K^{\beta+1} N_{11}}{N} \right), \quad jk \neq 11.$$

It is easy to check that in the case in which the four groups have identical preferences, i.e., $\lambda = \lambda_{jk}$ ($\xi = \xi_{jk}$ and $\delta = \delta_{jk}$) for all j, k. Using

$$K^{\beta} = \frac{\lambda N^{\beta}}{\delta}, \quad K_{jk} = \left(\frac{\lambda}{\delta} \right)^{1/\beta} z_{jk} N_{jk},$$

we get

$$\frac{dK}{dz_{11}} = \left(\frac{\lambda}{\delta} \right)^{1/\beta} N_{11} > 0, \quad \frac{dK_{11}}{dz_{11}} = \frac{N_{11}}{z_{11}} > 0, \quad \frac{dK_{jk}}{dz_{11}} = 0.$$

That is, an improvement of group 11's human capital enlarges group 11's capital stocks and the world's capital stocks, but has no impact on the capital stocks of the other three groups. Taking derivatives of $Y_{jk} = \delta_{jk} K_{jk} / \lambda_{jk}$ and $C_{jk} = \xi_{jk} K_{jk} / \lambda_{jk}$ with respect to z_{11} respectively, we obtain

$$\frac{1}{C_{jk}} \frac{dC_{jk}}{dz_{11}} = \frac{1}{Y_{jk}} \frac{dY_{jk}}{dz_{11}} = \frac{1}{K_{jk}} \frac{dK_{jk}}{dz_{11}}$$

for all j, k. The sign of dC_{jk} / dz_{11} and dY_{jk} / dz_{11} is the same as that of dK_{jk} / dz_{11}.

10.4 A Two-Region Growth Model with Capital and Knowledge

The model studies relationships between regional growth and regional trade patterns.[14] Each region's production is similar to the standard one-sector growth model. Knowledge accumulation is through learning by doing. We consider an economic system consisting of two regions, indexed by 1 and 2, respectively.

It is assumed that climates and environment are homogeneous within each region, but they may be different between the two regions. We

[14] This section is referred to Chap. 9 in Zhang (2003). The modeling framework is proposed by Zhang (1991a, 1993b).

select commodity to serve as numeraire. The amenity levels are assumed to be regionally fixed. Since people's locational choice is affected by regional environmental conditions and the temporary equilibrium condition for labor movement is that people achieve the same level of utility in two regions, we see that wage rates between the two regions may not be equal.

To describe the model, we introduce

N	= the given population of the economy;
$K(t)$	= the total capital stocks of the economy at time t;
$Z(t)$	= the level of knowledge stock at time t;
$F_j(t)$	= the output levels of region j's production sector;
$K_j(t)$ and $N_j(t)$	= the levels of capital stocks and labor force employed by region j's production sector;
$c_j(t)$ and $s_j(t)$	= the consumption level of and level of saving per capita in region j;
$y_j(t)$	= the net income per capita in region j;
$r(t)$	= the rate of interest; and
$w_j(t)$	= region j's wage rate.

Production functions of the two regions are

$$F_j(t) = Z^{m_j} K_j^{\alpha} N_j^{\beta}, \quad m_j \geq 0,$$

$$\alpha, \beta > 0, \quad \alpha + \beta = 1, \quad j = 1, 2, \tag{10.4.1}$$

in which m_j are region j's knowledge utilization efficiency parameters. They measure how effectively each region utilizes the knowledge reservoir of the economy. The marginal conditions are given by

$$r = \frac{\alpha F_j}{K_j}, \quad w_j = \frac{\beta F_j}{N_j}. \tag{10.4.2}$$

The rate of interest is identical in the whole economy and the wage rates, w_j, may be different between the two regions.

If we denote $k_j(t)$ the level of capital stocks per capita in region j, the interest payment per capita is given by $r(t)k_j(t)$. The current income per capita, $y_j(t)$ in region j consists of the wage income, $w_j(t)$, and interest payment, $r(t)k_j(t)$. That is

$$y_j(t) = w_j(t) + r(t)k_j(t), \ \ j = 1, 2 . \tag{10.4.3}$$

It is assumed that a typical person's utility level, $U_j(t)$, in region j is dependent on the person's consumption level, $c_j(t)$, of community, and the net saving, $s_j(t)$. The utility functions are specified as follows

$$U_j(t) = A_j c_j^{\xi_j} s_j^{\lambda_j}, \ \ \xi_j, \lambda_j > 0, \ \ \xi_j + \lambda_j = 1, \ \ j = 1, 2, \tag{10.4.4}$$

in which ξ_j and λ_j are respectively region j's propensities to consume commodity and to hold wealth.

A household's current income is distributed between consumption and saving. The budget constraints are given by

$$c_j + s_j = y_j + k_j - \delta_k k_j, \ \ j = 1, 2 .$$

The optimal problem has the following unique solution

$$c_j = \xi_j \Omega_j, \ \ s_j = \lambda_j \Omega_j , \tag{10.4.5}$$

where

$$\Omega_j \equiv y_j + dk_j, \ \ d \equiv 1 - \delta_k .$$

The wealth accumulation of a typical person in region j is given by

$$\dot{k}_j = s_j - k_j .$$

Substituting s_j in Eqs. (10.4.5) into these two equations yields

$$\dot{k}_j = \lambda_j \Omega_j - k_j . \tag{10.4.6}$$

As people are freely mobile between the two regions, the utility level of people should be equal, irrespective of which region they live. That is

$$U_1(t) = U_2(t). \tag{10.4.7}$$

This equation is the temporary equilibrium condition for interregional labor force markets.

Knowledge accumulation

In this section, we only take account of learning by doing in formulating knowledge accumulation.

We propose the following possible dynamics of knowledge

$$\frac{dZ}{dt} = \frac{\tau_1 F_1}{Z^{\varepsilon_1}} + \frac{\tau_2 F_2}{Z^{\varepsilon_2}} - \delta_z Z, \tag{10.4.8}$$

in which τ_j (≥ 0), ε_j, and δ_z (≥ 0) are parameters. Here, we interpret $\tau_j F_j / Z^{\varepsilon_j}$ as region j's contribution to knowledge accumulation through learning by doing. We assume that the contribution to the knowledge creation of region j's labor force is positively and linearly related to the region production scale, F_j. The term, $1/Z^{\varepsilon_j}$, implies that region j's knowledge accumulation exhibits return to scale effects. The parameters, ε_j, measure return to scale effects of knowledge in knowledge accumulation by region j's labor force. We say that the contribution to knowledge growth of region j exhibits increasing (decreasing) scale effects when $\varepsilon_j < (>) \, 0$. We interpret τ_j as a measurement of knowledge accumulation efficiency. It should be noted that it is conceptually not difficult to introduce other ways, such as research institutions and education, of accumulating knowledge.

By the definitions of K, k_j and N_j, we have

$$K = k_1 N_1 + k_2 N_2. \tag{10.4.9}$$

The above equation tells that the total capital stocks of the economy is equal to the sum of the capital stocks owned by the two regions. The assumption that labor force and capital stocks are fully employed is represented by

$$N_1 + N_2 = N, \quad K_1 + K_2 = K.$$

We complete the construction of the basic model. We are now concerned with conditions for the existence of equilibria of the dynamic system. By Eqs. (10.4.8) and (10.4.6), at equilibrium we have

$$\frac{\tau_1 F_1}{Z^{\varepsilon_1}} + \frac{\tau_2 F_2}{Z^{\varepsilon_2}} = \delta_z Z, \ \lambda_j \Omega_j = k_j, \ \ j = 1, 2. \tag{10.4.10}$$

Substituting $\lambda_j \Omega_j = k_j$ from Eqs. (10.4.10) into Eqs. (10.4.5) yields

$$c_j = \frac{\xi_j k_j}{\lambda_j}, \quad s_j = dk_j. \tag{10.4.11}$$

These equations tell that at a steady state the level of consumption per capita in region j is proportional to the capital stocks per capita.

Substituting Eqs. (10.4.11) into the utility functions (10.4.4) and then using Eqs. (10.4.7), we get

$$\frac{k_1}{k_2} = A \equiv \frac{A_2}{A_1}\left(\frac{\xi_2}{\lambda_2}\right)^{\xi_2}\left(\frac{\xi_1}{\lambda_1}\right)^{-\xi_1}. \tag{10.4.12}$$

The balance of demand and supply is given by

$$\delta_1 k_1 N_1 + \delta_2 k_2 N_2 = F_1 + F_2, \tag{10.4.13}$$

where

$$\delta_j \equiv \frac{\xi_j}{\lambda_j} + \delta_k .$$

By $\lambda_j \Omega_j = k_j$ in Eqs. (10.4.10) and (10.4.3) and the definitions of Ω_j, we get

$$w_j = (\delta_j - r)k_j, \quad j = 1, 2. \tag{10.4.14}$$

By Eqs. (10.4.1) and (10.4.2), we get

$$\Lambda \equiv \frac{K_1}{N_1 Z^{m_1/\beta}} = \frac{K_2}{N_2 Z^{m_2/\beta}}, \quad r = \frac{\alpha}{\Lambda^\beta}, \quad w_j = \beta Z^{m_j/\beta}\Lambda^\alpha. \tag{10.4.15}$$

Substituting

$$F_j = \frac{rK_j}{\alpha} = \frac{K_j}{\Lambda^\beta}$$

into Eq. (10.4.13) yields

$$\delta_1 k_1 N_1 + \delta_2 k_2 N_2 = \frac{K}{\Lambda^\beta}, \tag{10.4.16}$$

where we use $K = K_1 + K_2$. Substituting Eqs. (10.4.9) and (10.4.12) into Eq. (10.4.16) yields

$$\Gamma = \frac{\delta_2 \Lambda^\beta - 1}{(1 - \delta_1 \Lambda^\beta)A}, \tag{10.4.17}$$

where $\Gamma \equiv N_1 / N_2$. By $N_1 + N_2 = N$ and $\Gamma = N_1 / N_2$, we have

$$N_1 = \frac{\Gamma N}{1 + \Gamma}, \quad N_2 = \frac{N}{1 + \Gamma}. \tag{10.4.18}$$

Dividing the first equation in Eqs. (10.4.14) by the second one and then using Eq. (10.4.12) and Eqs. (10.4.15), we obtain

$$Z^m = \frac{\delta_1 \Lambda^\beta - \alpha}{\delta_2 \Lambda^\beta - \alpha}, \qquad (10.4.19)$$

where

$$m \equiv \frac{m_1 - m_2}{\beta}.$$

In the case of $m = 0$, we solve Λ and Γ respectively by Eqs. (10.4.19) and (10.4.17). If without special mention, $m \neq 0$ holds below.

For simplicity of discussion, we require:

$$\frac{\delta_2}{\alpha} > \delta_1 > \delta_2 .$$

We interpret this requirement later on. By Eqs. (10.4.14) and $r = \alpha / \Lambda^b$ in Eqs. (10.4.15), we see that $w_j > 0$ are guaranteed if

$$\Lambda^\beta > \max\left\{\frac{\alpha}{\delta_j},\ j = 1, 2\right\} = \frac{\alpha}{\delta_2}.$$

By Eq. (10.4.19) we see that this condition also guarantees $Z^m > 0$. By Eq. (10.4.17) $\Gamma > 0$ is guaranteed if $1/\delta_2 > \Gamma > 1/\delta_1$. It is thus sufficient for us to require

$$\frac{1}{\delta_2} > \Gamma > \max\left\{\frac{1}{\delta_1},\ \frac{\alpha}{\delta_2}\right\} = \frac{1}{\delta_1}. \qquad (10.4.20)$$

By the first equation in Eqs. (10.4.15) and $F_j = K_j / \Lambda^\beta$, we have

$$K_j = \Lambda N_j Z^{m_j/\beta}, \quad F_j = \Lambda^\alpha N_j Z^{m_j/\beta}. \qquad (10.4.21)$$

Substituting F_j from Eqs. (10.4.21) into the first equation in Eqs. (10.4.10), we get

$$\Phi(\Lambda) \equiv \Phi_1(\Lambda) + \Phi_2(\Lambda) - d = 0, \quad \frac{1}{\delta_2} > \Lambda^\beta > \frac{1}{\delta_1}, \qquad (10.4.22)$$

where we use Eq. (10.4.17) and

$$\Phi_1(\Lambda) \equiv \frac{\tau_1 N \Gamma Z^{x_1} \Lambda^{\alpha}}{1+\Gamma}, \quad \Phi_2(\Lambda) \equiv \frac{\tau_2 N Z^{x_2} \Lambda^{\alpha}}{1+\Gamma},$$

$$x_j \equiv \frac{m_j}{\beta} - \varepsilon_j - 1, \quad j = 1, 2.$$

In Eq. (10.4.22), Γ and Z are functions of Λ respectively defined by Eq. (10.4.17) and Eq. (10.4.19). By Eqs. (10.4.22), (10.4.17), and (10.4.19), we conclude that the function, $\Phi(\Lambda)$, has the following properties

$$\Phi\left(\frac{1}{\delta_2^{1/\beta}}\right) = \frac{\{(\delta_1/\delta_2 - \alpha)A/\beta\}^{x_2/\beta}}{\delta_2^{\alpha/\beta}} \tau_2 N - d,$$

$$\Phi\left(\frac{1}{\delta_1^{1/\beta}}\right) = \frac{\{\beta A/(\delta_2/\delta_1 - \alpha)\}^{x_1/\beta}}{\delta_1^{\alpha/\beta}} \tau_1 N - d,$$

$$\frac{d\Phi}{d\Lambda} = \frac{x_1\Phi_1 + x_2\Phi_2}{Z}\frac{dZ}{d\Lambda} + \frac{\Phi_1/\Gamma - \Phi_2}{1+\Gamma}\frac{d\Gamma}{d\Lambda} + \alpha\frac{\Phi_1 + \Phi_2}{\Lambda^{\beta}},$$

in which

$$m\frac{dZ}{d\Lambda} = \frac{\delta_2 - \delta_1}{(\delta_2\Lambda^{\beta} - \alpha)(\delta_1\Lambda^{\beta} - \alpha)}\frac{\alpha\beta Z}{\Lambda^{\alpha}} < 0,$$

$$\frac{\Lambda^{\alpha}}{\beta\Gamma}\frac{d\Gamma}{d\Lambda} = \frac{\delta_2}{\delta_2\Lambda^{\beta} - \alpha} + \frac{\delta_1}{\delta_1\Lambda^{\beta} - \alpha} > 0.$$

As it is difficult to explicitly judge the signs of $\Phi(1/\delta_1^{1/\beta})$, $\Phi(1/\delta_2^{1/\beta})$ and $d\Phi/d\Lambda$, we see that it is difficult to judge whether or not the equation, $\Phi(\Lambda) = 0$, has solutions for $1/\delta_2 > \Lambda^{\beta} > 1/\delta_1$. We may solve the equation properly specifying values of some parameters. For instance, if $\Phi(1/\delta_2^{1/\beta}) > 0$ (which is guaranteed, for instance, if τ_2 is large), $\Phi(1/\delta_2^{1/\beta}) < 0$ (which is guaranteed, for instance, if τ_1 is small), and $d\Phi/d\Lambda > 0$ (which is guaranteed, for instance, if $m > 0$, $x_1 < 0$ and $x_2 < 0$), then the dynamic system has a unique equilibrium. Summarizing the above discussion, we obtain the following proposition.

Proposition 10.4.1 We assume $m \geq 0$ and $\delta_2 / \alpha > \delta_1 > \delta_2$. If Eq. (10.4.22) has solutions for

$$\frac{1}{\delta_2} > \Lambda^\beta > \frac{1}{\delta_1},$$

for any given solution, the dynamic system has a unique equilibrium. The number of economic equilibrium is equal to the number of solutions of Eq. (10.4.22).

Since it is difficult to interpret the conditions that Eq. (10.4.22) has a meaningful solution, for illustration we now examine the case of $m = 0$. We still require: $\delta_2 / \alpha > \delta_1 > \delta_2$. By Eq. (10.4.19), we solve Λ^β as follows

$$\Lambda^\beta = \frac{\alpha(A-1)}{\delta_1 A - \delta_2}. \tag{10.4.23}$$

The unique solution, Λ^β, is positive and satisfies (10.4.20) if

$$\frac{\delta_2 / \delta_1 - \alpha}{\beta} < A < \min\left\{\frac{\delta_2}{\delta_1}, \frac{\beta}{\delta_2 / \delta_1 - \alpha}\right\} < 1. \tag{10.4.24}$$

Under (10.4.24), we solve Λ by Eq. (10.4.23). By Eq. (10.4.17), we solve Γ. It should be noted that the values of Γ and Λ are independent of Z. Similarly to Eq. (10.4.22), we get

$$\Phi_0(Z) \equiv \Phi_{01}(Z) + \Phi_{02}(Z) - d = 0, \quad \infty > Z > 0, \tag{10.4.25}$$

where

$$\Phi_{01}(Z) \equiv \frac{\tau_1 N \Gamma Z^{x_1} \Lambda^\alpha}{1 + \Gamma}, \quad \Phi_{02}(Z) \equiv \frac{\tau_2 N Z^{x_2} \Lambda^\alpha}{1 + \Gamma}.$$

It should be noted that in Eq. (10.4.25) Λ and Γ are treated as parameters. The equation includes a single variable, Z.

We obtain that $\Phi_0(Z)$ has the following properties: 1) in the case of $x_j < 0$, we have

$$\Phi_0(Z) > 0, \quad \Phi_0(\infty) < 0, \quad \frac{d\Phi_0}{dZ} < 0, \quad \infty > Z > 0,$$

2) in the case of $x_j > 0$, we have

$$\Phi_0(Z) < 0, \ \Phi_0(\infty) > 0, \ \frac{d\Phi_0}{dZ} > 0, \ \infty > Z > 0 .$$

3) in the case of $x_1 > 0$ and $x_2 < 0$ ($x_1 < 0$ and $x_2 > 0$), we have

$$\Phi_0(Z) > 0, \ \Phi_0(\infty) > 0,$$

$$\frac{d\Phi_0}{dZ} = 0 \text{ has a unique solution, } \infty > Z > 0 .$$

From these properties of $\Phi_0(Z)$, we have the following results.

Corollary 10.4.1 If $m = 0$, $\delta_2 / \alpha > \delta_1 > \delta_2$, and (10.4.24) holds, then, we have: (1) If $x_j < 0$, $j = 1, 2$, the system has a unique equilibrium; 2) If $x_j > 0$, $j = 1, 2$, the system has a unique equilibrium; 3) If $x_1 > 0$ and $x_2 < 0$ ($x_1 < 0$ and $x_2 > 0$), the system has either two equilibrium or no equilibrium point. Moreover, the equilibrium values of the variables are given by the following procedure: Λ by Eq. (10.4.23) $\rightarrow \Gamma$ by Eq. (10.4.17) $\rightarrow N_j$, $j = 1, 2$, by Eqs. (10.4.18) $\rightarrow Z$ by Eq. (10.4.25) $\rightarrow r$, w_j and K_j by Eqs. (10.4.15) $\rightarrow K = K_1 + K_2 \rightarrow k_j$ by Eqs. (10.4.14) $\rightarrow c_j$ and s_j by Eqs. (10.4.11) $\rightarrow F_j$ by Eqs. (10.4.1) $\rightarrow$ y_j by Eqs. (10.4.3) $\rightarrow U_j$ by (10.4.4).

The requirement, $m = 0$, means that the two regions have the same level of knowledge utilization efficiency. The requirement, $\delta_1 > \delta_2$ (i.e., $\lambda_2 > \lambda_1$), implies that region 2's propensity to hold wealth is higher than region 1. The condition, $\delta_2 / \alpha > \delta_1 > \delta_2$, implies that the preference difference between the two regions is not large. The condition (10.4.24) is guaranteed if the amenity ratio between the two regions is properly bound. As m_j is region j's knowledge utilization efficiency parameter and ε_j denotes the return to scale effects of knowledge in knowledge accumulation, we may interpret x_j as the measurement of return to scale effects of knowledge in region j. The condition, $x_j < (>)\ 0$, $j = 1, 2$, may be interpreted as that the two regions exhibit weak (strong) return to scale effects of knowledge.

In the remainder of this chapter, we examine effects of changes in some parameters on the economic structure. For simplicity, we assume that the assumptions in Corollary 10.4.1 are satisfied. Moreover, we are concerned with the two cases, $x_j > 0$ and $x_j < 0$, $j = 1, 2$, that guarantee the existence of a unique equilibrium in the dynamic system.

We now examine the impact of changes in the parameter, A. By the definition of A, A is changed either due to changes in A_j or λ_j. Here, we assume that a change in A is due to changes in amenity levels. An increase in A means an increase in A_2 or decrease in A_1. Taking derivatives of Eqs. (10.4.17), (10.4.18), $r = \alpha / \Lambda^\beta$ in (10.4.15), and (10.4.23) with respect to A yields

$$\frac{\beta}{\alpha\Lambda^\alpha}\frac{d\Lambda}{dA} = \frac{\delta_1 - \delta_2}{(\delta_1 A - \delta_2)^2} > 0,$$

$$\frac{dr}{dA} = -\frac{\alpha\beta}{\Lambda^{\beta+1}}\frac{d\Lambda}{dA} < 0,$$

$$\frac{d\Gamma}{dA} = \frac{\delta_1 - \delta_2}{(1 - \delta_2\Lambda^\beta)(1 - \delta_1\Lambda^\beta)}\frac{\beta\Gamma}{\Lambda^\alpha}\frac{d\Lambda}{dA} - \frac{\Gamma}{A} < 0,$$

$$\frac{1}{N_1}\frac{dN_1}{dA} = \frac{1}{(1+\Gamma)\Gamma}\frac{d\Gamma}{dA} < 0,$$

$$\frac{1}{N_2}\frac{dN_2}{dA} = \frac{-1}{1+\Gamma}\frac{d\Gamma}{dA} > 0.$$

As region 2's amenity level is increased, the rate of interest is reduced and some of region 1's population migrate to region 2. By Eq. (10.4.25), we have

$$-N\frac{d\Phi_0}{dZ}\frac{dZ}{dA} = \frac{\alpha d}{\Lambda^\beta}\frac{d\Lambda}{dA} + \left(\frac{\Phi_{01}}{N_1} - \frac{\Phi_{02}}{N_2}\right)\frac{N_2}{1+\Gamma}\frac{d\Gamma}{dA}. \quad (10.4.26)$$

Since $d\Phi / dZ < (>) \ 0$ in the case of $x_j < (>) \ 0$, we conclude that if we further require $\Phi_{02} / N_2 \geq \Phi_{01} / N_1$ then $dZ / dA > (<) \ 0$ in the case of $x_j < (>) \ 0$ In the case that the contribution, Φ_{02} / N_2, to knowledge accumulation by region 2's per capita is larger than the contribution, Φ_{01} / N_1, by region 1's per capita, if the two regions exhibit strong (weak) increasing return to scale effects in knowledge utilization and accumulation, then an improvement in region 2's amenity will reduce (increase) the equilibrium value, Z, of the knowledge stock. In the case of

$$\frac{\Phi_{02}}{N_2} < \frac{\Phi_{01}}{N_1},$$

it is difficult to determine the sign of dZ / dA.

By Eqs. (10.4.15) and $F_j = K_j / \Lambda^\beta$, we get the impact on K_j and w_j as follows

$$\frac{1}{w_j}\frac{dw_j}{dA} = \frac{m_1}{\beta Z}\frac{dZ}{dA} + \frac{\alpha}{\Lambda}\frac{d\Lambda}{dA}, \quad j = 1, 2,$$

$$\frac{1}{K_j}\frac{dK_j}{dA} = \frac{1}{N_j}\frac{dN_j}{dA} + \frac{1}{\Lambda}\frac{d\Lambda}{dA} + \frac{m_j}{\beta Z}\frac{dZ}{dA},$$

$$\frac{1}{F_j}\frac{dF_j}{dA} = \frac{1}{K_j}\frac{dK_j}{dA} - \frac{\beta}{\Lambda}\frac{d\Lambda}{dA}. \tag{10.4.27}$$

In the case of $dZ / dA > 0$, the sign of dw_j / dA is the same as that of dZ / dA. We see that the two regions' wage rates may be either increased or reduced in the case of $dZ / dA < 0$. From Eqs. (10.4.26) and (10.4.27), we see that it is complicated to explicitly determine the signs of dK_j / dA and dF_j / dA.

By Eqs. (10.4.14), (10.4.11) and $y_j = \delta_j k_j$, we have

$$\frac{1}{k_j}\frac{dk_j}{dA} = \frac{m_1}{\beta Z}\frac{dZ}{dA} + \frac{\alpha\delta_j - r}{(\delta_j - r)\Lambda}\frac{d\Lambda}{dA},$$

$$\frac{1}{y_j}\frac{dy_j}{dA} = \frac{1}{c_j}\frac{dc_j}{dA} = \frac{1}{k_j}\frac{dk_j}{dA}.$$

As

$$\frac{d\Lambda}{dA} > 0, \ \delta_j - r > 0, \ \alpha\delta_1 - r > 0, \ \alpha\delta_2 - r > 0,$$

we conclude that $dk_1 / dA > 0$ ($dk_2 / dA < 0$) in the case of $dZ / dA > (<)\, 0$. The signs of dk_2 / dA (dk_1 / dA) may be either negative or positive in the case of $dZ / dA > (<)\, 0$.

10.5 Money and Economic Growth

In the process of exchange and division of labor, money plays an essential role in modern economy.[15] We provide the neoclassical growth model with money – the generalized Tobin model. This section is based on Chap. 6 in Burmeister and Dobell and in Zhang.[16] The model in this section is developed and extended within the framework of Tobin's analysis.[17] We assume the presence of a paper currency in addition to a single capital good; wealth may be held in either of these two forms. The production side is identical to the OSG model. Money, issued by the government without cost, is assumed to serve as numeraire. Money is desired for current investment and transaction purposes. We assume that the per capita demand for money is a function of per capita money income, per capita money wealth, and the expected yield on capital. The money market is assumed to be always in equilibrium, that is, money demand per capita m^d always equals the actual money supply per capita

$$m^d = G(y, w, r) = m ,$$

where m $(= M / N$, where M is money stock, N labor force) is per capita money stock supply, G is a continuous function with regard to the arguments, y per capita value of output, w $(= pk + m)$ per capita money wealth; r is the expected money yield on the capital good. We have

$$r = f'(k) - \delta_k + E\left[\frac{dp / dt}{p}\right],$$

where p is price of output in terms of money as numeraire, F, $f(k)$, k, and δ_k are defined as in the OSG model. In the above formation, $E[dp / dt / p]$ is the expected inflation rate.

---

[15] See Hahn (1969), Gale (1983), Grandmont (1983, 1985), Orphanides and Solow (1990), Stadler (1990).

[16] See Burmeister and Dobell (1970) and Sec. 3.3 in Zhang (1990).

[17] See Tobin (1955, 1965). Extensions of this model to the Keynesian economics are referred to Chiarella and Flaschel (2000). Franke and Asada (1994) construct a nonlinear dynamic IS-LM model in four state variables representing real balances, inflation, income distribution, and a so-called state of confidence. The system exhibits time-dependent behavior.

We assume the absence of "money illusion", which means that $G(y,w,r)$ is homogeneous of degree one in its first two arguments. The real wealth W and real disposable income $Y_d(t)$ are defined as

$$W(t) = K(t) + \psi(t), \;\; Y_d(t) = F\big(K(t), N(t)\big) - \delta_k K(t) + \dot{\psi}(t),$$

where

$$\psi(t) \equiv \frac{M(t)}{p(t)}.$$

As

$$F(t) = C(t) + \delta_k K(t) + \dot{K}(t),$$

where $C(t)$ is the consumption, we have

$$Y_d(t) = C(t) + \dot{W}(t).$$

That is, real net disposable income is equal to the change in real wealth plus real consumption. It is assumed that real consumption is always a constant fraction of real net disposable income

$$C(t) = cY_d(t),$$

where c is the propensity to consume. This is a flow equilibrium condition satisfied at all points of time. From $Y_d = C + dW/dt$ and $C = cY_d$, we have

$$\dot{W}(t) = sY_d(t),$$

where $s \equiv 1 - c$. This implies

$$\dot{K} = s\big(Y + \dot{\psi} - \delta_k K\big) - \dot{\psi} = s(y - \delta_k k) - (1 - s)\dot{\psi}, \quad (10.5.1)$$

which is referred to as Tobin's fundamental equation.

Taking the derivative of $x = \psi / N$ yields

$$g_x = z - n - g_p,$$

where n is the fixed population growth rate and z is the constant proportional rate of increase in the nominal stock of money. The parameter z is fixed by the government. Let us denote

$$q \equiv E[g_p] = g_p.$$

By $y = f(k)$ and omitting depreciation factors (i.e., $\delta_k = 0$), it is obvious to show that we may write (10.5.1) and $g_x = z - n - g_p$ as follows

$$\dot{k}(t) = sf(k(t)) - c(z - q(t))x(t) - nk(t),$$

$$\frac{\dot{x}(t)}{x(t)} = z - n - \frac{\dot{p}(t)}{p(t)}.$$

To complete the system, we have to specify dp/dt. We permit the market for goods and services and the money market to be out of equilibrium and the actual rate of inflation to be different from the expected one. We adopt the Walrasian view that when there is excess demand the price rises, and when there is excess supply the price falls. Without taking account of expected inflation, we propose the following dynamics

$$\dot{p} = \alpha p[x - g(.)],$$

where α is a positive constant parameter, and the function g is to be specified. It is assumed that the expected inflation rate may be different from the actual inflation rate. The dynamics is specified as

$$\dot{q} = \beta\left[\frac{\dot{p}}{p} - q\right],$$

where β is the so-called "expectation coefficient". This is the "adaptive expectation" equation initially introduced by Cagan.[18] It states that expectations change a constant proportion of the "error" between the actual rate of inflation and the expected one. If $\beta \to \infty$, we again have the perfect foresight equation.

We now come to the problem of specifying the demand function for money. In the case in which the demand for real balances is only for asset purposes, then it is a function of the opportunity cost of holding them, $f'(k) + q$. We now examine the cases in which the two assets of our model, capital and real cash balances are perfect and imperfect substitutes. In the first case, the yields of both assets have to be the same. Otherwise, when $f'(k) + q > 0$ only capital is demanded, and when

---

[18] Cagan (1956).

$f'(k) + q < 0$ only real cash balances are demanded. In this case we have $g'(.) \to -\infty$, where the derivative is related to $f'(k) + q$. In the second case, we have $f'(k) + q > 0$ because of the obvious superiority of real cash balances. The demand function is negatively related to $f'(k) + q$. Tobin attributed this difference to a risk element involved in asset capital as compared to the risklessness of real cash balances, while Friedman *et al.* attributed this difference to what they call the utility yield of real cash balances. We also have to take account of money which is demanded for transaction purposes. The proxy for per capita transaction demand usually found in the literature is the per capita output $f(k)$; the higher the per capita output, the higher the per capita transaction demand. We can thus generally write g as,

$$g = G^*\{f(k), \ f' + q\}, \text{ or } g = g(k,q).$$

In the case of perfect substitutability, we have $g_k = +\infty$ and $g_q = -\infty$; in the case of imperfect substitutability, $g_k > 0$ and $g_q < 0$.

Summarizing the above discussions, we obtain the following generalized Tobin model

$$\dot{k} = sf(k) - c(z - q)x - nk,$$

$$\frac{\dot{x}}{x} = z - n - \alpha\{x - g(k,q)\},$$

$$\dot{q} = \beta[\alpha\{x - g(k,q)\} - q]. \tag{10.5.2}$$

A positive long-run equilibrium is determined as a solution of

$$sf(k) - c(z - q)x - nk = 0,$$

$$z - n - \alpha\{x - g(k,q)\} = 0,$$

$$\alpha\{x - g(k,q)\} = q.$$

First we have

$$x = (sf - nk)/cn,$$

which exhibits the non-neutrality of money in the sense that the capital-labor ratio of the monetary model is lower than that of the non-monetary one. If $x = 0$, then one has $sf = nk$, which is identical to the solution of the Solow model. If x_0 is positive, then $f/k > n/s$. The *non-neutrality* follows. As shown by Zhang,[19] the following theorem is held for the Tobin model.

Theorem 10.5.1 (i) If both α and $\beta \to \infty$, then the long-run model is locally unstable. (ii) Even if either α or $\beta \to \infty$, if money is a perfect substitute for capital, the long-run model is locally unstable.

As shown from the comparative analysis by Hadjimichalakis,[20] if the equilibrium is stable, an increase in the rate of change of the nominal quantity of money increases the long run capital intensity and the expected rate of inflation. The following theorem is proved by Benhabib and Miyao.[21]

Theorem 10.5.2 The equilibrium point is locally asymptotically stable if the following conditions are satisfied

$$\left\{\frac{1}{\alpha x} - \frac{q_0 g_q}{(g-n)g}\right\}\beta + \frac{(1-s)n}{\alpha k} \le 1, \quad \frac{k g_k}{g} \ge 1.$$

The theorem says that the smaller the value of β or the greater the value of α, the more likely it is stability. Also, the smaller the elasticity of the money demand function with respect to q, or the greater the elasticity with respect to k, the more likely it is stability.

The above two theorems imply that the equilibrium may be either stable or unstable. For instance, if we move from adaptive expectations towards perfect foresight, saddle-point instability may appear as it does in the Tobin model. For the sake of illustration, consider an increase in the stock of money at the equilibrium. The immediate impact of this is to increase the price level while the real money stock tends to fall back to its original level; but the initial increase in the stock of money also tends to increase price expectations and reduce the capital stock. The latter two effects reinforce the decrease in the money supply and may cause the money stock to overshoot its long-run equilibrium. As the money supply

[19] Zhang (1990).
[20] Hadjimichalakis (1971a, 1971b).
[21] Benhabib and Miyao (1981).

keeps falling beyond its equilibrium level, the effects on the two variables are reversed: the capital stock rises and expectations fall. Combined with the direct effect of the money stock on the accumulation of money, the fall of the money stock will now be reversed. This discussion hints at the possibility of oscillations in the long run. In fact, the existence of Hopf bifurcations in the generalized Tobin model has been identified by Benhabib and Miyao.[22] Their results can be summarized as follows.

Theorem 10.5.3 If there exist a set of parameter values which guarantee the stability of the equilibrium, we can find a value of β, denoted by β_0, such that the Jacobian of the system has a pair of purely imaginary eigenvalues. Moreover, there exists a continuous function $v(\varepsilon)$ ($v(0) = 0$) of a parameter ε such that when ε is sufficiently small, the generalized Tobin model has a continuous family of periodic solutions $(k(t,\varepsilon), x(t,\varepsilon), q(t,\varepsilon))^T$, which collapse to the equilibrium point as $\varepsilon \to \infty$.

This theorem identifies the existence of regular oscillations in the system. Such oscillations will continue permanently if the stability of the cycles can be identified. Non-equilibrium economic development is no longer a short-run phenomenon. This theorem shows that the loss of stability that occurs as expectations adjust is associated with the emergence of bounded, persistent oscillations in prices, output and expectations. This holds no matter how quickly prices adjust since there always exists a value of β at which the stability of the equilibrium is lost. The generalized Tobin model can be applied to explain business cycles. Zhang actually improved the results obtained by Benhabib and Miyao in the following aspects: (i) to find stability conditions of the cycle; (ii) to explicitly interpret the parameter h; (iii) to find the explicit expression of the cycle; and (iv) to discuss whether the Hopf bifurcation is supercritical or subcritical. We will not introduce Zhang's results for technical reasons. Supercritical bifurcations mean that if the bifurcation parameter β is increased, the system is stabilized, while if it is

---

[22] Benhabib and Miyao (1981).

decreased, the system becomes unstable and bifurcations may take place. Zhang's results give a complete description of the Hopf bifurcations near the equilibrium. At the equilibrium point, the system is very sensitive to changes of the parameter β. Even when perturbations in the parameter are sufficiently small, structural changes take place, resulting in limit cycles.

10.6 Limit Cycles and Aperiodic Behavior in the Multi-Sector Optimal Growth Model[23]

We now introduce the traditional optimal growth model with two or more capital goods. There are many studies on behavior of the multi-sector growth models.[24] Zhang's works are mainly concerned with identifying existence of periodic and aperiodic solutions of the nonlinear dynamic model.

Let there be m production sectors in the economy and the population grow at a fixed positive growth rate, n. The optimal growth problem is defined, in terms of per capita variables, as

$$Max \int_0^\infty U[T(y,k)]e^{-(r-n)t}\, dt\,,$$

$$\dot{k}_i(t) = y_i(t) - n\dot{k}_i(t), \quad i = 1, 2, \ldots, \quad m\,, \qquad (10.6.1)$$

where vectors y and k stand for output per capita and stocks of capital goods respectively, the consumption per capita is given by

$$c(t) = T(y,k)$$

while $U(T)$ is the utility derived from consumption, and r is the rate of interest. We require: $r - n \geq 0$. We apply the maximum principle to solve the optimal problem.

Economic interpretations of the following six assumptions are referred to Benhabib and Nishimura[25].

---

[23] This part is based on Zhang (1988c, 1989).

[24] See Cass and Shell (1976), Brock and Scheinkman (1976), Araujo and Scheinkman (1977), and Magill (1977).

[25] Benhabib and Nishimura (1979).

(A1) All goods are produced non-jointly with homogeneous production functions of degree one, strictly quasi-concave for non-negative inputs, and twice differentiable for positive inputs;

(A2) If we denote by (K_{ij}) the set of inputs used in the production of good j, then the j th goods cannot be produced without (K_{ij}). The maximum principle yields

$$\dot{k}_i(t) = y_i(t) - n\dot{k}_i(t),$$

$$\dot{q}_i(t) = -U'(t)w_i(t) + r(t)q_i(t),$$

$$q_i(t) = U'(t)p_i(t), \quad p_i(t) = -T_{y_i}(y,k),$$

$$w_i(t) = T_{k_i}(y,k), \quad i = 1, 2, \ldots, m, \tag{10.6.2}$$

where $T_{y_i}(y,k)$ and $T_{k_i}(y,k)$ represent the partial derivatives of $T(y,k)$ with regard to $y_i(t)$ and $k_i(t)$, respectively. Here, p_i and w_i are the price and the rental of the i th goods in terms of the price of the consumption goods. It can be shown that the above system has a unique solution for $r \in (n, \bar{r})$, where $\bar{r}$ may be positively infinite. The steady state of the system, denoted by (k_0, q_0), is a solution of $\dot{k} = 0$ and $\dot{q} = 0$ satisfying Eqs. (10.6.2). It can be shown that the system has a stationary state.

To examine whether the stationary state is locally stable, we write the dynamics of the system in a local form near the equilibrium. To do this, we make the following assumptions.

(A3) At the steady state, the capital coefficient matrix is indecomposable.

(A4) At the steady state, direct labor and at least one capital input is required in the production of consumption good.

(A5) Near the steady state, marginal utility of consumption is constant, i.e., $U'' = 0$ and $U' = 1$.

(A6) The input coefficient matrix is non-singular near the steady state.

It can be proved that if (A1)-(A6) hold, then we have

(i) $T(y,k)$ is twice differentiable;

(ii) The dynamics near the steady state are given by

$$\dot{k}_i(t) = y_i(k, p) - nk_i(t),$$

$$\dot{p}(t) = -\, w_i(k, p) + r(t)p_i(t), \quad i = 1, \ldots, \; m, \qquad (10.6.3)$$

where $y_i(k, p)$ and $w_i(k, p)$ are differentiable.

By standard analytical methods, we can analyze behavior of the system.[26] We are now interested in nonlinear phenomena of the dynamic system. We now add another assumption.

(A7) Let there be a value of r denoted with r_0 such that the Jacobian at the equilibrium has one pair of conjugate eigenvalues

$$z_{1,2} = \alpha(r) \pm i\beta(r),$$

which satisfy

$$\alpha(r_0) = 0, \quad i\beta(r_0) \neq 0, \quad \left.\frac{d\alpha}{dr}\right|_{r = r_0} \neq 0.$$

We now consider r as a bifurcation parameter with critical value r_0. We express perturbation in the rate of interest from r with x, i.e., $x \equiv r - r_0$. Let us denote the Jacobian by $L(x)$ at r. At the critical point, the Jacobian is $L(0)$.

(A8) Let $\pm\, i\beta_0$ ($\beta_0 = \beta(r_0)$) be simple, isolated eigenvalues of $L(0)$ and all real parts of other eigenvalues except $z_{1,2}(r)$ of $L(0)$ are negative. We also require that the strict loss of stability condition can be guaranteed, i.e.

$$\mathrm{Re}\left\{ \sum_{i = m+1}^{2m} Y_i \overline{Y}_i^* \right\} > 0.$$

Here, Y and Y^* are solutions of

$$L(0)Y = i\beta_0 Y, \quad L^T(0)Y^* = -\, i\beta Y^*,$$

subject to $\langle Y, Y^* \rangle = 1$ and $\langle Y, \overline{Y}^* \rangle = 0$ where $\langle , \rangle$ is product operator in C^{2m}.

[26] See Araujo and Scheinkman (1977), and Magill (1977).

The following theorem is proved by Zhang.[27]

Theorem 10.6.1 (Existence of limit cycles) Let the optimal problem (10.6.1) satisfy Assumptions (A1) - (A8), $y(k,p)$ and $w(k,p)$ be C^{ρ}, $\rho \geq 3$. Then there exist limit cycles bifurcated from the equilibrium (k_0, p_0) with the bifurcation parameter r of critical value being r_0. The bifurcated cycles, with period $2\pi / \omega(h)$, are explicitly expressed as

$$\begin{bmatrix} k(t,h) \\ p(t,h) \end{bmatrix} = \begin{bmatrix} k_0 \\ p_0 \end{bmatrix} + 2h[con\{\omega(h)t\}\mathrm{Re}(Y) - \sin\{\omega(h)t\}\mathrm{Im}(Y)]$$

$$+ \frac{h^2 U^2\{\omega(h)t\}}{2} + O(|h|^3),$$

where h is an expansion amplitude parameter, $\omega(h)$, $x(h)$, and U^2 are explicitly determined as functions of h in Zhang.[28]

We now introduce another dynamic property of the model, basing on Zhang.[29]

We are now interested in the existence of aperiodic time-dependent solutions in the optimal growth model with three sectors. We show that endogenous oscillations appear when stability is lost by virtue of two pairs of complex conjugate eigenvalues of the linearized system simultaneously crossing the imaginary axis.

In the remainder of this section, we assume $m = 2$. If Assumptions (A1) - (A6) hold, then the system is locally governed by

$$\dot{k}_i(t) = y_i(k,p) - nk_i(t),$$

$$\dot{p}(t) = -w_i(k,p) + r(t)p_i(t), \quad i = 1, 2. \tag{10.6.4}$$

In addition to Assumptions (A1)-(A6), we make the following assumption.

---

[27] As mentioned in Zhang (1988c), although Benhabib and Nishimura (1979) identified existence of limit cycles in the multisector optimal growth model, Zhang provide explicit — only in an approximate sense — solutions to the problem.

[28] See Zhang (1988c).

[29] Zhang (1989).

(A9) Let the system (10.6.3) possess two pairs of simple complex conjugate eigenvalues denoted by $z_{1,2}(r)$ and $z_{3,4}(r)$ respectively. Here,

$$z_{1,2}(r) = \alpha_1(r) + i\beta_1(r),$$

$$z_{3,4}(r) = \alpha_2(r) + i\beta_2(r),$$

where $\alpha_j(r)$ and $\beta_j(r)$ are real numbers. Assume that there exists a value of r, denoted by r_0, such that

$$\alpha_1(r_0) = \alpha_2(r_0) = 0, \quad \alpha_1(r) = \alpha_2(r) > 0, \quad r \neq r_0, \quad \left.\frac{d\alpha_j}{dr}\right|_{r = r_0} \neq 0,$$

$$\beta_j(r_0) \neq 0, \quad j = 1, 2.$$

As demonstrated by Benhabib and Nishimura,[30] all the assumptions made so far are economically acceptable. Let us introduce an expansion amplitude parameter as follows

$$h^2 = \begin{cases} x, & if\ \alpha_1'(r_0) > 0 \\ -x, & if\ \alpha_1'(r_0) < 0. \end{cases}$$

According to Zhang,[31] the following theorem holds.

Theorem 10.6.2 (Existence of aperiodic solutions) Let the optimal problem (10.6.1) satisfy Assumptions (A1)-(A6) and (A9), $y(k,p)$ and $w(k,p)$ be C^ρ, $\rho \geq 3$. Then if

$$|\beta_1 - 2\alpha_2|, \ |\beta_1 - \alpha_2|, \text{ and } |2\beta_1 - \alpha_2|$$

are all $O(1)$ with regard to h, we have

$$\begin{bmatrix} k(t,h) \\ p(t,h) \end{bmatrix} = \begin{bmatrix} k_0 \\ p_0 \end{bmatrix} + h[C_1 R(\Theta)\sin\Omega(\Theta,t) + C_2 R(\Theta)\cos\Omega(\Theta,t)$$

$$+ D_1 S(\Theta)\sin\Gamma(\Theta,t) + D_2 S(\Theta)\cos\Gamma(\Theta,t) + O(h^2),$$

where

$$\Theta \equiv h^2 t,$$

---

[30] See Benhabib and Nishimura (1979).

[31] Zhang (1989).

and C_i and D_i $(i = 1, 2)$ are constant four-dimensional vectors, and $\Gamma(\Theta, t)$, $\Omega(\Theta, t)$, $R(\Theta)$ and $S(\Theta)$ are scalar functions given by Zhang in 1989. Moreover, stability of the aperiodic solution is determined by the asymptotic behavior of $R(\Theta)$ and $S(\Theta)$ as $\Theta \to \infty$. If they approach constant values or oscillate, then the bifurcated solution is stable.

Calculating the parameters and determining the functions in the theorem are tedious; hence we omit them. The theorem describes the oscillations bifurcated from the equilibrium. In sharp contrast to the case of the bifurcation at a pair of simple complex eigenvalues as described by Theorem 10.6.2, it is possible for the time dependent solution not to be periodic. The superposition of harmonics in $\Gamma(\Theta, t)$ and $\Omega(\Theta, t)$ is not periodic if β_1 and β_2 are incommensurate.

10.7 Theoretical Insight into China's Modern Economic Development

Inspired by China's history and modernization of overseas Chinese, in 1990 Zhang attempted to explain possible paths of interdependence between China's economic reform and political development.[32] We consider an economic system consisting of two, agricultural and industrial, sectors. It is assumed that the agricultural sector produces goods such as corn, rice and vegetables, which are only for consumption. The industrial sector produces commodities for investment and consumption. Industrial commodity is selected to serve as numeriare.

Behavior of production sectors and households

We denote $K(t)$, $r(t)$, and $p_a(t)$, the total capital, the rate of interest and price of agricultural commodity. We define the following indexes and variables

a, i = subscripts denoting agriculture and industry;

L and N = the fixed land and total labor force of the economy;

[32] This chapter is based on a model built by Zhang (1990, 1992) to reveal possible paths of Mainland China's contemporary economic dynamics. The Chinese history does prove that the mathematical model "economizes" the thinking processes for revealing the essence of the dynamic processes of Chinese societies.

$N_j(t)$ and $K_j(t)$ = the labor force and capital stocks employed by sector j ($j = a, i$);

$L_a(t)$ = the land used by the agricultural sector;

$S(t)$ and $L_h(t)$ = the total saving and land used for housing;

$F_j(t)$ and $C_j(t)$ = sector j 's output and consumption levels of product j; and

$w_j(t)$ and $R(t)$ = sector j 's wage rate and land rent.

We now describe the basic model. We assume that production processes of each sector can be described by some aggregate production functions. We assume that agricultural production is a process of combining land, labor force and capital. For simplicity, production function of the agricultural sector is specified as follows

$$F_a(t) = K_a^{\alpha_a} (H^{m_q/\beta_a} N_a)^{\beta_a} L_a^{\varsigma},$$

$$m_a \geq 0, \ \alpha + \beta_a + \varsigma = 1, \ \alpha_a, \beta_a, \varsigma > 0 .$$

Here, the term $H^{m_a/\beta_a} N_a$ is the qualified labor input. The parameter m_a / β_a describes how effectively the agricultural sector utilizes human capital. The marginal conditions for the agricultural sector are given by

$$r = \frac{\alpha_a p_a F_a}{K_a}, \ w = \frac{\beta_a p_a F_a}{N_a}, \ R = \frac{\varsigma p_a F_a}{L_a} . \qquad (10.7.1)$$

We now describe the industrial sector. Production function of the industrial sector is specified as follows

$$F_i(t) = N_i^{\alpha_i} (H^{m_i/\beta_i} N_i)^{\beta_i}, \ m_i \geq 0, \ \alpha_i + \beta_i = 1, \ \alpha_i, \beta_i > 0 .$$

Two, machines and the qualified labor force, inputs are taken into account by describing industrial production. It should be noted that possible land use by the industrial sector is omitted. The marginal conditions for the industrial sector are given by

$$r = \frac{\alpha_i F_i}{K_i}, \ w = \frac{\beta_i F_i}{N_i} . \qquad (10.7.2)$$

We have thus described behavior of the two production sectors.

Assume the public land ownership, which means that the revenue from land is equally shared among the population. As the urban land and rural land are respectively homogeneous in urban and rural land markets, each household gets identical land revenue from the land markets. The total land revenue is given by $R(t)L$. Let us denote $Y(t)$ the net income of the households. The net incomes consist of three parts: wage income, interest payment and revenue from land ownership, i.e.

$$Y(t) = rK + wN + RL .$$

It is assumed that utility level $U(t)$ of each household is dependent on consumption levels of industrial commodity and agricultural commodity, $C_i(t)$ and $C_a(t)$, housing conditions (measured in terms of lot size), $L_h(t)$, and saving, $S(t)$. The utility function is specified as follows

$$U(t) = C_a^{\mu} C_i^{\xi} L_h^{\eta} S^{\lambda}, \quad \mu, \xi, \eta, \lambda > 0, \quad \mu + \xi + \eta + \lambda = 1,$$

in which the parameters, μ, ξ and η, are respectively the propensities to consume agricultural good, industrial commodity, and housing, and the parameter λ is the propensity to hold wealth.

The budget constraint of households is given by

$$p_a(t)C_a(t) + C_i(t) + R(t)L_h(t) + S(t) = \hat{Y}(t),$$

where

$$\hat{Y}(t) = Y(t) + \delta K(t).$$

In the left-hand side of the budget constraint, $p_a(t)C_a(t)$ and $C_i(t)$ are spending on consumption of agricultural good and industrial commodity, respectively, $R(t)L_h(t)$ is payment for housing, and $S(t)$ is saving. Maximizing $U(t)$ subject to the above budget constraint yields

$$p_a C_a = \mu \hat{Y}, \quad C_i = \xi \hat{Y}, \quad RL_h = \eta \hat{Y}, \quad S = \lambda \hat{Y} . \qquad (10.7.3)$$

Substituting S from Eqs. (10.7.3) into

$$\dot{K} = S - K$$

yields

$$\dot{K}(t) = \lambda \hat{Y}(t) - K(t) .$$

The above equation determines capital accumulation of households.

We assume that capital, labor, and land are fully employed, i.e.

$$K_i + K_a = K, \quad N_i + N_a = N, \quad L_h + L_a = L.$$

Assume that industrial product is either consumed or invested. We have

$$C_i + S - K + \delta_k K = F_i.$$

The balance of demand of and supply for agricultural product is represented by

$$C_a = F_a.$$

In order to formulate the dynamics of human capital H, we first introduce another variable - a measurement of the degree of openness of a nation with respect to the rest of the world. It is supposed that international interactions may have significant impact upon economic development of the society under consideration. Developed economies are considered a stimulus and source of human capital for social progress and economic development. However, there are many deterministic as well as uncertain factors which affect the degree of openness of a national economy. In the case of China, its long history and vast size, virtual inaccessibility (both in terms of transportation and communication) to developed economies, low education, and other factors make China responsively inert with respect to events in the rest of the world. The current reforms may be considered as a policy designed to change the openness of the system. We model the dynamics of openness X as follows

$$\dot{X} = T_x[\varepsilon X - \theta X^3 + q(K, H)],$$

where T_x is a positive adjustment speed parameter. The term, $\varepsilon X - \theta X^3$, represents the political forces which affect the openness of the nation. The linear term εX expresses the strength of the 'reformers' who support scientific and technological development and extra-cultural learning effects; the nonlinear term θX^3 represents the power of the conservative fraction which tacitly or openly opposes increasing openness. The term εX implies that the more open the nation becomes, the greater the efforts of reformers toward even more openness; the term θX^3 states that when the country becomes more open, the opposition of conservatives toward even more openness increases.

In the above formula, the relative strength of the conservatives vis-à-vis the reformers increases very rapidly as the country becomes more open. The absolute value of the relative strength is dependent upon the values of ε and θ. The term $\varepsilon X - \theta X^3$ also shows that for any given ε and θ, the effects of the conservatives tend to be greater than those of the reformers with respect to the openness of the nation.

Indeed in order to explain the long-run dynamics of openness, one may necessarily treat political struggle parameters, ε and θ, as endogenous (slow-)variables. The function $q(K,H)$ specifies that human capital and living conditions affect economic openness. In general, we are not certain about the effects of these factors upon the openness. As living conditions are improved and the level of human capital is increased, the nation may become more open and eager to learn more from other cultures. But validity of this hypothesis obviously depends upon the culture to which it is applied. The adjustment speed, T_x, is much difficult to determine. During the whole period of the Cultural Revolution, this parameter was almost equal to zero because no force at that time was strong enough to open China.[33] In the present situation, the parameter is positive, but not infinitely large. It may be important for us to investigate what will happen to the system when T_x takes on different values.

Changes in the education system, openness policy, the freedom of communication among people, etc. may increase or decrease the level of human capital.[34] We propose that human capital is accumulated from three sources: learning by doing and learning from other cultures (a reflection of economic openness). We specify the following dynamics

$$\dot{H} = T_h \left(\frac{aX}{1 + bH} + \frac{\tau_a N_a F_a}{NH^{\varepsilon_a}} + \frac{\tau_i N_i F_i}{NH^{\varepsilon_i}} - \delta_h H \right),$$

where a, b, T_h, τ_a, τ_i, δ_h are parameters.

---

[33] See Zhang (1998).

[34] The two typical ways of knowledge are "learning by doing" (Arrow, 1962) and "learning through education" (Uzawa, 1965). There are many models of growth with endogenous knowledge (e.g., Becker and Burmeister, 1991, Breschger, 1999, Young, 1993, Grossman and Helpman, 1991, Zhang, 1996, Aghion and Howitt, 1998).

The term $\delta_h H$ in the above formula describes depreciation of human capital, where δ_h is the depreciation rate of human capital. The first term, $aX/(1 + bH)$, implies that as the economic system becomes more open, the level of human capital tends to increase. However, effects of international interactions upon human capital accumulation tend to decline if the society's human capital is already very high: if the level of human capital is already very high, a country tends to have less to learn from others. We interpret $\tau_i N_i F_i / NH^{\varepsilon_i}$ as effects of "learning by doing" of each worker in the industrial sector upon accumulation of human capital.

We have thus established the economic dynamics with endogenous economic structure, physical capital and human capital, and openness. The system has 18 endogenous variables, $K(t)$, $H(t)$, $X(t)$, $K_a(t)$, $K_i(t)$, $N_a(t)$, $N_i(t)$, $L_h(t)$, $L_a(t)$, $F_a(t)$, $C_a(t)$, $C_i(t)$, $S(t)$, $U(t)$, $r(t)$, $w(t)$ and $p_a(t)$. We now examine dynamic properties of the system.

Since this system is highly dimensional with a complicated internal economic structure, it is difficult to explicitly determine all possible behavior of the dynamic system. In order to analyze dynamic behavior of the system, it is necessary to show that the 3-dimensional systems are governed by the dynamics of three variables. The following lemma guarantees this.

Lemma 10.7.1 For any given $K(t) > 0$, $H(t) > 0$ and $X(t) > 0$ at any point of time, the other variables in the system are uniquely determined as functions of $K(t)$ and $H(t)$ by the following procedure: $\Omega(t)$ by Eq. (10.A.1.7) $\rightarrow$ $K_a(t)$ and $K_i(t)$ by Eqs. (10.A.1.5) $\rightarrow$ $N_a(t)$ and $N_i(t)$ by Eqs. (10.A.1.6) $\rightarrow$ $L_h(t)$ and $L_a(t)$ by Eqs. (10.A.1.2) $\rightarrow$ $R(t)$ by Eq. (10.A.1.2) $\rightarrow$ $F_a(t) = K_a^{\alpha_a}(H^{m_q/\beta_a} N_a)^{\beta_a} L_a^{\varsigma} \rightarrow$ $p_a(t)$ by Eqs. (10.7.1) $\rightarrow$ $F_i(t) = N_i^{\alpha_i}(H^{m_i/\beta_i} N_i)^{\beta_i}$ $\rightarrow$ $r(t)$ and $w(t)$ by Eqs. (10.7.2) $\rightarrow$ $C_a(t)$, $C_i(t)$, and $S(t)$ by Eqs. (10.7.3) $\rightarrow$ $U(t) = C_a^{\mu} C_i^{\xi} L_h^{\eta} S^{\lambda}$. Moreover, the motion of $K(t)$, $H(t)$, and $X(t)$ are determined by the following three-dimensional differential equations

$$\dot{K} = \lambda \hat{Y}(K, H) - K,$$

$$\dot{X} = T_x[\varepsilon X - \theta X^3 + q(K,H)],$$

$$\dot{H} = T_h\left[\frac{aX}{1+bH} + \frac{\tau_a N_a F_a}{NH^{\varepsilon_a}} + \frac{\tau_i N_i F_i}{NH^{\varepsilon_i}} - \delta_h H\right].$$

We prove the above lemma in the Appendix. The lemma guarantees that if we know the dynamics of $K(t)$, $H(t)$, and $X(t)$, then all the other economic variables are determined as unique functions of these three variables at any point of time. Hence, it is sufficient to be concerned with the above differential equations. This result is important for carrying out stability analysis.

For simplicity, we first set $b = 0$. Equilibrium of the dynamics is defined by follows

$$\lambda\hat{Y}(K,H) = K, \ \varepsilon X - \theta X^3 + q(K,H) = 0,$$

$$aX + \frac{\tau_a N_a F_a}{NH^{\varepsilon_a}} + \frac{\tau_i N_i F_i}{NH^{\varepsilon_i}} = \delta_h H. \tag{10.7.4}$$

We now show how we can determine equilibrium. First, substituting $\lambda\hat{Y} = K$ into Eq. (10.A.1.7), we solve

$$K = \alpha_h H^{m_i/\beta_i}, \tag{10.7.5}$$

where

$$\alpha_h \equiv \left(\frac{\alpha_i}{\alpha_1/\lambda - \delta_1}\right)^{\alpha_i/\beta_i} \frac{\beta_i N}{\alpha_2/\lambda - \delta_2} > 0.$$

By $\lambda\hat{Y} = K$, and Eqs. (10.A.1.5) and (10.A.1.6), we have

$$K_a = \alpha_a^* K, \ K_i = \alpha_i^* K,$$

$$N_a = \frac{\beta_a \mu}{\alpha_2 - \delta_2\lambda} N, \ N_i = \frac{\beta_i \xi_k}{\alpha_2 - \delta_2\lambda} N, \tag{10.7.6}$$

in which

$$\alpha_a^* \equiv \frac{\alpha_a \mu}{\alpha_1 - \delta_1\lambda}, \ \alpha_i^* \equiv \frac{\alpha_i \xi_k}{\alpha_1} - \delta_1\lambda.$$

We see that N_a and N_i are already determined. By Eqs. (10.7.6) and (10.A.1.2), we see that F_a and F_i are given as functions of K and H as follows

$$F_a = (\alpha_a^* \alpha_h)^{\alpha_a} N_a^{\beta_a} L_a^{\varsigma} H^{m_q + \alpha_a m_i / \beta_i},$$

$$F_i = (\alpha_i^* \alpha_h)^{\alpha_i} N_i^{\beta_i} H^{m_i + \alpha_i m_i / \beta_i}. \qquad (10.7.7)$$

Substituting Eqs. (10.7.7) into the last two equations in Eqs. (10.7.4), we have

$$\Phi_x(X,H) \equiv \varepsilon X - \theta X^3 + q(\alpha_h H^{m_i/\beta_i}, H) = 0,$$

$$\Phi_h(X,H) \equiv \frac{aX}{H} + \Phi_a(H) + \Phi_i(H) - \delta_h = 0, \qquad (10.7.8)$$

in which

$$\Phi_a \equiv \frac{(\alpha_a^* \alpha_h)^{\alpha_a} N_a^{1+\beta_a} L_a^{\varsigma}}{N} H^{x_a}, \quad \Phi_i \equiv \frac{(\alpha_i^* \alpha_h)^{\alpha_i} N_i^{\beta_i}}{N} H^{x_i},$$

$$x_a \equiv m_q + \frac{\alpha_a m_i}{\beta_i} - \varepsilon_a - 1, \quad x_i \equiv m_i + \frac{\alpha_i m_i}{\beta_i} - \varepsilon_i - 1.$$

Since K is uniquely determined by H, we see that the number of equilibria is equal to the number of solutions of the two equations (10.7.8). It is straightforward to show that for any fixed $H > 0$, $\Phi_x = 0$ may have one or three solutions. When $a = 0$, the equation $\Phi_h(H) = 0$ has two positive solutions in the case of $x_a > 0$ and $x_i < 0$ (or $x_a < 0$ and $x_i > 0$) and has a unique positive solution in the case of $x_a > 0$ and $x_i > 0$ (or $x_a < 0$ and $x_i < 0$). This implies that the dynamic system may have multiple equilibria.

It is straightforward to prove that when we neglect human capital accumulation ($H(t) = 1$) and omit the impact of openness on the system ($X(t) = 0$), then the economic system consisting of a one-dimensional differential equation for capital accumulation has a unique stable equilibrium. As stationary states are independent of T_x and T_h, it can be seen that the system is locally stable under certain constraints, when T_x and T_h are sufficiently small. If the human capital and opening policies adapt to new situations very slowly, the economic system tends to be stable. Since T_x and T_h, in particular T_x, may be rather large, the system is faced with possible instabilities.

The isolated economy with human capital accumulation

We omit any possible impact of openness on the dynamics, i.e., $X(t) = 0$. By Lemma 10.7.1, the dynamics are given by

$$\dot{K} = \lambda\Omega(K,H) - K,$$

$$\dot{H} = T_h\left[\frac{\tau_a N_a F_a}{NH^{\varepsilon_a}} + \frac{\tau_i N_i F_i}{NH^{\varepsilon_i}} - \delta_h H\right].$$

Equilibrium of the dynamics is given by

$$\lambda\Omega(K,H) = K,$$

$$\frac{\tau_a N_a F_a}{NH^{\varepsilon_a}} + \frac{\tau_i N_i F_i}{NH^{\varepsilon_i}} = \delta_h H.$$

The variables, K, K_a, K_i, N_a, and N_i are still given by Eqs. (10.7.5) and (10.7.6) as functions of H. Hence, if we find H, we determine all the variables in the system. By Eq. (10.7.8), we see that H is determined by

$$\Phi_h(H) \equiv \Phi_a(H) + \Phi_i(H) - \delta_h = 0,$$

in which Φ_a and Φ_i are defined in the previous section.

We omit the case of $x_a = x_i = 0$. Equilibrium of the system is given by a positive H such that $\Phi(H) = 0$. When $x_a > 0$ and $x_i > 0$, $\Phi(H) = 0$ has a unique positive solution as $\Phi' > 0$ for any positive H, $\Phi(H) < 0$ and $\Phi(\infty) > 0$. Similarly, if $x_a < 0$ and $x_i < 0$, $\Phi(H) = 0$ has a unique positive solution. It is easy to check that if either $x_a = 0$ or $x_i = 0$, then the system has a unique positive solution under certain conditions. We now prove that if $x_a > 0$ and $x_i < 0$ (or $x_a < 0$ or $x_i > 0$), then the system has either two solutions or no solution. It is sufficient for us to examine one case, for instance, of $x_a > 0$ and $X(t)$. As $\Phi(H) > 0$, $\Phi(\infty) > 0$, we see that $\Phi(H) = 0$ cannot have a unique solution. That is, $\Phi(H) = 0$ has either multiple solutions or no solution. On the other hand, as $\Phi'(H) = 0$ has a unique positive solution, we conclude that $\Phi(H) = 0$ has two solutions if $\Phi(H)$ has solutions. The necessary and sufficient condition for the existence of two solutions is that there exists a positive value H_1, of H such that $\Phi(H_1) < 0$ and $\Phi'(H_1) = 0$.

By calculating the Jacobian and eigenvalues at each equilibrium, it is straightforward to prove that if the term

$$\Phi^* \equiv -\frac{x_a \Phi_a + x_i \Phi_i}{H},$$

is negative, then the equilibrium is unstable; if it is positive, then the equilibrium is stable. Moreover, when the system has two equilibria, Φ^* is positive (negative) at the equilibrium with the low (high) value of H. Summarizing the above analytical results, we prove the following proposition.

Proposition 10.7.1 If $x_a < 0$ and $x_i < 0$, the system has a unique stable equilibrium. If $x_a > 0$ and $x_i > 0$, the system has a unique unstable positive equilibrium. If $x_a < 0$ and $x_i < 0$ ($x_a < 0$ and $x_i > 0$), the system has two equilibria. The one with higher values of K and H is unstable; the other one is stable.

By the definitions of x_a and x_i, we see that we may interpret x_a and x_i respectively as measurements of return to scales of the agricultural and industrial sectors in the dynamic system. When x_a (>) 0, we say that the agricultural sector displays decreasing (increasing) returns to scale in the dynamic economy. Hence, the above proposition tells us that if the two sectors display decreasing (increasing) returns, then the dynamic system has a unique stable (unstable) equilibrium; if one sector displays decreasing (increasing) returns and the other sector exhibits increasing (decreasing), the system has two equilibria. When the system has two possible equilibria, it may be located in the stabilized situation with low creativity and low living standard.

Slow opening and catastrophes

We now consider the case that the adjustment speed of the openness X is much slower than the adjustment speeds of human capital and capital. That is, $T_h = 1$ and T_x is sufficiently small. For simplicity of analysis, we specify the functional form of $q(K, H)$ as

$$q(K,H) = b_0\left\{c^* - \frac{Y(K,H)}{N}\right\}, \qquad (10.7.9)$$

where c^* is the average net income per capita in the developed economies which have great influences upon China, b_0 is a given non-negative parameter. The nations which affect China may include the USA, Japan and some developed economies in Europe. In Eq. (10.7.9), Y/N is the net income per capita.

Introducing $T^* = tT_x$, we can rewrite the dynamics in the following form

$$T_x \frac{dK}{dT^*} = \lambda \hat{Y}(K,H) - K,$$

$$\frac{dX}{dT^*} = \varepsilon X - \theta X^3 + q(K,H),$$

$$T_x \frac{dH}{dT^*} = aX + \frac{\tau_i N_i F_i}{N} - \delta_h H,$$

in which we require

$$b = \tau_a = \varepsilon_i = 0, \quad m_i = \beta_i .$$

We assume that T_x is so small that we can safely let

$$\frac{T_x dK}{dT^*} = 0, \quad \frac{T_x dH}{dT^*} = 0$$

in the dynamic analysis. It should be noted that this assumption is valid only under some conditions. As the system is subjected to instabilities, its behavior may be rather sensitive to the small parameter of the system even from a qualitative point of view.

The long-run dynamics are thus approximately given by

$$\frac{dX}{dT^*} = \varepsilon X - \theta X^3 + q(K,H), \qquad (10.7.10)$$

subject to

$$\lambda \Omega(K,H) = K,$$

$$aX + \frac{\tau_i N_i F_i}{N} = \delta_h H. \qquad (10.7.11)$$

Although the dynamic behavior appears to be dependent only on Eq. (10.7.10), the equations expressing capital accumulation and change in human capital affect the actual paths of X since they determine the term $q(K,H)$ in Eq. (10.7.10).

From the first equation in Eq. (10.7.11) and the definition of Ω, we have $Y = (1/\lambda - 1 + \delta_k)K$. By the second equation in Eqs. (10.7.11), (10.7.5) and (10.7.7), we solve

$$H = a_1 X, \ \ Y = a_2 X,$$

where

$$a_1 \equiv \left(\delta_h - \frac{\tau_i (\alpha_i^* \alpha_h)^{\alpha_i} N_i^{1+\beta_i}}{N} \right)^{-1} a, \ \ a_2 \equiv \left(\frac{1}{\lambda} - 1 + \delta_k \right) \alpha_h a_1 .$$

By $H = a_1 X$, $Y = a_2 X$ and Eq. (10.7.9), we rewrite Eq. (10.7.10) as follows:

$$\frac{dX}{dT^*} = - \theta X^3 + (\varepsilon - b_2) X + b_1 ,$$

where $b_1 \equiv b_0 c^*$ and $b_2 \equiv b_0 a_2 / N$. Here, we interpret the terms $(\varepsilon - b_2)$ and b_1 as measurements of progressive forces for China. The parameter b_1 measures how the foreign economic and technological conditions push the government to open the nation; and the parameter b_2 measures how improved living conditions may slow down the rate of change of openness. The term $(\varepsilon - b_2)$ is an aggregated measurement of the progressive forces of the reformers and people's attitudes toward other cultures. According to these interpretations of the parameters, we see that openness is determined by dynamic interactions of the conservative and progressive forces upon the system.

The stationary values are given by

$$X^3 + r_1 X + r_2 = 0,$$

in which

$$r_1 \equiv \frac{\varepsilon - b_2}{\theta}, \ \ r_2 \equiv -\frac{b_1}{\theta} .$$

This equation has either one or three real roots. If

$$\left(-\frac{r_1}{3}\right)^3 > \left(\frac{r_2}{2}\right)^2,$$

then the equation has three roots.

As the left term is always positive, a necessary condition for the inequality is that

$$r_1 < 0, \text{ i.e., } \varepsilon > b_2.$$

This is the case only when the reformers have strong political influence with respect to anti-foreign attitudes. Otherwise, there is only one equilibrium in the system. This discussion shows that the existence of a unique X is dependent on the power of the reformers.

The boundaries of the region for single and multiple solutions are determined as follows

$$4r_1^3 + 27r_2^2 = 0.$$

This produces the cusp-shaped curves on the control manifold - the (r_1, r_2) plane. As shown in Fig. 10.7.1, outside the cusp-shaped region there is only one root and this is a minimum of the corresponding potential of the system

$$F^* = \frac{X^4}{4} + \frac{r_1 X^2}{2} + r_2 X.$$

The unique equilibrium is stable. Inside the region, there are three real roots — one maximum (an unstable state) and two minima (stable states). The shaded region is the catastrophe set and the boundary is the bifurcation set. The r_1-axis, for $r_1 < 0$ (i.e., the reformers are strong), represents the conflict set: here there are two stable states of equal value. In the case of cusp catastrophe, r_1 is termed as the 'splitting factor' and r_2 the normal factor.[35] The reason is that it is the value of r_1 which determines whether a trajectory is in a region where the surface is folded. If $r_1 > 0$, the surface is single valued; if $r_1 < 0$, the surface is double valued. In the case of the normal factor r_2, the variable X changes monotonically as r_2 changes, and continuously except for jumps at the bifurcation points.

[35] Refer Zeeman (1977) for catastrophe theory.

Proposition 10.7.2 If capital and human capital accumulation are fast variables and openness is a slow variable, the long-run dynamics may be described by a single dynamic equation of the slow variable. In other words, capital and human capital are 'enslaved' by the change in openness.[36] And sudden structural changes in the long-run dynamic evolution may exist, depending upon the whole structure of dynamic interactions of economic development, human capital growth, and political struggles.

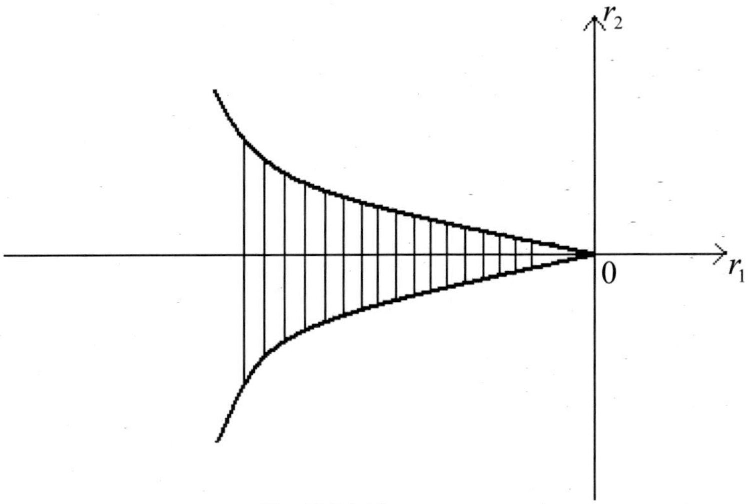

Fig. 10.7.1 The cusp catastrophe

We can find three types of behavior which are not familiar in the traditional comparative analysis. They are: (i) a sudden jump (or catastrophe); (ii) hysteresis - a reverse path to some point not being the same as the original; and (iii) divergence - a small difference approach towards a cusp point leads the system to the upper or lower surface and hence to a very different state.

We are particularly interested in the parameters ε and θ. As $r_1 \equiv (\varepsilon - b_2)/\theta$, we see that for a (positive) fixed b_2, when the reformers are not politically strong, there is a unique stationary state. As

[36] The "Slaving principle" is introduced as a key concept in synergetics developed by Haken (1977, 1983).

the power of the reformers increases to such a degree that r_1 becomes negative, the situation becomes more complicated. There are multiple equilibria in the system. Figure 11.5.2 illustrates the relationship between the opening policy and the power of the conservatives.

Depending on the power of the conservatives, there may be sudden changes the openness. Outside the interval $[\theta_1, \theta_2]$, there is a unique X for each value of θ. However, if $\theta_1 < \theta < \theta_2$, there are two stable and one unstable equilibria. Consider a possible case of the dynamics. When the economy is just opened, the power of the conservatives begins to increase, i.e., θ increases toward θ_1 from the right. The economy becomes smoothly more open as θ continuously changes. When θ arrives at the critical point θ_1, there are sudden increases in communication and trades between the economy under consideration and the rest of the world. Near such a point, there are structural changes. The consumption per capita, capital per capita, and average human capital in the society increase during a very short period.

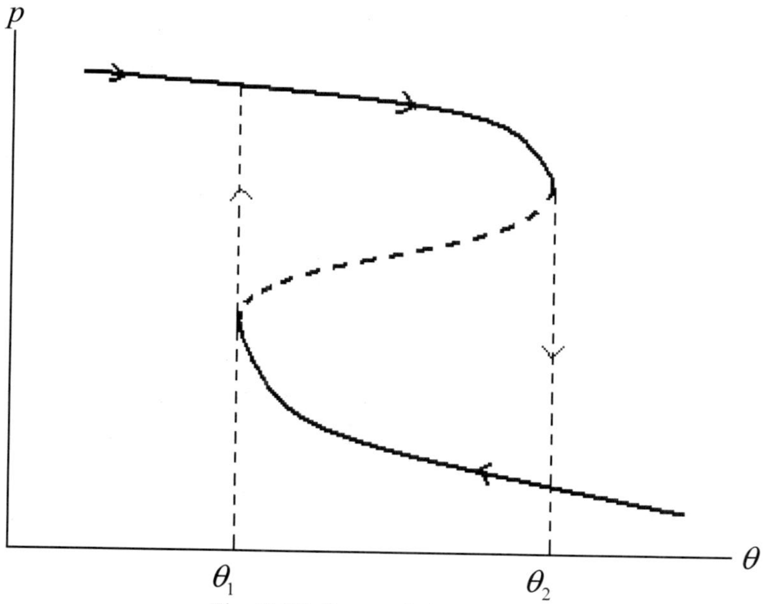

Fig. 10.7.2 Structural changes.

However, as such increased external communication and improved living conditions may cause officials to be corrupt and may introduce some 'undesirable elements' from other countries, the conservatives' power may either increase or decrease, which is, however, uncertain. If the conservatives continuously become weaker, the country will become more open and there are no further sudden changes in the system. However, if the conservatives become stronger after the sudden change, the country becomes more isolated. When the conservatives wield so much power that the parameter value reaches θ_2, there is a sudden change again. The nation suddenly becomes more isolated and it is impossible for scientists and entrepreneurs to interact with other countries. Near such points, there are sudden decreases in consumption, capital per capita, and the average human capital.

Fast capital accumulation and social cycles

The learning processes have been very slow, as observed, for instance, during the period from the Opium War to the fall of Qing Dynasty in 1911 or the period from the establishment of New China to the start of the economic reforms. It may be mentioned that during the later period the society 'forgot' more than it learned. It is meaningful to examine what will happen to the system if both the openness and learning processes are very slow, i.e., T_x and T_h are very small. For simplicity, let $T_x = T_h$.

We introduce $T^* = tT_x$. Like the preceding section, we still require: $b = \tau_a = \varepsilon_i = 0$ and $m_i = \beta_i$. We still assume Eq. (10.7.9). Under these requirements we can rewrite the system as follows:

$$T_x \frac{dK}{dT^*} = \lambda\Omega(K,H) - K,$$

$$\frac{dX}{dT^*} = \varepsilon X - \theta X^3 + b_0\left\{c^* - \frac{Y(K,H)}{N}\right\},$$

$$\frac{dH}{dT^*} = aX + \frac{\tau_i N_i F_i}{N} - \delta_h H. \qquad (10.7.12)$$

From $dK/dT^* = 0$, we have $\lambda\Omega(K,H) = K$. From this equation and the definition of Ω, we have

$$Y = \left(\frac{1}{\lambda} / - 1 + \delta_k \right) K .$$

By this equation and Eq. (10.7.5), we solve $Y = a_h H$, where

$$a_h \equiv \frac{1/\lambda - 1 + \delta_k}{\alpha_h} .$$

By $Y = a_h H$ and Eq. (10.7.7), we rewrite Eqs. (10.7.12) as

$$\frac{dX}{dT^*} = \varepsilon X - \theta X^3 - b_2 H + b_1 ,$$

$$\frac{dH}{dT^*} = aX + a_0 H ,$$

where

$$b_1 \equiv b_0 c^* , \quad b_2 \equiv \frac{a_h b_0}{N} , \quad a_0 \equiv \frac{\tau_i (\alpha_i^* \alpha_h)^{\alpha_i} N_i^{1+\beta_i}}{N} - \delta_h .$$

It is necessary to require that a_0 be negative. The stationary values of X are determined by

$$X^3 + r_1 X + r_2 = 0 ,$$

in which

$$r_1 \equiv \frac{\varepsilon - b_2}{\theta} , \quad r_2 \equiv - \frac{b_1}{\theta} .$$

This equation has either one or three real roots. In this section we still choose the power of the conservatives as a bifurcation parameter. From the previous section, we know that outside the interval $[\theta_1, \theta_2]$, there is a unique level of openness for each value of θ. We now consider the case in which $\theta > \theta_2$. That is, during the study period under consideration the conservatives are politically very strong. As $r_1 > 0$, this requirement is equal to saying that the reformers are rather weak.

The equilibrium is denoted by (X_0, H_0). The two eigenvalues are given by

$$\phi_{1,2} = \frac{a_0 + \varepsilon - 3\theta X^2}{2}$$

$$\pm\left\{\left(\frac{a_0 + \varepsilon - 3\theta X^2}{2}\right)^2 - a_0(\varepsilon - 3\theta X^2) - ab_2\right\}^{1/2}.$$

As θ is rather large, r_1 is small and $r_2 < \phi(0)$ is large. It is not difficult to see that the term $(a_0 + \varepsilon - 3\theta X^2)$ is negative when ε is rather small and θ is large, i.e., the conservatives are strong and the reformers are weak. As ε increases and θ decreases, the term tends to increase. It is also not difficult to see that for sufficient large ε and appropriately small θ, the term becomes positive. This means that we can appropriately choose the combinations of ε and θ such that

$$a_0 + \varepsilon - 3\theta X^2 = 0,$$

which defines a critical point of the system. We let ε_0 be the value of ε for which the term is equal to zero. In what follows, we denote small perturbations of ε from ε_0 by x, i.e.

$$x = \varepsilon - \varepsilon_0.$$

When x is positive, the power of the reformers increases, and when x is negative, the reformers' power becomes weaker. At $x = 0$, we have

$$\phi_{1,2}(0) = \pm iy,$$

where

$$y \equiv \{a_0(\varepsilon - 3\theta X^2) + ab_2\}^{1/2} > 0.$$

That is, at $x = 0$ we have a pair of purely imaginary eigenvalues. As the eigenvalues are continuous functions of x, for the neighborhood of $x = 0$, $\phi(x)$ denotes the eigenvalue which equals iy at $x = 0$. As

$$\phi_x(0) = \frac{1}{2} - \frac{ia_0}{2y},$$

the real part of the derivative of the eigenvalue ϕ is positive when x becomes positive. Hence, when x crosses its critical value, the system becomes unstable. That is, an increase in the power of the reformers may result in instabilities.

According to the Hopf bifurcation theorem, we know that when x becomes positive, limit cycles appear around the stationary state. The results can be summarized in the following theorem.

Proposition 10.7.3 Near the critical state, when the power of the reformers increases, social cycles appear in the system. The cycles are approximately (the first order) given by

$$X(t,h) = X_0 + 2ha^* \cos(yt) + O(h^2),$$

$$H(t,h) = H_0 + 2ha\{(\varepsilon - 3\theta X^2)\cos(yt) + y\sin(yt)\} + O(h^2),$$

where h is a small expansion parameter and $a^* \equiv Y / X$.

The proposition can be proved by applying the bifurcation method by Iooss and Joseph.[37] What should be noted, however, is that the eigenvectors W and adjoint eigenvectors W^* are given by

$$W = \left(aa^*,\ a\left(\varepsilon - 3\theta X^2 - iy\right)\right)^T,$$

$$W^* = \left(x_1 + ix_2,\ \left(\varepsilon - 3\theta X^2\right)x_1 - x_2 y + i\left(x_1 y + x_2\varepsilon - 3x_2\theta X^2\right)\right)^T,$$

where

$$x_1 \equiv \frac{m}{m^2 + 4y^2(\varepsilon - 3\theta X^2)^2},\quad x_2 \equiv \frac{2y(\varepsilon - 3\theta X^2)}{m^2 + 4y^2(\varepsilon - 3\theta X^2)^2},$$

$$m \equiv aa^* - (\varepsilon - 3\theta X^2)^2 + y^2.$$

Moreover, the stability conditions and more accurate expressions of the limit cycles can be given by using the eigenvectors and the adjoint eigenvectors. As these expressions are too complicated to deepen our insight into the problem, we do not explicitly calculate them here. The cycles can be illustrated as in Fig. 10.7.3.

The system oscillates around the stationary state: (X_0, H_0). Let us begin the movement at point A. Human capital tends to increase near A. Since the level of human capital is increased, the nation tends to become more open, which results in further expansion of human capital. When the system arrives at B, the conservatives become so strong that it is impossible to increase the openness of the economy; thus the nation becomes more isolated. The situation is continued. When the conservatives have increased their power, the level of human capital does

[37] Iooss and Joseph (1980). This method has been applied to some economic systems by Zhang (1991).

not seem to decrease rapidly. Instead, human capital is increased until the system arrives at C. This also implies that just after the nation becomes isolated, production and consumption will not decrease. The economic conditions are further improved because of the improved human capital. After C, the level of human capital decreases and the nation continues to be further isolated. During the period $C - D$, the nation may assume a very pessimistic outlook. However, after the social conditions tend to worsen, the effects of the conservatives begin to weaken. The nation becomes open again. However, even when the nation is open, human capital cannot be increased very rapidly. It will still take a long time for the effects of the opening policy to be recognized. It is after A that the masses may become a little more optimistic because everything seems to be improved from now on.

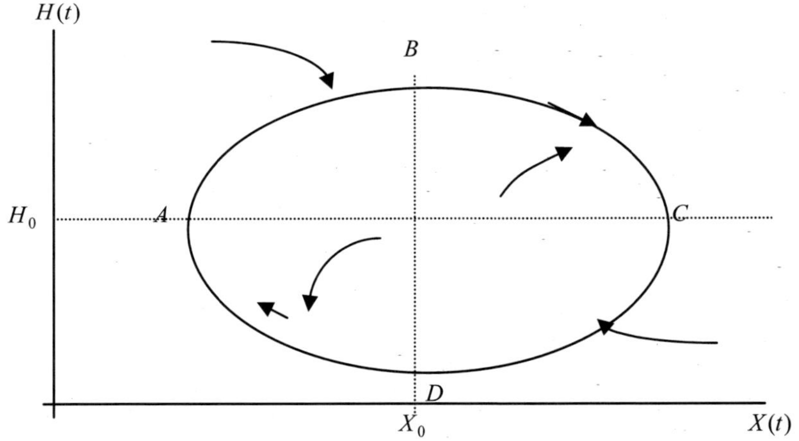

Fig. 10.7.3 The social cycles.

Appendix: Proving Lemma 10.7.1

First, by Eqs. (10.7.1) and (10.7.2) we have

$$\frac{r}{w} = \frac{\alpha_a N_a}{\beta_a K_a} = \frac{\alpha_i N_i}{\beta_i K_i}. \tag{10.A.1.1}$$

Substituting $p_a C_a = \mu \hat{Y}$ into $C_a = F_a$ yields

$$p_a F_a = \mu \hat{Y}.$$

Substituting the above equation into $R = \varsigma p_a F_a / L_a$ we get

$$R = \frac{\varsigma \mu \hat{Y}}{L_a}.$$

From $RL_h = \eta \hat{Y}$ from Eqs. (10.7.3), $R = \varsigma \mu \hat{Y} / L_a$ and $L_h + L_a = L$, we solve

$$R = \frac{(\varsigma \mu + \eta)\hat{Y}}{L}, \quad L_h = \frac{\eta L}{\varsigma \mu + \eta} \quad L_a = \frac{\varsigma \mu L}{\varsigma \mu + \eta}. \quad (10.A.1.2)$$

Substituting C_i and S in Eqs. (10.7.3) into

$$C_i + S - K + \delta_k K = F_i,$$

we get

$$F_i = (\xi + \lambda)\hat{Y} - (1 - \delta_k)K. \quad (10.A.1.3)$$

From

$$r = \frac{\alpha_a p_a F_a}{K_a} = \frac{\alpha_i F_i}{K_i},$$

we get:

$$K_i = \frac{\alpha_i K_a F_i}{\alpha_a p_a F_a}.$$

Substituting $p_a F_a = \mu \hat{Y}$ and Eqs. (10.A.1.3) into this equation yields

$$\frac{K_i}{K_a} = \frac{(\xi + \lambda)\hat{Y} - (1 - \delta_k)K}{\alpha_a \mu \hat{Y}} \alpha_i. \quad (10.A.1.4)$$

By Eq. (10.A.1.4) and $K_i + K_a = K$, we solve

$$K_a = \frac{\alpha_a \mu \hat{Y}}{\alpha_1 \hat{Y} - \delta_1 K} K, \quad K_i = \frac{(\xi + \lambda)\hat{Y} - \delta K}{\alpha_1 \hat{Y} - \delta_1 K} \alpha_i K, \quad (10.A.1.5)$$

where

$$\alpha_1 \equiv \alpha_a \mu + \alpha_i \xi + \alpha_i \lambda, \quad \delta_1 \equiv \delta \alpha_i.$$

In order that $0 < K_i$, $K_a < K$ are held, it is sufficient to require: $\hat{Y} / K > \delta_1 / \alpha_1$ for any $K > 0$ and $\hat{Y} > 0$.

By Eqs. (10.A.1.1) and (10.A.1.4), we get

$$\frac{N_i}{N_a} = \frac{(\xi + \lambda)\hat{Y} - \delta K}{\beta_a \mu \hat{Y}} \beta_i .$$

By this equation and $N_i + N_a = N$, we solve

$$N_a = \frac{\beta_a \mu \hat{Y}}{\alpha_2 \hat{Y} - \delta_2 K} N, \quad N_i = \frac{(\xi + \lambda)\hat{Y} - \delta K}{\alpha_2 \hat{Y} - \delta_2 K} \beta_i N , \quad (10.A.1.6)$$

where

$$\alpha_2 \equiv \beta_a \mu + \beta_i \xi + \beta_i \lambda , \quad \delta_2 \equiv \delta \beta_i .$$

In order that $0 < N_i, \; N_a < N$ are held, it is sufficient to require: $\hat{Y} / K > \delta_2 / \alpha_2$ for any $K > 0$ and $\hat{Y} > 0$. We see that the capital and labor distribution can be expressed as functions of K and $\hat{Y}$ at any point of time if

$$\frac{\hat{Y}}{K} > \delta \equiv \min\left\{ \frac{\delta_1}{\alpha_1}, \frac{\delta_2}{\alpha_2} \right\}.$$

Substituting K_i in Eqs. (10.A.1.5) and N_i in Eqs. (10.A.1.6) into

$$F_i(t) = N_i^{\alpha_i} (H^{m_i / \beta_i} N_i)^{\beta_i}$$

yields

$$F_i(t) = H^{m_i} \left\{ \frac{(\xi + \lambda)\hat{Y} - \delta K}{\alpha_1 \hat{Y} - \delta_1 K} \alpha_i K \right\}^{\alpha_i} \left\{ \frac{(\xi + \lambda)\hat{Y} - \delta K}{\alpha_2 \hat{Y} - \delta_2 K} \beta_i N \right\}^{\beta_i} .$$

By this equation and Eq. (10.A.1.3), we have

$$\Phi_i(\hat{Y}; K, H) \equiv H^{m_i} \left(\frac{\alpha_i K}{\alpha_1 \hat{Y} - \delta_1 K} \right)^{\alpha_i} \left(\frac{\beta_i K}{\alpha_2 \hat{Y} - \delta_2 K} \right)^{\beta_i} - 1 = 0 .$$

(10.A.1.7)

We now show that for any given positive K and H, the equation $\Phi_i(\hat{Y}; K, H) = 0$ has a unique solution $\hat{Y}(K, H)$ satisfying $\infty > \hat{Y} > \delta^* K$. It is straightforward to check that the function $\Phi_i(\hat{Y})$ has the following properties

$$\Phi_i(\delta^* K) > 0, \quad \Phi_i(+\infty) < 0, \quad \Phi_i' < 0 \text{ for } \infty > \hat{Y} > \delta^* K .$$

Hence, for any positive K and H, the equation $\Phi_i(\hat{Y}; K, H) = 0$ has a unique solution. We thus proved Lemma 10.7.1.

Chapter 11

Epilogue: Economic Evolution with Changeable Speeds and Structures

We have filled many pages with differential equations and their applications to economics. Mathematically, we have left many theorems unproved and have been unable to mention many new developments in dynamical systems. Economically, we have provided examples to illustrate applications of analytical techniques. Those numerous economic models (each of which appears to be quite reasonable), which are supposed to deal with the same economic system might have caused the reader to wonder whether economics should provide a consistent theory which treats those ideas within the same framework. We now would like to offer in these closing sentences a "general" vision about economic theories for explaining economic reality. We are now concerned with two issues which are important but rarely addressed in economics.

Time scales and changeable speeds

As time passes, economic issues with which economists are concerned have shifted. Even since the time of Adam Smith, the economic variables that economists have dealt with appear to have been invariant. But the ways in which these variables are combined and the speeds at which they change have constantly varied and the dominant economic doctrines have shifted over time and space. The complexity of economic reality is constantly increasing in modern time. This is partially because of the expanded capital and knowledge stocks of mankind. Knowledge, in fields of philosophy, arts, literature, music, technology and sciences, expands man's imagination and extends possibilities of human action, not to mention that the knowledge reservoir can directly satisfy the desires of an unlimited number of people at the same time. Knowledge is not only power and sources of money, but also the most durable capital goods for human mind. Increases in machines, housing and infrastructures has enriched human environment, increased accessibility to various locations, and enlarged variety of human behavior. The explosion of knowledge and capital in this century has resulted in far more complicated human action fields than anyone could have imagined in the last century.

Time is at the main difficulty of almost every economic problem. The role of time in decision-makings and action is becoming increasingly complicated as variety of action and social networks are expanding. It is a difficult issue to decide the length of time which affects a special decision making since each kind of human decisions are made with different time scales and two persons may have different time scales with regard to the same kind of decision making. Because of the high variety of human behavior and time scales, in order to analyze a single person's economic behavior as a whole we have to conduct the analysis within a framework with varied time scales. Human behavior are connected in direct or indirect ways in human action fields; but we may miss interdependence between some elements if we do not properly recognize the role of time.

If we examine the complexity of economic evolution from a historical perspective, we may argue that mankind has experienced three economic structural transformations - from hunting society to agricultural one, agricultural society to industrial one, industrial society to information/knowledge-based one. These transformations are still occurring in different nations in different forms at different speeds. Each of these economic systems has certain corresponding dominant ideologies such as religions, socialism and capitalism. At each turning point there tend to be great conflicts among different social classes, though forms of conflicts are affected by geographical conditions, cultural traditions, international environment, and other factors. As an illustration of applying the concept of speeds of changes in analyzing economic structural changes, we may select three basic variables, the population, capital and knowledge. As shown Table 1, these three variables may be roughly considered to be changeable at different speeds in different societies.

Table 11.1 Change speeds of the economic variables.

Society / Variables	Agricultural	Industrial	Knowledge
Population	Fast	fast/slow	slow
Capital	Slow	slow/fast	fast
Knowledge	Slow	slow/fast	fast

It may be argued that if we are interested in examining agricultural economies, we may concentrate on studying population (and power struggle) dynamics. But the analytical conclusions about agricultural economies cannot be applied to explain economic dynamics of industrialized economies, as capital is the dominant variable of industrial economies. Similarly the analytical conclusions about capital-based societies cannot be applied to explain economic phenomena of knowledge-based societies. In fact, from the studies of history of economic analysis[1], it is clear that many economic ideas were created at the time when the societies were faced with new economic problems (such as structural transformation) and thus required new ideas to solve those problems. We may thus expect certain correspondence between creation of economists' ideas and historical conditions.

Another dimension in analysis is space. Man, action, capital, knowledge and time can become culturally and socially meaningful only if we locate them over space. Each human being is born into a unique existence and each piece of land has its unique attributes in affecting human action. Space means individual characteristics and accordingly requires refined classification. This is particularly important in analyzing modern economies. Fast technological changes, richness of material living conditions, complicated international interactions, and many other modern phenomena have increased complexity of spatial economies. The

[1] See Schumpeter (1954), Blaug (1985), Negishi (1989).

subsystems such as ecological, economical and social subsystems, which could be once decomposable as separate elements in analyzing the social system at least in short terms over a homogenous space, have to be treated as a part of the whole system. Some economic relations cannot be recognized if we don't explicitly introduce spatial and temporal dimensions. It will take some time for what is happening in a scientific lab to affect economic reality. Without spatial dimension, we can hardly analyze actual processes of, for instance, how Japanese economy may actually affect the world economy. In fact, the choice of spatiotemporal scale is a delicate and obligatory process and must be made before actual study of any special economic problem. The explicit awareness of this necessity is very important for understanding both economic reality and structure of economics. For instance, for human life what is good to one's taste (assessment on a short timescale) may be harmful to one's health (assessment on a longer timescale). One can hardly explain differences between Keynes and Schumpeter's economic visions without differentiating their temporary scales. Temporal scales in the economist's vision have complicated interdependent relationships with actual analyses and abstraction of reality.

We are in an era of high economic complexity. This implies that economic decisions have to be made within a large context in which internal structures of each subsystem and connections of different subsystems have to be taken into account within a genuinely dynamic framework. The bringing-up of children, lower and higher education, family structure, and family values are all connected in a subtle and complicated way in economic networks. We have to consider reciprocal relations of different aspects of social and economic factors rather than considering these facts in isolation. Simple one-sector growth models without economic structures will hardly provide any useful information about the complexity of modern economies. We need to enlarge analytical frameworks to handle multiple hierarchical levels, multiple space degrees and multiple time scales.

Dynamic economic structures

An economic system is composed of many people, and the psychology of people and the relations (which are reflected in values, institutions and customs) among people are constantly changing. The difficult task is to find out whether or not there are durable (if not permanent) patterns or orders in human behavior and in human societies and to explicitly construct descriptions (usually, models) for these orders if they exist. In order to construct a comprehensive theory it is necessary to understand general patterns of people's behavior in a society over time and space. The difficult task is how to construct such a comprehensive economic theory.

It is significant to examine economic systems with a spatiotemporal structural vision. The key words are space, time, and structure. It is hard to give a precise concept of structure. Here, a structure means a sum of elements and relationships between those elements. In other words, structure stands for the way the elements and constituent parts of a whole are arranged with respect to each other. Structure represents a whole in which each element depends on the others by virtue of its relation with them. According to Thom,[2] structure is defined as a spatiotemporal morphology described by significant spatial discontinuities and by the syntax that determines how these sets of discontinuities form into relatively stable systems. In evolving structures relations depend on time. The structure includes properties, which are properties of the whole rather than only properties of its component parts. Any change in one element or one relationship will cause a modification in other elements or relationships. By means of the cooperation of the individual parts of different subsystems new properties may emerge that are not present in the subsystems. Economic evolution involves not only changes in variable levels and functions but also in organizational structures that concern the way elements are connected within subsystems, the way subsystems are embedded in large ones, and the way that organizational structures emerge or disappear. As mentioned in the introduction, advances in theory of complex systems provide promising ways for understanding the dynamics of structural changes in socio-economic

---

[2] Thom (1977).

systems. Theory of complex systems provides many deep insights into structural evolution. The modern study of economic chaos permits the discovery of chaotic economic structures disguised by very complicated fluctuations. The concept of structural stability in theory of complex systems is essentially significant in the study of structural evolution.

Hierarchy is a main character of economic structures. Economic systems consist of a hierarchical structuring among the component parts. Hierarchy here means, following Herbert Simon[3], a set of Chinese boxes of a particular kind. Opening any given box discloses a whole small set of boxes; and opening any one of these component boxes discloses a new set in turn. Power distribution is an important indicator of this structure. For instance, in common situations the state organizes the regions within the country, and the regional governments in turn organize the lower-levels within them. Each society is characterized by its own hierarchical structure. In social evolution, these structures may be either stable or unstable, depending on material, affective, cognitive, and spiritual, factors. In the traditional societies economic structures often remained quite stable over many generations; in modern societies structural changes may occur several times within a short period of time.

Hierarchy is not only the character of human societies; even sciences exhibit hierarchical structures. Dawkins sees scientific theories and areas as a hierarchical structure, on different levels, corresponding to levels of description of phenomena[4]. Philosophers and some scientists have sought ultimate reality in the structure of matter at increasingly finer scales in order to provide the most elementary explanation, while astronomers have sought the structure of the universe in increasingly wider domains. In natural science, the complexities of ecosystems are explained by examining those of organisms, organisms are explained by referring them back to the growth of spatially organized proteins and other macromolecule, the complex organization of organisms is explained back to the linear complexity of their DNA code, the complexity of DNA is referred back to combinations of simpler atoms, and so on. We should have national macroeconomics based on regional economics; regional

---

[3] Simon (1973).
[4] Dawkins (1986).

economics should be referred back to urban and rural economics; spatial economics referred back to family-level and company-level economics. In a broader perspective, psychology and behavior sciences should be the starting point of microeconomics. Chemists will to explain psychological processes in terms of natural laws. The processes can be further going on. Darwin's remarks that it is not necessary to refer every phenomenon back down this chain of reductions in order to understand it. In natural sciences, chemistry can be considered as a 'fixed parameter' for the purpose of understanding DNA. In economics, macroeconomics can be considered as 'given' for labor economics and family economics. It is obviously important to construct a grand theory, which connects all the levels within a compact framework.

Connections between levels in a social hierarchy are usually not simple. An economic hierarchical system may operate on different scales. Its variables and substructures may operate or change in different process rates. Since higher levels usually strongly and quickly affect low levels in the hierarchical structure, higher levels usually tend to be changed in lower frequencies. But this asymmetry in change speeds is not always held. To study the hierarchical nature of complex systems, we have to accept a different perspective - a different spatiotemporal scale used. There are gaps between any two levels of social hierarchy. For instance, we may have a reasonable understanding of single male or female behavior and we know how men and women get married and form families. But the functioning of families is far more complicated. Micro level phenomena such as family ties have significant implications for macro economies. An economic theory without endogenous family structure can hardly explain modern economic reality since on one hand family structures have been affected by economic development, on the other hand economic development is the consequences of cooperative (and competitive) behavior among family members.

All these intrinsic difficulties related to economic structures heavily affect the efficiency of modeling economic systems. Multiple levels have to be described in long-term studies. This requires economic theory to have internal structures to represent the complexity of subsystems and connections of the subsystems. Such structural models will eventually turn out to be very complicated. Indeed, we may find out some special characteristics of the system under consideration and thus are able to simplify the analysis. For instance, some hierarchical systems are decomposable, at least in short timescales. This means that it is possible to effectively isolate and describe a part of the system for a given timescale. We may analyze behavior of the independent subsystem in isolation from the rest of the hierarchy to which it belongs. A study of dynamics of a particular process on a particular level can thus be conducted by taking behavior of higher levels as fixed and 'enslaving' behavior of the low levels as structurally determined flows. In other words, for the chosen time scale the behavior of higher levels are so slow that they can be effectively negligible and the behavior of lower levels are so fast that perturbations generated by the behavior of lower levels can also be effectively neglected. For instance, an economic analysis may be conducted in a time scale short enough to assume changes in ecological processes negligible and long enough to average out noise from processes occurring at individual levels. It should be remarked that this method might be invalid especially in 'revolutionary' periods. At such critical points, neither the dynamics of higher levels nor the perturbations generated by the behavior of lower levels are negligible. The model used to describe the dynamic interaction of the chosen subsystem is no longer able to provide reliable information about possible behavior of the subsystem.

An important feature of economic structures is that they are intrinsically complicated at each level. Individuals, groups or clubs, regions and nations, even as they develop under practically similar conditions, are never exactly the same. Detailed studies of their evolution have provided many examples of an intrinsic complexity. For instance, random fluctuations in tastes may affect microeconomic evolutionary processes on a large scale. The economic structure represents the values and principles of the economic organization. The system may be analyzed by dividing the whole system into different levels, each representing a subsystem, which consists of relatively uniform elements that interact with each other either in simple or complicated ways. To find and describe these interactions are the key elements for analyzing order and disorder at any given level. Economists have sought structural invariants on macro, meso and micro levels. The construction of a theory with structure is not arbitrary and gratuitous. We first have to determine issues under examination, scales (both of variables, time and space) and domains, and analytical methods. Here, when assuming the habitual three-dimensional representation of space, with time as a fourth dimension, scale is defined as the smallest volume within the interior of which it is agreed not to try to distinguish the nonuniformity of a property being measured and as the shortest interval of time during which it is agreed not to try to distinguish variations of a given property. The domain is defined as the greatest volume and the longest time interval over which the study will be extended. For instance, the whole economy can be studied by employing several scales. The variables used at one scale may be treated as a coarser scale, macroscopic in comparison with the first by taking averages of larger volumes and longer intervals of time. In building a sophisticated economic theory, one has to construct, without making any mistakes, a long chains of assertions, has to be aware of what one is doing at each step of the construction process, and has to speculate about where one is going. The constructor has be to able to guess what is true and what is false at each level and be able to judge what is useful and what is not in the whole framework.

Appendix

A.1 Matrix Theory

We present some important concepts and theorems from linear algebra and matrix theory. Some elementary concepts, such as identity matrices and null matrices, matrix operations, and proofs of theorems are omitted.[1]

Let vectors A be a nonempty set of vectors in R^n. A vector x in R^n is *linearly dependent* on the set A if there exist vectors $y_1, y_2, ..., y_m$ and scalars $a_1, a_2, ..., a_m$ such that

$$x = \sum_{j=1}^{m} a_j y_j .$$

For any nonempty set A of vectors in R^n, $\langle A \rangle$ is the set of all vectors in R^n that are dependent on A. $\langle A \rangle$ is a subspace of R^n. A vector of the form

$$\sum_{j=1}^{m} a_j y_j$$

is called a *linear combination.* A set A of vectors is a *basis* of the subspace U if (i) A "spans" U and (ii) A is linearly independent. If U is any subspace of R^n, the number of vectors in a basis of U is called the *dimension* of U and is abbreviated as $\dim(U)$. The dimension of R^n is n.

[1] This part on matrix theory is based on Gilbert and Gilbert (1970). See also Chiang (1984), Berman and Plemons (1979), and Peterson and Sochachi (2002).

Let $U = \{U_1, U_2, ..., U_r\}$ be a set of vectors in R^n and $V = \{V_1, V_2, ..., V_s\}$ be a set of vectors in $\langle U \rangle$, a *matrix of transition* from U to V is a matrix $A = \lfloor a_{ij} \rfloor_{rxs}$ such that

$$V_j = \sum_{i=1}^{r} a_{ij} U_i, \quad j = 1, 2, ..., s.$$

Definition A.1.1 A square matrix $A = \lfloor a_{ij} \rfloor_{nxn}$ is *nonsingular* if and only if A is a matrix of transition from one basis of R^n to another basis of R^n. A square matrix that is not nonsingular is called *singular*.

We denote the identity matrix by I_n ($\equiv \lfloor \delta_{ij} \rfloor_{n \times n}$ where δ_{ij} is the Kronecker delta).

For any $m \times n$ rectangular matrix, if the maximum of linearly independent rows that can be found in such a matrix is r, the matrix is said to be of *rank* r, denoted by $Rank(A)$ or $RankA$. The rank also tells us the maximum number of linearly independent columns in the same matrix. As a square matrix has n linearly independent rows (or columns), it must be of rank of n. If A is $m \times n$ matrix over R and P is any invertible $n \times n$, then we have $Rank(A) = Rank(AP)$.

Definition A.1.2 An $n \times n$ matrix B is an *inverse* of the $n \times n$ matrix $A = \lfloor a_{ij} \rfloor_{nxn}$ if $AB = I_n = BA$. Furthermore a square matrix is called *invertible* if it has an inverse.

Theorem A.1.1 An $n \times n$ matrix A is invertible if and only if A is nonsingular. The inverse of an invertible matrix is unique.

If $A = \lfloor a_{ij} \rfloor_{nxn}$ is invertible, its unique inverse is denoted by A^{-1}. If $A_1, A_2, ..., A_m$ are square invertible matrices of order n over R^n, then $A_1 A_2 \cdots A_m$ is invertible and

$$(A_1 A_2 \; A_m)^{-1} = A_m^{-1} \cdots A_2^{-1} A_1^{-1}.$$

For any $m \times n$ matrix A, the transpose of A is denoted by A^T. If a square matrix A is invertible, then A^T is also invertible and $(A^T)^{-1} = (A^{-1})^T$. The following concept is only referred to square matrices.

Definition A.1.3 The *determinant* of the square matrix $A = \lfloor a_{ij} \rfloor_{n\times n}$ is the scalar $\det(A)$ defined by

$$\det(A) = \sum_j (-1)^t a_{1j_1} a_{1j_2} \cdots a_{1j_n},$$

where $\sum_j$ denotes the sum of all terms of the form $(-1)^t a_{1j_1} a_{1j_2} \cdots a_{1j_n}$ as $j_1, j_2, \ldots, j_n$ assumes all possible permutations of the numbers of the numbers $1, 2, \ldots n$, and the exponent t is the number of interchanges used to carry $j_1, j_2 \ldots, j_n$ into the natural order $1, 2, \ldots, n$.

The notations $\det A$ and $|A|$ are used interchangeably with $\det(A)$. When $n = 2$ and $n = 3$, we have

$$|A_{2\times2}| = \begin{vmatrix} a_{11} & a_{12} \\ a_{21} & a_{22} \end{vmatrix} = a_{11}a_{22} - a_{21}a_{12},$$

$$|A_{3\times3}| = \begin{vmatrix} a_{11} & a_{12} & a_{13} \\ a_{21} & a_{22} & a_{23} \\ a_{31} & a_{32} & a_{33} \end{vmatrix} = a_{11}a_{22}a_{22} + a_{12}a_{23}a_{31} + a_{13}a_{32}a_{21},$$

$$- a_{31}a_{22}a_{13} - a_{32}a_{23}a_{11} - a_{33}a_{12}a_{21}. \tag{A.1.1}$$

Definition A.1.4 The *minor* of the element a_{ij} in $A = \lfloor a_{ij} \rfloor_{n\times n}$ is the determinant M_{ij} of the $(n-1)\times(n-1)$ submatrix of A obtained by deleting row i and column j of A. The *cofactor*, denoted by A_{ij}, of a_{ij} in $A = \lfloor a_{ij} \rfloor_{n\times n}$ is the product of $(-1)^{(i+j)}$ and M_{ij}, that is,

$$A_{ij} = (-1)^{(i+j)} M_{ij}.$$

The *adjoint* of A, denoted by $adj(A)$, is given by

$$adj(A) = [A_{ij}]^T_{n\times n}.$$

Theorem A.1.2 If $A = \lfloor a_{ij} \rfloor_{n\times n}$, then

$$a_{i1}A_{k1} + a_{i2}A_{k2} \cdots + a_{in}A_{kn} = \delta_{ik} \det(A), \quad i, k = 1, 2, \ldots, n,$$

$$a_{1j}A_{1k} + a_{2j}A_{2k} \cdots + a_{nj}A_{nk} = \delta_{jk} \det(A), \quad j, k = 1, 2, \ldots, n.$$

By the above formula and (A.1.1) with $\delta_{ii} = 1$, we can calculate $\det(A)$ of any dimension, in principle. For instance, when $n = 4$

$$\det(A_{4\times 4}) = a_{11}A_{11} + a_{21}A_{21} + a_{31}A_{31} + a_{41}A_{41},$$

where A_{i1} are calculated from the corresponding 3×3 matrices.

It can be shown that if $A = \lfloor a_{ij} \rfloor_{n\times n}$ is invertible, then

$$A^{-1} = \frac{1}{\det(A)} adj(A).$$

Definition A.1.5 If A is $n\times n$ matrix, an *eigenvector* of A is a nonzero column vector v in R^n such that $Av = \rho v$ for some scalar ρ; the scalar ρ is called an *eigenvalue* of A.

Theorem A.1.3 If A is an $n\times n$ matrix, a number ρ is an eigenvalue of A if and only if $\det(\rho I_{n\times n} - A) = 0$.

The equation $\det(\rho I_{n\times n} - A) = 0$ is called the *characteristic equation* of the matrix A. Upon expanding the determinant $\det(\rho I_{n\times n} - A)$ we will have a polynomial of degree n in ρ called the *characteristic polynomial* of A.

An $n\times n$ matrix B is said to be *similar* to the $n\times n$ matrix A if there is an invertible $n\times n$ matrix P such that $B = P^{-1}AP$. A square matrix is said to be *diagonalizable* if it is similar to a diagonal matrix. It can be proved that a square matrix A is diagonalizable if and only if there is a basis for R^n consisting of eigenvectors of A. A non-square matrix is not diagonalizable. There is something close to diagonal form called the *Jordan canonical form* of a square matrix. A *basic Jordan block* associated with a value ρ is expressed

$$\begin{bmatrix} \rho & 1 & 0 & \cdots & 0 & 0 \\ 0 & \rho & 1 & \cdots & 0 & 0 \\ \vdots & \vdots & \vdots & & \vdots & \vdots \\ 0 & 0 & 0 & \cdots & \rho & 1 \\ 0 & 0 & 0 & \cdots & 0 & \rho \end{bmatrix}.$$

The Jordan canonical form of a square matrix is compromised of such Jordan blocks.

Theorem A.1.4 Suppose that A is an $n \times n$ matrix and suppose that

$$\det(\rho I - A) = (\rho - r_1)^{m_1} (\rho - r_2)^{m_2} \cdots (\rho - r_k)^{m_k},$$

where $\rho_1, \rho_2, \cdots, \rho_k$ are distinct roots of the characteristic polynomial of A. Then A is similar to a matrix of the form

$$\begin{bmatrix} B_1 & 0 & \dots & 0 \\ 0 & B_2 & \cdots & 0 \\ \vdots & \vdots & & \vdots \\ 0 & 0 & \cdots & B_k \end{bmatrix},$$

where each B_i is an $m_i \times m_i$ matrix of the form

$$\begin{bmatrix} J_{i_1} & 0 & \dots & 0 \\ 0 & J_{i_2} & \cdots & 0 \\ \vdots & \vdots & & \vdots \\ 0 & 0 & \cdots & J_{i_l} \end{bmatrix},$$

and each J_{i_j} is a basic Jordan block associated with r_i.

A.2 Systems of Linear Equations

A system of linear equations is

$$a_{i1}x_1 + a_{i2}x_2 + \cdots + a_{in}x_n = b_j, \quad j = 1, 2, ..., m,$$

or in the matrix form

$$A_{m\times n} x_{n\times 1} = b_{m\times 1}. \tag{A.2.1}$$

A solution of the system is a set of values of x that satisfies $Ax = b$. In this system, A is called the *coefficient matrix*, x the *matrix of unknowns*, and b the *matrix of constants*. The matrix $[A, b]$ is called the *augmented matrix* of the system.

Theorem A.2.1 The system $Ax = b$ has a solution if and only if $Rank([A,b]) = Rank(A)$.

Theorem A.2.2 If $Rank([A,b]) = Rank(A) = r$, then the solution to $A_{m\times n}x_{1\times n} = b_{1\times n}$ can be expressed in terms of $n - r$ parameters.

Theorem A.2.3 Consider a system of linear equations

$$A_{n\times n}x_{n\times 1} = b_{n\times 1}.$$

If $\det(A) \neq 0$, then the unique solution is given by $x = A^{-1}b$.

Theorem A.2.4 Consider a system of linear equations $A_{n\times n}x_{n\times 1} = b_{n\times 1}$. If $\det(A) \neq 0$, then the unique solution of the system is given by

$$x_j = \frac{\sum_{k=1}^{n} b_k A_{kj}}{\det(A)}, \quad j = 1, 2, \cdots, n,$$

where A_{ij} are cofactors of A.

The above formula is called *Cramer's Rule*. We note that

$$\sum_{k=1}^{n} b_k A_{kj}$$

is the determinant of the matrix obtained by replacing the jth column of A by the column of constants b.

A.3 Properties of Functions and the Implicit Function Theorem

First we state a few theorems from analysis.

Definition A.3.1 Suppose that V_1 and V_2 are two normalized linear spaces with respective norms $\| \|_1$ and $\| \|_2$. Then $F : V_1 \to V_2$ is *continuous* at $x_0 \in V_1$ if for all $\varepsilon > 0$ there exists a $\delta > 0$ such that $x \in V_1$ and $\|x - x_0\|_1 \in \delta$ implies that $\|F(x) - F(x_0)\|_2 \leq \varepsilon$. And F is said to be continuous on the set $U \in V_1$ if it is continuous at each point $x \in U$, and we write $F \in C(U)$.

Theorem A.3.1 (The intermediate-value theorem) If the function $f:[a, b] \to R$ is continuous and

$$f(a) < 0 < f(b),$$

then there exists a point $c \in (a, b)$ such that $f(c) = 0$.

Definition A.3.2 The function $f: R^n \to R^n$ is *differentiable* at $x_0 \in R^n$ if there is a linear transformation $Df(x_0)$ that satisfies

$$\lim_{|h| \to 0} = \frac{|f(x_0 + h) - f(x_0) - Df(x_0)h|}{|h|} = 0.$$

The linear transformation v is called the *derivative* of f at x_0.

The following theorem gives us a method for computing the derivative in coordinates.

Theorem A.3.2 If the function $f: R^n \to R^n$ is differentiable at $x_0 \in R^n$, then the partial derivatives $\partial f_i / \partial x_j$ all exist at x_0 and for all $x \in R^n$

$$Df(x_0)x = \sum_{j=1}^{n} \frac{\partial f}{\partial x_j}(x_0)x_j .$$

Thus if f is a differentiable function, the derivative Df is given by the $n \times n$ Jacobian matrix

$$Df = \left[\frac{\partial f_i}{\partial x_j}\right].$$

For U an open subset of R^n, the higher order derivatives $D^k f(x_0)$ are defined in a similar way.

Definition A.3.3 Suppose that $f: U \to R^n$ is *differentiable* on U. Then $f \in C^1(U)$ if the derivative Df is continuous on U.

We can define

$$f \in C^k(U),\ k = 2, 3, \cdots$$

in a similar manner.

Theorem A.3.3 Suppose that U is an open subset of R^n and that $f: U \to R^n$. Then $f \in C^1(U)$ if and only if the partial derivatives $\partial f_i / \partial x_j$ all exist and are continuous on U.

It can be shown that $f \in C^k(U)$ if and only if the partial derivatives

$$\frac{\partial^k f_i}{\partial x_{j_1} \cdots \partial x_{j_k}},$$

with $i, j_1, \cdots, j_k = 1, \cdots, k$ exist and are continuous on U.

Theorem A.3.4 (Inverse function theorem) Let U be an open set in R^n and $f: U \to R^n$ be a C^k function with $k \geq 1$. If a point $\bar{x} \in U$ is such that the $n \times n$ matrix $Df(\bar{x})$ is invertible, then there is an open neighborhood V of $\bar{x}$ in U such that $f: V \to f(V)$ is invertible with a C^k inverse.

The inverse function theorem implies that if the matrix $f_x(0)$ is nonsingular, then there is a smooth locally defined function

$$x = g(y), \quad g: R^n \to R^n$$

such that

$$f(g(y)) = y,$$

for all y in some neighborhood of the origin of R^n. The function g is called the inverse function for f and is denoted by $g = f^{-1}$.

If $y = g(x)$, $g: R^n \to R^m$ and $z = f(y)$, $f: R^m \to R^k$ are two maps, then their *superposition*

$$h = f \circ g$$

is a map $z = h(x)$, $R^n \to R^k$, defined by the formula $h(x) = f(g(x))$. Let $f_y(y)$ denote the Jacobian matrix f evaluated at a point $y \in R^m$

$$f_y(y) = \left(\frac{\partial f_i(y)}{\partial y_j} \right)_{k \times m}, \quad i = 1, 2, ..., k, \; j = 1, 2, ..., m.$$

We similarly define $h_x(x)$ as

$$h_x(x) = \left(\frac{\partial f_i(y)}{\partial y_j}\right)\left(\frac{\partial y_p}{\partial x_q}\right).$$

We consider a map

$$(x, y) \mapsto F(x, y),$$

where $F: R^n \times R^m \to R^n$ is a smooth map defined in a neighborhood of $(x, y) = (0, 0)$ and $F(0,0) = 0$. Let $F_x(0,0)$ denote the matrix of first partial derivatives of F with respect to x evaluated at $(0, 0)$

$$F_x(0,0) = \left(\frac{\partial F_i(x, y)}{\partial x_j}\right)_{(x,y)=(0,0)}.$$

Theorem A.3.4 (The implicit function theorem) If the matrix $F_x(0,0)$ is nonsingular, then there is a smooth locally defined function $y = f(x)$, $f: R^n \to R^m$ such that

$$F(x, f(x)) = 0,$$

for all x in some neighborhood of the origin of R^n. Moreover

$$f_x(0) = -[F_x(0,0)]^{-1} F_y(0,0).$$

The degree of smoothness of the function f is the same as that of F.

Theorem A.3.5 (The submanifold theorem) Let U be an open set in R^n and let $f: U \to R^p$ be a differentiable function such that $Df(x)$ has rank p whenever $f(x) = 0$. Then $f^{-1}(0)$ is an $(n - p)$-dimensional manifold in R^n.

Lemma A.3.1 (The Morse lemma) Let $f: R^n \to R$ be a sufficiently differentiable function. If x^* is a *nondegenerate critical point* of f, that is, $Df(x^*) = 0$ and the *Hessian matrix* $\left[\partial^2 f(x^*)/\partial x_i \partial x_j\right]$ is nonsingular, then there is a local coordinate system $(y_1, \cdots, y_n)$ in a neighborhood U of x^* with $y_i(x^*) = 0$ for all i, such that

$$f(y) = f(x^*) - \sum_{i=1}^{k} y_i^2 + \sum_{i=1+k}^{n} y_i^2,$$

for all $y \in U$. The integer k is the number of negative eigenvalues of the Hessian matrix.

Sard's Theorem Let U be an open set in R^n and let $f : U \to R^p$ be a differentiable function. Let C be the set of critical points of f, that is, the set of all $x \in U$ with $rank Df(x) < p$. Then $f(C)$ has measure zero in R^p.

A.4 Taylor Expansion and Linearization

Given a successively differentiable one-variable function $f(x)$, the *Taylor expansion* around a point x^* gives the series

$$f(x) = f(x^*) + f'(x^*)(x - x^*) + \frac{f''(x^*)}{2!}(x - x^*)^2 + \ldots$$

$$+ \frac{f^{(n)}(x^*)}{n!}(x - x^*)^n + R(x),$$

where a polynomial involving higher powers (than n) of $(x - x_0)$ appears on the right. For a two-variable function, $f(x, y)$, the Taylor expansion around a point (x^*, y^*) is given by

$$f(x, y) = f(x^*, y^*) + f_x(x^*, y^*)(x - x^*) + f_y(x^*, y^*)(y - y^*)$$

$$+ \frac{1}{2!}\Big[f_{xx}(x^*, y^*)(x - x^*)^2 + 2f_{xy}(x^*, y^*)(x - x^*)(y - y^*)$$

$$+ f_{yy}(x^*, y^*)(y - y^*)^2\Big] + \ldots + R(x, y).$$

Linearization of a function is obtained by simply dropping all terms of order higher than one from the Taylor series of the function. For instance, the linear approximation of a one-variable function $f(x)$ gives

$$f(x) = f(x^*) + f'(x^*)(x - x^*).$$

In the case of two variables

$$f(x, y) = f(x^*, y^*) + f_x(x^*, y^*)(x - x^*) + f_y(x^*, y^*)(y - y^*).$$

We now give the Taylor expansion for any dimension around the origin. Let U be a region in R^n containing the origin $x = 0$. We denote the set of all continuous functions $f : U \to R^m$ by $C^0(U, R^m)$ and the set of all differentiable functions with continuous first derivatives by $C^1(U, R^m)$. Analogously, we will use $C^k(U, R^m)$ to indicate the functions with continuous derivatives up to order k. If $f \in C^k(U, R^m)$ with a sufficiently large k, the function f is called *smooth*. A C^∞ function has continuous partial derivatives of any order. Any function $f \in C^k(U, R^m)$ can be represented near $x = 0$ in the Taylor expansion

$$f(x) = \sum_{|i|}^{k} \frac{1}{i_1! i_2! \ldots i_n!} \frac{\partial^{|i|} f(x)}{\partial x_1^{i_1} \partial x_2^{i_2} \ldots \partial x_n^{i_n}}\bigg|_{x=0} x_1^{i_1} x_2^{i_2} \ldots x_n^{i_n} + R(x),$$

where

$$|i| \equiv i_1 + i_2 + \ldots + i_n,$$

and

$$R(x) = O\left(\|x\|^{k+1}\right) = o\left(\|x\|^k\right),$$

namely

$$\frac{\|R(x)\|}{\|x\|^k} \to 0 \quad as \ \|x\| \to 0,$$

in which $\|x\| = \sqrt{x^T x}$. Here, we give precise definitions of O and o. Let f and g be two given functions. We say that

$$f(x) = O(g(x)), \quad as \ x \to 0,$$

if there are constants $\alpha > 0$ and $A > 0$ such that $|f(x)| \le A|g(x)|$ for $|x| < \alpha$. We say that

$$f(x) = o(g(x)), \quad as \ x \to 0,$$

if for any $\varepsilon > 0$ there is a $\delta > 0$ such that $|f(x)| \le \varepsilon |g(x)|$ for $|x| < \delta$.

A C^∞-function f is called *analytical* near the origin if the corresponding *Taylor series*

$$\sum_{|i|}^{\infty} \frac{1}{i_1! i_2! \ldots i_n!} \frac{\partial^{|i|} f(x)}{\partial x_1^{i_1} \partial x_2^{i_2} \ldots \partial x_n^{i_n}}\bigg|_{x=0} x_1^{i_1} x_2^{i_2} \ldots x_n^{i_n},$$

converges to $f(x)$ at any point x sufficiently close to $x = 0$.

A.5 Structural Stability[2]

The stability concepts introduced so far are related to the way in which a dynamic system reacts to perturbations in initial conditions. However, sometimes we are interested in properties of functional forms. For instance, it is important to know about the stability of the money supply function itself. The concept of structural stability is related to qualitative properties of functions.

Models we build for describing reality may be very sensitive to small changes. In such cases, an arbitrarily small change in a model leads to another model with essentially different properties. To explain the concept of structural stability, let us consider the differential equation $\dot{x} = f(x)$, for a given vector, f, on the manifold, M.

Definition A.5.1 Two systems are said to be *topologically orbitally equivalent* if there exists a homeomorphism of the phase space of the first system onto the phase space of the second, conserving oriented phase curves of the first system onto oriented phase curve of the second. No coordination of the motion on the corresponding phase curve is required.

Definition A.5.2 Let M be a compact manifold (of class C^{k-1}, $k = 1$). Let f be a vector field of class k (if M has a boundary, then it is assumed that f is not tangent to it). The system (M, f) is said to be *structurally stable* if there exists a neighborhood of f in the space C^1 such that every vector field in this neighborhood defines a system topologically orbitally equivalent to the initial one, and the homeomorphism of the equivalence is close to the identity homeomorphism.

A system which is not structurally stable is defined to be *structurally unstable*. The following predator-prey system is structurally unstable

$$\dot{x}(t) = \alpha(y_0 - y(t))x(t),$$

[2] This section is based on Sec. 3.5 in Zhang (1991).

$$\dot{y}(t) = \beta(x(t) - x_0)y(t),$$

where $x(t)$ and $y(t)$ are respectively the population of preys and predators, and α, β, x_0 and y_0 are parameters. We can define the concept more simply, which is enough for our purpose. A system $\dot{x} = f(x)$ is structurally stable if there exists a homeomorphism from the orbits of $\dot{x} = f(x)$ to the orbits of

$$\dot{x} = f(x) + p(x)$$

for sufficiently small perturbations $p(x)$.

Let M denote the interior of a closed curve without contact to any of the vector fields to be considered and let G be the set of all such C^k vector fields. We have the following theorem. The theorem provides a necessary and sufficient condition for identifying structural stability of a dynamic system. However, the results are not so easy to apply because it is difficult to check the conditions for real problems.

Theorem A.5.1[3] A function $f(x)$ in G is structurally stable if and only if every equilibrium point and every periodic orbit is hyperbolic and there are no connections between saddle points. Also, the set of structurally stable system is open and dense in G.

A.6 Optimal Control Theory[4]

The *optimal control problem* is one to obtain the trajectory $x(t)$ by choosing a function $v(t)$ to maximize or minimize a certain objective. The theory for such a problem is called *optimal control theory*. In general, $x(t)$ is an n-dimensional vector function and $v(t)$ is an m-dimensional vector function. The basic result of the Pontryagin and his associates is called *Pontryagin's maximum principle*. The principle gives the necessary conditions for optimality. We now introduce this principle.

Consider the following system of n first-order differential equations

[3] The concepts introduced above and Theorem A.5.1 are referred to Chow and Hale (1982).

[4] Different extensions of the maximum principle and various economic applications of the theory are referred to Takayama (1996) and Seierstad and Sydsaeter (1987). The theorems listed here are referred to Takayama (1994).

$$\dot{x}_i(t) = f_i[x(t), v(t), t], \quad i = 1, 2, \cdots, n, \tag{A.6.1}$$

where

$$x(t) = (x_1(t), x_2(t), \cdots, x_n(t)),$$

$$v(t) = (v_1(t), v_2(t), \cdots, v_m(t)),$$

where f_i, x_j and v_k are all real-valued functions. The boundary conditions for the differential equations (A.6.1) are specified as $x(t_0) = x_0$. If we specify the path of $v(t)$, then we can determine the trajectory $x(t; x_0, t_0)$. The existence theorem in Chap. 3 provides sufficient conditions for the local existence and the uniqueness of a solution. The optimal control theory is to find the trajectory $v(t)$ $(\in \Omega)$ which maximizes an objective function. We now consider an objective function

$$S = \sum_{i=1}^{n} c_i x_i(T), \tag{A.6.2}$$

where c_i and T are constant. Our problem is

$$\textit{Maximize}\, S,$$

$$\text{Subject to (A.6.1)}, \ v(t) \in \Omega, \ t_0 \le t \le T, \ x(t_0) = x_0. \tag{A.6.3}$$

In optimal control theory, $x(t)$, which are assumed to be continuous in t, are called the *state variables*, and $v(t)$ are called the *control variables*. Here, Ω is the *set of admissible controls*. When $v(t) \in \Omega$, $v(t)$ is called an *admissible control function*. We assume that Ω is restricted to the set where $v(t)$ is "piecewise continuous."[5] Here, we assume that f_i's are continuous in each x_i, v_j, and t, and possess continuous partial derivatives with respect to each x_i and t. The range of $x(t)$ is denoted by Π, where Π is an open connected subset of R^n.[6]

---

[5] By *piecewise continuous*, here we mean that a function is continuous except possibly at a finite number of points and that the discontinuity is limited to the *first kind* (which means that the left-hand and right-hand limits are finite though they are not equal.

[6] A subject X of R^n is said to connected if it cannot be partitioned into two disjoint nonempty subsets of R^n which are open in X.

Theorem A.6.1 (Pontryagin's maximum principle) For the problem (A.6.3), in order that $v^*(t)$ is a solution of the problem with the corresponding state variable $x^*(t)$, it is necessary that there exists a continuous, vector-valued function

$$\lambda(t) = (\lambda_1(t),\ \lambda_2(t),\ \cdots,\ \lambda_n(t)),$$

not vanishing simultaneously for each t,[7] such that

(1) $\lambda(t)$ together with $x^*(t)$ and $v^*(t)$ solving the following Hamiltonian system

$$\dot{x}_i^* = \frac{\partial H}{\partial \lambda_i}\Big|_{(x^*, v^*, t, \lambda)},$$

$$\dot{\lambda}_i^* = -\frac{\partial H}{\partial x_i}\Big|_{(x^*, v^*, t, \lambda)}, \quad i = 1,\ 2,\ \cdots,\ n, \qquad \text{(A.6.4)}$$

where the *Hamiltonian* H is defined by

$$H(x(t),\ v(t),\ t,\ \lambda(t)) = \sum_{i=1}^{n} \lambda_i(t) f_i(x(t),\ v(t),\ t).$$

(2) The Hamiltonian is maximized with respect to $v(t)$, that is

$$H(x^*(t),\ v^*(t),\ t,\ \lambda(t)) \geq H(x^*(t),\ v(t),\ t,\ \lambda(t)), \quad \text{for all } v(t) \in \Omega.$$

(3) The following *traversality condition* holds

$$\lambda_i(T) = c_i, \quad i = 1,\ 2,\ \cdots,\ n.$$

(4) $x_i(t_0) = x_{i0}$, $i = 1,\ 2,\ \cdots,\ n.$

The theorem gives the necessary conditions for $v^*(t)$ to be optimal. It turns out that these conditions are also sufficient (for global optimum) if the f_i's are concave in x and t. The variables $\lambda(t)$ are called the *Portryagin multipliers*, the *auxiliary variables*, or the *costate variables*. The conditions (A.6.3) can be rewritten as

---

[7] The phrase "not vanishing simultaneously" means that $\lambda(t)$ is a non-zero vector. In the appendix, $\lambda(t)$, like $x(t)$, is continuous and has piecewise continuous derivatives on the interval $[t_0,\ T]$. The possible discontinuities of $\dot{\lambda}(t)$ and $\dot{x}(t)$ occur at the points of discontinuity of $v(t)$.

$$\dot{x}_i^* = f_i\left(x^*(t), v^*(t), t\right), \ \dot{\lambda}_i = -\sum_{j=1}^{n} \lambda_j \frac{\partial f_j}{\partial x_i}\bigg|_{\left(x^*(t), v^*(t), t\right)},$$

$$i = 1, 2, \cdots, n. \tag{A.6.5}$$

The system (A.6.5) consists of $2n$ first-order differential equations for $2n$ variables $x(t)$ and $\lambda(t)$. There are $2n$ boundary conditions, $x(t_0) = x_0$ and $\lambda(T) = c$ corresponding to these $2n$ equations. The condition (2) in the theorem determines $v(t)$ as functions of $x(t)$ and $\lambda(t)$. We call the pair $x^*(t)$ and $v^*(t)$ the *optimal pair* or the *solution pair*. The function $v^*(t)$ is called the *optimal control*, and $x^*(t)$ the *optimal trajectory*.

It should be noted that the objective function S in the condition (A.6.2) is more general than it appears. It includes frequently-used objective function

$$\Psi \equiv \int_{t_0}^{T} f_0\left(x(t), v(t), t\right) dt.$$

Introduce a new variable $x_0(t)$ by

$$\dot{x}_0(t) \equiv f_0\left(x(t), v(t), t\right), \ x_0(t_0) = 0.$$

Then $\Psi = x_0(T)$ is a special case of the condition (A.6.2). Hence the problem of maximizing Ψ subject to $\dot{x}_0(t) = f_0(x, v, t)$, the condition (A.6.1), and $x(t_0) = x_0$ can be converted to the problem of maximizing $x_0(T)$ subject to $\dot{x}_0(t) = f_0(x, v, t)$, the condition (A.6.1), and $x(t_0) = x_0$. We can then immediately apply Theorem A.6.1.

The above discussion shows that the target in the integral form can be converted to a summation form S. Conversely, we can convert S in the condition (A.6.2) to the integral form. To see this, we use

$$S = \sum_{i=1}^{n} c_i x_i(T) = \int_{t_0}^{T} \sum_{i=1}^{n} c_i \dot{x}_i(T) + \sum_{i=1}^{n} c_i x_i(t_0).$$

Hence, the maximization of S subject to the differential equations (A.6.1), and $x(t_0) = x_0$ is equivalent to the maximization of the integral

$$\int_{t_0}^{T} \sum_{i=1}^{n} c_i f_i \, dt,$$

subject to the differential equations (A.6.1), and $x(t_0) = x_0$.

Since $v_j(t)$ can be any piecewise continuous function, $v_j^*(t)$ may be such that

$$u_j^*(t) = 0, \quad t_0 \le t < \bar{t},$$

$$u_j^*(t) = 1, \quad \bar{t} \le t < T.$$

Such a control is called the *bang-bang* control.

Theorem A.6.1 is extended to varied forms of optimal problems. We now provide two cases which are frequently applied in economics. First, consider case of fixed time with variable right-hand end-points problem.

Theorem A.6.2 Consider the following problem

$$\underset{v(t)}{\text{Maximize}} \int_{t_0}^{T} f_0(x(t), v(t), t)\,dt\,,$$

$$\text{Subject to } \dot{x}_i(t) = f_i[x(t), v(t), t], \quad i = 1, 2, \cdots, n,$$

$$v(t) \in \Omega, \ t_0 \le t \le T, \ x(t_0) = x_0,$$

with T being fixed and $x_i(T)$ to be determined. In order that $v^*(t)$ is a solution of the problem with the corresponding state variable $x^*(t)$, it is necessary that there exists a continuous, vector-valued function

$$\lambda(t) = (\lambda_1(t), \lambda_2(t), \cdots, \lambda_n(t)),$$

not vanishing simultaneously for each t, such that

(1) $\lambda(t)$ together with $x^*(t)$ and $v^*(t)$ solving the following Hamiltonian system

$$\dot{x}_i^* = \frac{\partial H}{\partial \lambda_i}\bigg|_{(x^*, v^*, t, \lambda)},$$

$$\dot{\lambda}_i^* = -\frac{\partial H}{\partial x_i}\bigg|_{(x^*, v^*, t, \lambda)}, \quad i = 1, 2, \cdots, n, \qquad \text{(A.6.4)}$$

where the *Hamiltonian* H is defined by

$$H(x(t), v(t), t, \lambda(t)) = \sum_{i=1}^{n} \lambda_i(t) f_i(x(t), v(t), t).$$

(2) The Hamiltonian is maximized with respect to $v(t)$, that is

$$H\left(x^*(t),\, v^*(t),\, t,\, \lambda(t)\right) \geq H\left(x^*(t),\, v(t),\, t,\, \lambda(t)\right), \text{ for all } v(t) \in \Omega .$$

(3) $\lambda_i(T) = 0,\; i = 1, 2, \cdots, n$.

(4) $x_i(t_0) = x_{i0},\; i = 1, 2, \cdots, n$.

Theorem A.6.3 If $f_0(x(t), v(t), t)$ and $f_i(x(t), v(t), t)$ are all concave in x and v, then the set of necessary conditions stated in Theorems A.6.1 and A.6.2 are also sufficient for optimum for their respective problems. In addition, if f_0 is strictly concave in x and v, the optimal path is unique.

The following form of the principle is most frequently applied in economics.

Theorem A.6.4 Consider the following problem

$$\underset{v(t)}{\text{Maximize}} \int_0^{\infty} f_0(x(t), v(t), t) e^{-\rho t}\, dt$$

$$\text{Subject to } \dot{x}_i(t) = f_i[x(t), v(t), t],\; i = 1, 2, \cdots, n,$$

$$v_j(t) \geq 0,\; j = 1, 2, \cdots, m,\; x(0) = x_0,$$

where ρ is a positive constant. Assume that f_0 and f_i are continuously differentiable in the (x, v, t)-space. In order that $v^*(t)$ is a solution of the problem with the corresponding state variable $x^*(t)$, it is necessary that there exist multipliers

$$\lambda(t) = (\lambda_1(t), \lambda_2(t), \cdots, \lambda_n(t)),$$

such that

(1) $\lambda(t)$ together with $x^*(t)$ and $v^*(t)$ solve the following Hamiltonian system

$$\dot{x}_i^* = \frac{\partial H}{\partial \lambda_i}\bigg|_{(x^*, v^*, t, \lambda)},$$

$$\dot{\lambda}_i^* = \rho\lambda_i - \frac{\partial H}{\partial x_i}\Big|_{(x^*, v^*, t, \lambda)}, \quad i = 1, 2, \cdots, n$$

where the current value *Hamiltonian* H is defined by

$$H(x(t), v(t), t, \lambda(t)) = f_0(x(t), v(t), t) + \sum_{i=1}^{n} \lambda_i(t) f_i(x(t), v(t), t).$$

(2) H is maximized with respect to $v(t)$ subject to $v_j(t) \geq 0$, that is

$$\frac{\partial H}{\partial v_j} \leq 0,$$

$$v_j \frac{\partial H}{\partial v_j} = 0, \quad j = 1, 2, \cdots, m.$$

(3) $\lim_{t\to\infty} \lambda_i(t) e^{-\rho t} x_i(t) = 0, \quad i = 1, 2, \cdots, n.$

(4) $x(0) = x_{i0}, \quad i = 1, 2, \cdots, n.$

Theorem A.6.3 is also applicable to Theorem A.6.2.

Bibliography

Abe, N. (1995). Poverty trap and growth with public goods, Economic Letters, 47, pp. 361-66.

Abel, A.B. and Bernanke, B.S. (1998). Macroeconomics, 3rd edition. New York: Addison-Wesley.

Aghion, P. and Howitt, P. (1998). Endogenous Growth Theory. Mass., Cambridge: The MIT Press.

Alonso, W. (1964). Location and Land Use. Mass., Cambridge: Harvard University Press.

Araujo, A. and Scheinkman, H.A. (1977). Smoothness, comparative dynamics and turnpike property, Econometrica, 45, pp. 601-20.

Arnold, V.I. (1978). Ordinary Differential Equations, translated from the Russian by Silverman, R.A. Mass., Cambridge: MIT Press.

Arnott, R.J. (1979). Unpriced transport congestion, Journal of Economic Theory, 21, pp. 294-316.

Arrow, K.J. (1962). The economic implications of learning by doing, Review of Economic Studies, 29, 155-73.

Arrow, K.J., Block, H.D. and Hurwicz, L. (1959). On stability of competitive equilibrium, II, Econometrica, 27, pp. 82-109.

Arrow, K.J. and Hahn, F.H. (1971). General Competitive Analysis. San Francisco: Holden-Day, Inc.

Arrowsmith, D.K. and Place, C.M. (1990). An Introduction to Dynamic Systems. Cambridge: Cambridge University Press.

Ashenfelter, O. and Layard, R. (eds.) (1992). Handbook of Labor Economics, Volume I. Amsterdam: North-Holland.

Barro, R. (1990). Government spending in a simple model of endogenous growth, Journal of Political Economy, 98, pp. 103-125.

Barro, R.J. and Sala-i-Martin, X. (1995). Economic Growth. New York: McGraw-Hill, Inc.

Becker, G.S. (1957). The Economics of Discrimination. Chicago: The University of Chicago Press.

Becker, G. S. (1976). The Economic Approach to Human Behavior. Chicago: The University of Chicago Press.

Becker, G.S. (1985). Human capital, effort, and the sexy division of labor, Journal of Labor Economics 3, pp. 34-58.

Benhabib, J. and Miyao, T. (1981). Some new results of the dynamics of the generalized Tobin model, International Economic Review, 22, 589-96.

Becker, R. and Burmeister, E. (collected) (1991). Growth Theory, three volumes. Vermont: Edward Elgar.

Benhabib, J. and Nishimura, K. (1979). The Hopf bifurcation and the existence and stability of closed orbits in multisector models of optimal economic growth, Journal of Economic Theory, 21, 421-44.

Berman, A. and Plemons, R.J. (1979). Nonnegative Matrices in Mathematical Sciences. New York: Academic Press.

Blanchard, O. (1997). Macroeconomics. New Jersey: Prentice-Hall International.

Blaug, M. (1985). Economic Theory in Retrospect, fourth edition. Cambridge: Cambridge University Press.

Boyce, W.E. and DiPrima, R.C. (2001). Elementary Differential Equations. New York: John Wiley & Sons.

Bretschger, L. (1999). Growth Theory and Sustainable Development. Cheltenham: Edward Elgar.

Britton, N.F. (1986). Reaction-Diffusions and Their Applications to Biology. New York; Academic Press.

Brock, W.A. and Scheinkman, J.A. (1976). Global asymptotic stability of optimal control systems with applications to the theory of economic growth, Journal of Economic Theory, 12, pp. 164-90.

Brueckner, J.K. and Rabenau, B.V. (1981). Dynamics of land use for a closed city, Regional Science and Urban Economics, 11, pp. 1-17.

Buiter, W.H. (1981). Time preference and international lending and borrowing in an overlapping generations model, Journal of Political Economy, 89, pp. 769-97.

Burmeister, E. and Dobell, A.R. (1970). Mathematical Theories of Economic Growth. London: Collier Macmillan Publishers.

Cagan, P. (1956). The monetary dynamics of hyperinflation, in Friedman, M. (ed.) Studies in the Quantity Theory of Money. Chicago: University of Chicago Press.

Cain, G.G. (1986). The economic analysis of labor market discrimination: A Survey, in Ashenfelter, O.C. and Layard, R. (eds.) Handbook of Labor Economics. Amsterdam: North-Holland.

Carr, J. (1981). Applications of Center Manifold Theory. New York: Springer-Verlag.

Cass, D. and Shell, K. (1976). The structure and stability of competitive dynamical systems, Journal of Economic Theory, 12, pp. 31-70.

Cesari, L. (1971). Asymptotic Behavior and Stability Problems in Ordinary Differential Equations. Berlin: Springer.

Chang, W.W. and Smyth, D.J. (1971). The existence and persistence of cycles in a non-linear model: Kaldor's 1940 model re-examined, Review of Economic Studies, 38, pp. 37-44.

Chiang, A.C. (1984). Fundamental Methods of Mathematical Economics. London: McGraw-Hill Book Company.

Chiappori, P.A. (1988). Rational household labor supply, Econometrica, 51, pp. 63-89.

Chiappori, P.A. (1992). Collective labor supply and welfare, Journal of Political Economy, 100, pp. 438-67.

Chiarella, C. and Flaschel, P. (2000). The Dynamics of Keynesian Monetary Growth: Macro Foundations. Cambridge: Cambridge University Press.

Dai, L.S. (1981). Nonconstant periodic solutions in predator-prey systems with continuous time delay. Math. Biosci., 53, pp. 149-57.

Dendrinos, D.S. and Mullally, H. (1985). Urban Evolution: Studies in the Mathematical Ecology of Cities. Oxford: Oxford University Press.

Desai, M. (1973). Growth cycles and inflation in a model of the class struggle, Journal of Economic Theory, 6, pp. 527-45.

Devereux, M.B. and Shi, S. (1991). Capital accumulation and the current account in a two-country model, Journal of International Economics, 30, pp. 1-25.

Domar, E.V. (1946). Capital expansion, rate of growth and employment, Econometrica, 14, pp. 137-47.

Enns, R.H. and McGuire, G.C. (2001). Nonlinear Physics with Mathematica for Scientists and Engineers. Boston: Birkhäuser.

Farkas, M. (1984). Stable oscillations in a predator-prey with time lag. J. Math. Anal. Appl., 102, pp. 175-88.

Farkas, M. (1994). Periodic Motions. Berlin: Springer.

Farkas, M., Farkas, A. and Szabó, G. (1988). Multiparameter bifurcation diagrams in predator-prey models with time lag. J. Math. Biol., 26, pp. 93-103.

Fei, J.C.H., Ranis, G. and Kuo, S.W.Y. (1978). Growth and the family distribution of income by factor components, Quarterly Journal of Economics, XCII, pp. 17-53.

Ferguson, B.S. and Lim, G.C. (1998). Introduction to Dynamic Economic Models. Manchester: Manchester University Press.

Flaschel, P., Franke, R. and Semmler, W. (1997). Dynamic Macroeconomics. Mass., Cambridge: The MIT Press.

Frenkel, J. and Razin, A. (1987). Fiscal Policy and the World Economy. MA., Cambridge: MIT Press.

Folbre, N. (1986). Cleaning house - new perspectives on households and economic Development, Journal of Economic Development, 22, pp. 5-40.

Forster, B.A. (1977). On a one state variable optimal control problem; consumption-pollution trade-offs, in Pitchford, J.D. and Turnovsky, S.J. (eds.) Applications of Control Theory to Economic Analysis. Amsterdam: North Holland.

Franke R. and Asada, T. (1994). A keynes-goodwin model of the business cycle, Journal of Economic Behavior and Organization, 24, pp. 273-95.

Freedman, H.I. (1980). Deterministic Mathematical Models in Population Ecology. New York: Dekker.

Futagami, K., Morita, Y. and Shibata, A. (1993). Dynamic analysis of an endogenous growth model with public capital, Scandinavian Journal of Economics, 95, pp. 607-25.

Gabisch, G. and Lorenz, H.W. (1989). Business Cycle Theory, 2nd edition. Berlin: Springer-Verlag.

Gale, D. (1983). Money: In Disequilibrium. Cambridge: Cambridge University Press.

Gandolfo, G. (1996). Economic Dynamics, third edition. Berlin: Springer.

Gilbert J. and Gilbert, L. (1995). Linear Algebra and Matrix Theory. New York: Academic Press.

Goodwin, R.M. (1951). The non-linear accelerator and the persistence of business cycles, Econometrica, 19, pp. 1-17.

Goodwin, R.M. (1989). Essays in Nonlinear Economic Dynamics. Bern: Peter Lang.

Grandmont, J.M. (1983). Money and Value. Cambridge: Cambridge University Press.

Grandmont, J.M. (1985). On endogenous competitive business cycles, Econometrica, 53, pp. 535-72.

Greenwood, J. and Jovanovic, B. (1990). Financial development, growth, and distribution of income, Journal of Political Economy, 98, pp. 1076-07.

Grossman, G.M. and Helpman, E. (1991). Innovation and Growth in the Global Economy. Cambridge, Mass.: The MIT Press.

Haavelmo, T. (1954). A Study in the Theory of Economic Evolution. North-Holland: Amsterdam.

Hadjimichalakis, M. (1971a). Money, expectations, and dynamics - an alternative view, International Economic Review, 12, pp. 381-402.

Hadjimichalakis, M. (1971b). Equilibrium and disequilibrium growth with money - the Tobin models, Review of Economic Studies, 38, pp. 457-79.

Hahn, F.H. (1965). On two-sector growth models, Review of Economic Studies, 32, pp. 339-346.

Hahn, F. (1982). Stability, in Arrow, K.J. and Intriligator, M.D. (eds.) A Handbook of Mathematical Economics, vol. 2. Amsterdam: North-Holland.

Haken, H. (1977). Synergetics: An Introduction. Berlin: Springer-Verlag.

Haken, H. (1983). Advanced Synergetics - Instability Hierarchies of Self-Organizing Systems and Devices. Berlin: Springer.

Hale, J. and Koçak, H. (1991). Dynamics and Bifurcations. Berlin: Springer-Verlag.

Harrod, R. (1948). Towards a Dynamic Economics. London: Macmillan.

Hartman, P. (1964). Ordinary Differential Equations. Berlin: Springer.

Heckman, J.J. and Macurdy, T.E. (1980). A life cycle model of labor supply, Review of Economic Studies, 47, pp. 47-74.

Hirsh, M. and Smale, S. (1974). Differential Equations, Dynamical Systems, and Linear Algebra. New York: Academic Press.

Ikeda, S. and Ono, Y. (1992). Macroeconomic dynamics in a multi-country economy - a dynamic optimization approach, International Economic Review, 33, pp. 629-44.

Iooss, G. and Joseph, D.D. (1980). Elementary Stability and Bifurcation Theory. New York: Springer-Verlag.

Jordan, D.W. and Smith, P. (1999). Nonlinear Ordinary Differential Equations – An Introduction to Dynamic Systems. Oxford: Oxford University Press.

Kaldor, N. (1940). A model of the trade cycle, Economic Journal, 50, pp. 78-92.

Kaldor, N. (1957). A model of economic growth, Economic Journal, 67, pp. 591-624.

Kaldor, N. (1963). Capital accumulation and economic growth, in Lutz, F.A. and Hague, D.C. (eds.) Proceedings of a Conference held by the International Association. London: Macmillan.

Kalecki, M. (1937). A theory of the business cycle, Review of Economic Studies, 4, pp. 77-97.

Kalecki, M. (1939). A theory of the business cycle, in Kalecki, M. (1972). Essays in the Theory of Economic Fluctuation. London: Allen-Unwin.

King, R.G. and Rebelo, S.T. (1993). Transitional dynamics and economic growth in the neoclassical model, American Economic Review, 83, pp. 908-31.

Kuznetsov, Y.A. (1998). Elements of Applied Bifurcation Theory. Berlin: Springer.

Lancaster, K.J. (1966). A new approach to consumer theory, Journal of Political Economy, 74, pp. 132-57.

Lancaster, K. (1971). Consumer Demand. New York: Columbia University Press.

Levinson, N. and Smith, O.K. (1942). A general equation for relaxation oscillations, Duke Mathematical Journal, 9, pp. 382-403.

Lorenz, H.W. (1987). On the uniqueness of limit cycles in business cycle theory, Metroeconomica, 38, pp. 281-93.

Lorenz, H.W. (1993). Nonlinear Dynamic Economics and Chaotic Motion. Berlin: Springer-Verlag.

Lotka, A. (1924). Elements of Mathematical Biology, reprinted in 1956. New York: Dover.

MacDonald, N. (1978). Time Lags in Biological Models, Lecture Notes in Biomathematics 27. Berlin: Springer-Verlag.

Magill, M.J.P. (1977). Some new results on the local stability of the process of capital accumulation, Journal of Economic Theory, 15, pp. 174-210.

Mankiw, N.G. and Weil, D.N. (1989). The baby boom, the baby bust, and the housing market, Regional Science and Urban Economics, 19, pp. 235-58.

Marsden, J.E. and McCracken, M. (1976). The Hopf Bifurcation and Its Applications. New York: Springer-Verlag.

Mas-Colell, A. (1986). Notes on Price and Quantity Tâtonnement Dynamics, in Sonnenschein, H (ed.) Models of Economic Dynamics. New York: Springer.

Mas-Colell, A., Whinston, M.D. and Green, J.R. (1995). Microeconomic Theory. New York: Oxford University Press.

Mills, E.S. and Hamilton, B.W. (1985). Urban Economics. London: Foresman and Company.

Mincer, J. (1962). Labor force participation of married woman: a study of labor supply, in Lewis, H.G. (ed) Aspects of Labor Economics. Princeton: Princeton University Press.

Miyao, T. (1981). Dynamic Analysis of the Urban Economy. New York: Academic Press.

Morishima, M. (1959). A reconsideration of the Walras-Cassel-Leontief models of general equilibrium, in Arrow, K. (ed.) Mathematical Social Sciences. Stanford: Stanford University Press.

Morishima, M. (1977). Walras' Economics. Cambridge: Cambridge University Press.

Muth, R.F. (1973). A vintage model of the housing stock, Papers of the Regional Science Association, 30, pp. 141-156.

Negishi, T. (1989). History of Economic Theory. Amsterdam: North-Holland.

Obstfeld, M. and Rogoff, K. (1999). Foundations of International Macroeconomics. Mass., Cambridge: MIT Press.

Oniki, H. and Uzawa, H. (1965). Patterns of trade and investment in a dynamic model of international trade, Review of Economic Studies, 32, pp. 15-38.

Orphanides, A. and Solow, R.M. (1990). Money, Inflation and Growth, in Friedman, B.M. and Hahn, F.H. (eds.) Handbook of Monetary Economics, vols. I, II. Amsterdam: North-Holland.

Perko, L. (2001). Differential Equations and Dynamical Systems, third edition. Berlin: Springer.

Phillips, A.W. (1954). Stabilisation policy in a closed economy, Economic Journal, 64, pp. 290-323.

Pitchford, J.D. (1991). Optiumum response of the current account when income is uncertain, Journal of Economic Dynamics and Control, 15, pp. 285-96.

Peterson, G.L. and Sochacki, J.S. (2002). Linear Algebra and Differential Equations. Boston: Addison-Wesley Publishing Company.

v.d. Ploeg, F. (1983). Predator-prey and neo-classical models of cyclical growth, Journal of Economics, 43, pp. 235-56.

v.d. Ploeg, F. (1987). Growth cycles, induced technical change and perpetual conflict over the distribution of income, Journal of Macroeconomics, 9, pp. 1-12.

Prigogine, I. and Lefever, R. (1968). Symmetry breaking instabilities in dissipative systems II, Journal of Chemical Physics, 48.

Puu, T. (2000). Attractors, Bifurcations, and Chaos: Nonlinear Phenomena in Economics. Berlin: Springer-Verlag Telos.

Rauch, J.E. (1991). Comparative advantage, geographic advantage and the volume of trade, The Economic Journal, 101, pp. 1230-44.

Rapoport, A. (1960). Fights, Games and Debates. Ann Arbor: University of Michigan Press.
Richardson, L.F. (1960). Mathematics of War and Foreign Politics, in Arms and Insecurity: A Mathematical Study of the Causes and Origins of War. Pittsburgh: Boxwood.
Rodriguez, C.A. (1975). Brain drain and economic growth - a dynamic model, Journal of Development Economics, 2, pp. 223-47.
Ros, J. (2000). Development Theory and the Economics of Growth. Ann Arbor: The University of Michigan Press.
Rosser, J.B.Jr. (1991). From Catastrophe to Chaos: A General Theory of Economic Discontinuities, Boston: Kluwer Academic Publishers.
Rouche, N., Habets, P. and Laloy, M. (1977). Stability Theory by Liapunov's Direct Method. New York: Springer.
Ruffin, R.J. (1979). Growth and the long-run theory of international capital movements, American Economic Review, 69, pp. 832-42.
Sato, R. (1965). Stability conditions in two-sector models of economic growth, Journal of Economic Theory, 1, pp. 107-17.
Scarf, H. (1960). Some examples of global instability of the competitive equilibrium, International Economic Review, 1, 157-72.
Schultz, T.W. (1993). Origins of Increasing Returns. Oxford: Blackwell Publishers.
Schumpeter, J.A. (1954). History of Economic Analysis. New York: Oxford University Press.
Seierstad, A. and Sydsaeter, K. (1987). Optimal Control Theory with Economic Applications. Amsterdam: North-Holland.
Shah, A. and Desai, M. (1981). Growth cycles with induced technical change, The Economic Journal, 91, pp. 1006-10.
Shone, R. (2002). Economic Dynamics – Phase Diagrams and Their Economic Application. Cambridge: Cambridge University Press.
Simon, H.A. (1973). The organization of complex systems, in Pattee. H. (ed.) Hierarchy Theory: The Challenge of Complex Systems. New York: G. Braziller.
Solow, R. (1956). A contribution to the theory of growth, Quarterly Journal of Economics, 70, pp. 65-94.
Solow, R. (1961). Note on Uzawa's two-sector model of economic growth, Review of Economic Studies, 29, pp. 48-50.
Sparrow, C. (1982). The Lorenz Equations, Bifurcations, Chaos, and Strange Attractors. Berlin: Springer.
Stadler, G.W. (1990). Business cycle models with endogenous technology, The American Economic Review, 80, pp. 763-76.
Stevenson, A., Muscatelli, V. and Gregory, M. (1988). Macroeconomic Theory and Stabilisation Policy. Oxford: Philip Allan.
Swan, T.W. (1956). Economic growth and capital accumulation, Economic Record, 32, pp. 334-61.
Tabor, M. (1989). Chaos and Integrability in Nonlinear Dynamics – An Introduction. New York: John Wiley & Sons.
Takayama, A. (1965). On a two-sector model of economic growth with technological progress, Review of Economic Studies, 32, pp. 251-62.
Takayama, A. (1994). Analytical Methods in Economics. New York: Harvester Wheatsheaf.
Takayama, A. (1996). Mathematical Economics, 2nd edition, New York: Cambridge University Press.

Thom, R. (1977). Structural Stability and Morphogenesis. New York: Addison Wesley.

Tobin, J. (1955). A dynamic aggregative model, Journal of Political Economy, 63, pp. 103-15.

Tobin, J. (1965). Money and economic growth, Econometrica, 33, pp. 671-84.

Tobin, J. (1969). A general equilibrium approach to monetary theory, Journal of Money, Credit and Banking, 1, pp. 15-29.

Torre, V. (1977). Existence of limit cycles and control in complete Keynesian system by theory of bifurcations, Econometrica, 45, pp. 1457-66.

Turnovsky, S.J. (1997). International Macroeocnomic Dynamics. Mass., Cambridge: The MIT Press.

Turnovsky, S.J. (2000). Methods of Macroeconomic Dynamics. Mass., Cambridge: The MIT Press.

Uzawa, H. (1961). On a two-sector model of economic growth, Review of Economic Studies, 29, pp. 47-70.

Uzawa, H. (1965). Optimal technical change in an aggregative model of economic growth, International Economic Review, 6, pp. 18-31.

Velupillai, K. (1978). Some stability properties of Goodwin's growth cycle model, Journal of Economics, 39, pp. 245-57.

Volterra, V. (1931). *Lecons sur la Théorie Mathématique de la Lutte pour la Vie.* Paris: Gauthier-Villars.

Walker, D.A. (1987). Walras's theories of tâtonnement, Journal of Political Economy, 95, pp. 758-74.

Wang, J.Y. (1990). Growth, technology transfer, and the long-run theory of international capital movements, Journal of International Economics, 29, pp. 255-71.

Wang, P. (1993). Agglomeration in a linear city with heterogeneous households, Regional Science and Urban Economics, 23, pp. 291-306.

Weiss, Y. and Willis, R.J. (1985). Children as collective goods and divorce settlements, Journal of Labor Economics, 3, pp. 268-92.

Wiggins, S. (1990). Introduction to Applied Nonlinear Dynamical Systems and Chaos. Berlin: Springer.

Wilson, H.K. (1971). Ordinary Differential Equations. Reading: Addison-Wesley.

Wörz-Busekros, A. (1978). Global stability in ecological systems with continuous time delay, SIAM J. Appl. Math., 35, pp. 124-34.

Young, A. (1993). Invention and bounded learning by doing, Journal of Political Economy, 101, pp. 443-72.

Zeeman, E.C. (1977). Catastrophe Theory: Selected Papers 1972-1977. Reading, MA: Addison-Wesley.

Zhang, J.X. (2000jx). Public services, increasing returns, and equilibrium dynamics, Journal of Economic Dynamics & Control, 24, pp. 227-46.

Zhang, W.B. (1988). The Pattern formation of an urban system, Geographical Analysis, 20, pp. 75-84.

Zhang, W.B. (1988a). Limit cycles in van der Ploeg's model of economic growth and conflict over the distribution of income, Journal of Economics, 48, pp. 159-73.

Zhang, W.B. (1988b). Urbanizing processes with moving boundaries, Geographical Analysis, 20, pp. 328-39.

Zhang, W.B. (1988c). Hopf bifurcations in multisector models of optimal economic growth, Economics Letters, 26, pp. 329-34.

Zhang, W.B. (1989). Existence of aperiodic time-dependent solutions in optimal growth economy with three sectors, International Journal of Systems Science, 20, pp. 1943-53.

Zhang, W.B. (1990). Economic growth and technological change, International Journal of Systems Science, 21, pp. 1933-49.

Zhang, W.B. (1991). Synergetic Economics. Berlin: Springer.

Zhang, W.B. (1991a). Regional dynamics with creativity and knowledge diffusion, The Annals of Regional Science, 25, pp. 179-91.

Zhang, W.B. (1992). A development model of developing economies with capital and knowledge accumulation, Journal of Economics, 55, pp. 43-63.

Zhang, W.B. (1992a). A two-country growth model - knowledge accumulation with international interactions, Journal of Scientific & Industrial Research, 31, pp. 187-94.

Zhang, W.B. (1992b). Trade and world economic growth - differences in knowledge utilization and creativity, Economic Letters, 39, pp. 199-206.

Zhang, W.B. (1993). Knowledge, economic structure, and growth, Technological Forecasting and Social Change, 43, pp. 63-74.

Zhang, W.B. (1993a). Trade, knowledge and government research policy, International Review of Economics and Finance, 2, pp. 103-14.

Zhang, W.B. (1993b). Knowledge, economic geography and growth in a two-group island-economy, Technological Forecasting and Social Change, 44, pp. 405-18.

Zhang, W.B. (1994). A multi-country free trade model with capital accumulation, International Economic Journal 8, 53-66.

Zhang, W.B. (1995). Leisure time, savings and trade patterns - a two-country growth model, Economic Modelling, 12, pp. 425-34.

Zhang, W.B. (1996). Knowledge and economic growth with research and job amenity, The Interdisciplinary Information Sciences, 2, pp. 1-10.

Zhang, W.B. (1998). Japan versus China in the Industrial Race. London: Macmillan.

Zhang, W.B. (1999). Capital and Knowledge - Dynamics of Economic Structures with Non-constant Returns. Berlin: Springer.

Zhang, W.B. (2000). A Theory of International Trade – Capital, Knowledge and Economic Structures. Berlin: Springer.

Zhang, W.B. (2002). An Economic Theory of Cities – Spatial Models with Capital, Knowledge, and Structures. Berlin: Springer.

Zhang, W.B. (2003). A Theory of Interregional Dynamics – Spatial Models with Capital, Knowledge, and Structures. Berlin: Springer.

Zhang, W.B. (2005). Economic Growth Theory. London: Ashgate.

Zhang, Z.F. (1958zf). On the uniqueness of limit cycles of certain equations of nonlinear oscillations, Dokl. Akad. Nauk. SSSR, 19, pp. 659-62.

Zhang, Z.F. (1981zf). On the existence of exactly two limit cycles for the lienard equation, Acta Math. Sinica, 24, pp. 710-16.

Series on Advances in Mathematics for Applied Sciences

Series on Advances in Mathematics for Applied Sciences

Aims and Scope

This Series reports on new developments in mathematical research relating to methods, qualitative and numerical analysis, mathematical modeling in the applied and the technological sciences. Contributions rlated to constitutive theories, fluid dynamics, kinetic and transport theories, solid mechanics, system theory and mathematical methods for the applications are welcomed.

This Series includes books, lecture notes, proceedings, collections of research papers. Monograph collections on specialized topics of current interest are particularly encouraged. Both the proceedings and monograph collections will generally be edited by a Guest editor.

High quality, novelty of the content and potential for the applications to modern problems in applied science will be the guidelines for the selection of the content of this series.

Instructions for Authors

Submission of proposals should be addressed to the editors-in-charge or to any member of the editorial board. In the latter, the authors should also notify the proposal to one of the editors-in-charge. Acceptance of books and lecture notes will generally be based on the description of the general content and scope of the book or lecture notes as well as on sample of the parts judged to be more significantly by the authors.

Acceptance of proceedings will be based on relevance of the topics and of the lecturers contributing to the volume.

Acceptance of monograph collections will be based on relevance of the subject and of the authors contributing to the volume.

Authors are urged, in order to avoid re-typing, not to begin the final preparation of the text until they received the publisher's guidelines. They will receive from World Scientific the instructions for preparing camera-ready manuscript.

SERIES ON ADVANCES IN MATHEMATICS FOR APPLIED SCIENCES

Published:

Vol. 1 Mathematical Topics in Nonlinear Kinetic Theory II: The Enskog Equation
by N. Bellomo et al.

Vol. 2 Discrete Models of Fluid Dynamics
ed. A. S. Alves

Vol. 3 Fluid Dynamics Applications of the Discrete Boltzmann Equation
by R. Monaco and L. Preziosi

Vol. 4 Waves and Stability in Continuous Media
ed. S. Rionero

Vol. 5 A Theory of Latticed Plates and Shells
by G. I. Pshenichnov

Vol. 6 Recent Advances in Combustion Modelling
ed. B. Larrouturou

Vol. 7 Linear Kinetic Theory and Particle Transport in Stochastic Mixtures
by G. C. Pomraning

Vol. 8 Local Stabilizability of Nonlienar Control Systems
by A. Bacciotti

Vol. 9 Nonlinear Kinetic Theory and Mathematical Aspects of Hyperbolic Systems
ed. V. C. Boffi

Vol. 10 Nonlinear Evolution Equations. Kinetic Approach
by N. B. Maslova

Vol. 11 Mathematical Problems Relating to the Navier-Stokes Equation
ed. G. P. Galdi

Vol. 12 Thermodynamics and Kinetic Theory
eds. W. Kosiński et al.

Vol. 13 Themomechanics of Phase Transitions in Classical Field Theory
by A. Romano

Vol. 14 Applications of Padé Approximation Theory in Fluid Dynamics
by A. Pozzi

Vol. 15 Advances in Mathematical Modelling of Composite Materials
ed. K. Z. Markov

Vol. 16 Advanced Mathematical Tools in Metrology
eds. P. Ciarlini et al.

SERIES ON ADVANCES IN MATHEMATICS FOR APPLIED SCIENCES

Vol. 35 Mechanics of Mixtures
by K. R. Rajagopal and L. Tao

Vol. 36 Dynamical Mechanical Systems Under Random Impulses
by R. Iwankiewicz

Vol. 37 Oscillations in Planar Dynamic Systems
by R. E. Mickens

Vol. 38 Mathematical Problems in Elasticity
ed. R. Russo

Vol. 39 On the Way to Understanding the Time Phenomenon: The Constructions of Time in Natural Science — Part 2. The "Active" Properties of Time According to N. A. Kozyrev
ed. A. P. Levich

Vol. 40 Advanced Mathematical Tools in Metrology II
eds. P. Ciarlini et al.

Vol. 41 Mathematical Methods in Electromagnetism Linear Theory and Applications
by M. Cessenat

Vol. 42 Constrained Control Problems of Discrete Processes
by V. N. Phat

Vol. 43 Motor Vehicle Dynamics: Modeling and Simulation
by G. Genta

Vol. 44 Microscopic Theory of Condensation in Gases and Plasma
by A. L. Itkin and E. G. Kolesnichenko

Vol. 45 Advanced Mathematical Tools in Metrology III
eds. P. Cianlini et al.

Vol. 46 Mathematical Topics in Neutron Transport Theory — New Aspects
by M. Mokhtar-Kharroubi

Vol. 47 Theory of the Navier–Stokes Equations
eds. J. G. Heywood et al.

Vol. 48 Advances in Nonlinear Partial Differential Equations and Stochastics
eds. S. Kawashima and T. Yanagisawa

Vol. 49 Propagation and Reflection of Shock Waves
by F. V. Shugaev and L. S. Shtemenko

Vol. 50 Homogenization
eds. V. Berdichevsky, V. Jikov and G. Papanicolaou

Vol. 51 Lecture Notes on the Mathematical Theory of Generalized Boltzmann Models
by N. Bellomo and M. Lo Schiavo

SERIES ON ADVANCES IN MATHEMATICS FOR APPLIED SCIENCES

Vol. 52 Plates, Laminates and Shells: Asymptotic Analysis and Homogenization
by T. Lewiński and J. J. Telega

Vol. 53 Advanced Mathematical and Computational Tools in Metrology IV
eds. P. Ciarlini et al.

Vol. 54 Differential Models and Neutral Systems for Controlling the Wealth of Nations
by E. N. Chukwu

Vol. 55 Mesomechanical Constitutive Modeling
by V. Kafka

Vol. 56 High-Dimensional Nonlinear Diffusion Stochastic Processes — Modelling for Engineering Applications
by Y. Mamontov and M. Willander

Vol. 57 Advanced Mathematical and Computational Tools in Metrology V
eds. P. Ciarlini et al.

Vol. 58 Mechanical and Thermodynamical Modeling of Fluid Interfaces
by R. Gatignol and R. Prud'homme

Vol. 59 Numerical Methods for Viscosity Solutions and Applications
eds. M. Falcone and Ch. Makridakis

Vol. 60 Stability and Time-Optimal Control of Hereditary Systems — With Application to the Economic Dynamics of the US (2nd Edition)
by E. N. Chukwu

Vol. 61 Evolution Equations and Approximations
by K. Ito and F. Kappel

Vol. 62 Mathematical Models and Methods for Smart Materials
eds. M. Fabrizio, B. Lazzari and A. Morro

Vol. 63 Lecture Notes on the Discretization of the Boltzmann Equation
eds. N. Bellomo and R. Gatignol

Vol. 64 Generalized Kinetic Models in Applied Sciences — Lecture Notes on Mathematical Problems
by L. Arlotti et al.

Vol. 65 Mathematical Methods for the Natural and Engineering Sciences
by R. E. Mickens

Vol. 66 Advanced Mathematical and Computational Tools in Metrology VI
eds. P. Ciarlini et al.

Vol. 67 Computational Methods for PDE in Mechanics
by B. D'Acunto

SERIES ON ADVANCES IN MATHEMATICS FOR APPLIED SCIENCES

Vol. 68 Differential Equations, Bifurcations, and Chaos in Economics
by W. B. Zhang

Vol. 69 Applied and Industrial Mathematics in Italy
eds. M. Primicerio, R. Spigler and V. Valente